Frontiers in Space

Human
Spaceflight

JOSEPH A. ANGELO, JR.

Facts On File
An imprint of Infobase Publishing

To all hardworking elementary school teachers—
like my wife, Joan—whose daily efforts help transform the
childhood dreams of today into the reality of tomorrow.

✧

Facts On File, Inc.
An imprint of Infobase Publishing
132 West 31st Street
New York NY 10001

Library of Congress Cataloging-in-Publication Data
Angelo, Joseph A.
   Human spaceflight / Joseph A. Angelo, Jr.
     p. cm. — (Frontiers in space)
   Includes bibliographical references and index.
   ISBN 10: 0-8160-5775-3
   ISBN 13: 978-0-8160-5775-7
   1. Manned space flight—Juvenile literature. 2. Outer space—Exploration—Juvenile literature. 3. Astronautics—History—Juvenile literature. 4. Space colonies—Juvenile literature. I. Title.
   TL793.A54 2007
629.45—dc22 2006029488

Contents

✧ 12 Large Space Settlements—Hallmark of a Solar System Civilization   280

✧ 13 Conclusions   293

Preface

Frontiers in Space is a comprehensive multivolume set that explores the scientific principles, technical applications, and impacts of space technology on modern society. Space technology is a multidisciplinary endeavor, which involves the launch vehicles that harness the principles of rocket propulsion and provide access to outer space, the spacecraft that operate in space or on a variety of interesting new worlds, and many different types of payloads (including human crews) that perform various functions and objectives in support of a wide variety of missions. This set presents the people, events, discoveries, collaborations, and important experiments that made the rocket the enabling technology of the space age. The set also describes how rocket propulsion systems support a variety of fascinating space exploration and application missions—missions that have changed and continue to change the trajectory of human civilization.

The story of space technology is interwoven with the history of astronomy and humankind's interest in flight and space travel. Many ancient peoples developed enduring myths about the curious lights in the night sky. The ancient Greek legend of Icarus and Daedalus, for example, portrays the age-old human desire to fly and to be free from the gravitational bonds of Earth. Since the dawn of civilization, early peoples, including the Babylonians, Mayans, Chinese, and Egyptians, have studied the sky and recorded the motions of the Sun, the Moon, the observable planets, and the so-called fixed stars. Transient celestial phenomena, such as a passing comet, a solar eclipse, or a supernova explosion, would often cause a great deal of social commotion—if not outright panic and fear—because these events were unpredictable, unexplainable, and appeared threatening.

It was the ancient Greeks and their geocentric (Earth-centered) cosmology that had the largest impact on early astronomy and the emergence of Western Civilization. Beginning in about the fourth century B.C.E., Greek philosophers, mathematicians, and astronomers articulated a geocentric model of the universe that placed Earth at its center with everything else revolving about it. This model of cosmology, polished and refined in about 150 C.E. by Ptolemy (the last of the great early Greek astronomers), shaped and molded Western thinking for hundreds of years until displaced in the 16th century by Nicholaus Copernicus and a heliocentric (Sun-centered) model of the solar system. In the early 17th century, Galileo Galilei and Johannes Kepler used astronomical observations to validate heliocentric cosmology and, in the process, laid the foundations of the Scientific Revolution. Later that century, the incomparable Sir Isaac Newton completed this revolution when he codified the fundamental principles that explained how objects moved in the "mechanical" universe in his great work *The Principia*.

The continued growth of science over the 18th and 19th centuries set the stage for the arrival of space technology in the middle of the 20th century. As discussed in this multivolume set, the advent of space technology dramatically altered the course of human history. On the one hand, modern military rockets with their nuclear warheads redefined the nature of strategic warfare. For the first time in history, the human race developed a weapon system with which it could actually commit suicide. On the other hand, modern rockets and space technology allowed scientists to send smart robot exploring machines to all the major planets in the solar system (save for tiny Pluto), making those previously distant and unknown worlds almost as familiar as the surface of the Moon. Space technology also supported the greatest technical accomplishment of the human race, the Apollo Project lunar landing missions. Early in the 20th century, the Russian space travel visionary Konstantin E. Tsiolkovsky boldly predicted that humankind would not remain tied to Earth forever. When astronauts Neil Armstrong and Edwin (Buzz) Aldrin stepped on the Moon's surface on July 20, 1969, they left human footprints on another world. After millions of years of patient evolution, intelligent life was able to migrate from one world to another. Was this the first time such an event has happened in the history of the 14-billion-year-old universe? Or, as some exobiologists now suggest, perhaps the spread of intelligent life from one world to world is a rather common occurrence within the galaxy. At present, most scientists are simply not sure. But, space technology is now helping them search for life beyond Earth. Most exciting of all, space technology offers the universe as both a destination and a destiny to the human race.

Each volume within the Frontiers in Space set includes an index, a chronology of notable events, a glossary of significant terms and concepts, a helpful list of Internet resources, and an array of historical and current print sources for further research. Based upon the current principles and standards in teaching mathematics and science, the Frontiers in Space set is essential for young readers who require information on relevant topics in space technology, modern astronomy, and space exploration.

Acknowledgments

I wish to thank the public information specialists at the National Aeronautics and Space Administration (NASA), the National Oceanic and Atmospheric Administration (NOAA), the United States Air Force (USAF), the Department of Defense (DOD), the Department of Energy (DOE), the National Reconnaissance Office (NRO), the European Space Agency (ESA), and the Japanese Aerospace Exploration Agency (JAXA), who generously provided much of the technical materials used in the preparation of this series. Acknowledgment is made here for the efforts of Frank Darmstadt and other members of the editorial staff at Facts On File whose diligent attention to detail helped transform an interesting concept into a series of publishable works. The support of two other special people merits public recognition here. The first individual is my physician, Dr. Charles S. Stewart III, M.D., whose medical skills allowed me to successfully complete the series. The second individual is my wife, Joan, who, as she has for the past 40 years, provided the loving spiritual and emotional environment so essential in the successful completion of any undertaking in life, including the production of this series.

Introduction

Human Spaceflight is a volume that describes the epoch journeys of human beings as they first ventured beyond Earth's atmosphere, starting in the early 1960s, and traveled through outer space. The epiphany of these daring ventures occurred on July 20, 1969, when the American astronaut Neil Armstrong became the first human being to walk on another world. As he descended from the last step of the lunar excursion module's ladder and made contact with the Moon's surface, he uttered these famous words: "That's one small step for (a) man, one giant leap for mankind." His simple statement most eloquently summarized a major milestone in the evolution of conscious intelligence beyond the confines of our tiny planetary biosphere. Here on Earth, the last such major evolutionary unfolding occurred some 350 million years ago, when prehistoric fish, called crossopterygians, first left the ancient seas and crawled upon the land. Scientists consider these early "explorers" to be the ancestors of all terrestrial animals with backbones and four limbs. Perhaps some future galactic historian will note how life emerged out of Earth's ancient oceans, paused briefly on the land, and then boldly ventured forth to the stars.

Minutes later, on that historic day in July 1969, astronaut Edwin (Buzz) Aldrin joined Armstrong on the surface of the Moon. While they explored features of the lunar surface near the lunar excursion module, their fellow astronaut, Michael Collins, orbited overhead in the Apollo Command module. Back on Earth, more than 500 million people watched this momentous event through live television broadcasts. As predicted decades earlier by the Russian astronautics pioneer Konstantin Tsiolkovsky, the rocket and the complementary technologies needed to support human spaceflight would liberate us from the planetary cradle of Earth and help our species come of age in a vast and beautiful universe. *Human Spaceflight* describes how this marvelous and important application of space technology came about in just a few years after the start of the Space Age.

Any discussion of human spaceflight and its overall significance in history must pay homage to the political decision and technical efforts that allowed American astronauts to walk on another world. In addition to Armstrong and Aldrin, 10 other astronauts became "Moon walkers" as part of NASA's Apollo Project—a daring technical effort born out of political necessity during the cold war.

In May 1961, President John F. Kennedy made a bold decision to send American astronauts to the Moon and to return them safely to Earth before the decade was out. He made this decision to thwart the global political impact of numerous space technology achievements by the former Soviet Union during the cold war. What is often lost in the glare of the successful Moon landings is that Kennedy's decision was made *before* an American astronaut had even successfully orbited Earth in a space capsule. Aerospace engineers certainly recognized the incredible challenge inherent in Kennedy's decision. Before any astronaut could walk on the Moon, there were many basic technical questions that needed to be answered. Could humans survive in space? Could a spacecraft be designed to keep them alive while they traveled in orbit around Earth? Could astronauts survive the fiery, high-speed reentry into the atmosphere and return safely to Earth's surface? *Human Spaceflight* describes how NASA's Mercury Project answered these fundamental questions and many other challenging technical issues.

For example, Mercury Project engineers had to devise a space vehicle that could protect a human being from the temperature extremes, vacuum, and newly discovered radiation of space. Added to those demands was the need to keep an astronaut cool during the astronaut vehicle's fiery, high-speed reentry into the atmosphere. The vehicle that best fit these demanding requirements was a wingless capsule designed for ballistic reentry. The Mercury capsule had an ablative heat shield that burned off as the spacecraft made its fiery plunge through the atmosphere. Following a stepwise conservative engineering development philosophy, before human beings flew into space in the Mercury space capsule, NASA sent two chimps (first Ham and then Enos) into space on test missions that demonstrated the integrity of the spacecraft's design.

The American human spaceflight program experienced not only an increase in the numbers of people traveling into orbit but also a marked improvement in the spacecraft that supported these missions. Each successive spacecraft from the Mercury Project through the Apollo Project, followed by the space shuttle, has been larger, more comfortable, and more capable. Some spaceflight activities produced stunning firsts, while others, such as *Skylab* and most recently the *International Space Station* (*ISS*), systematically advanced capabilities by extending the range and sophistication of human operations in space.

The current NASA human spaceflight vision involves a return to the Moon, followed by a crewed expedition to Mars, using a new crew vehicle that resembles the Apollo capsule but is significantly larger. NASA's next-generation spacecraft and launch system will be capable of delivering crew and supplies to the *ISS,* carrying four astronauts to the Moon, and supporting up to six crewmembers on future missions to Mars.

Human Spaceflight describes the historic events, scientific principles, and technical breakthroughs that allowed people to live and work in space. This book also presents some exciting future human spaceflight activities—including a return to the Moon to establish permanent lunar surface bases, human expeditions to Mars, and even the creation of large space settlements in orbit around Earth and at other strategic locations throughout the solar system.

This book contains a special collection of illustrations that depict historic, contemporary, and future human spaceflight activities. The illustrations allow readers to appreciate the tremendous technical progress that has occurred since the early 1960s and what lies ahead. A generous number of sidebars are strategically positioned throughout the book to provide expanded discussions of fundamental physical concepts, engineering choices, and life support techniques. There are also capsule biographies of prominent scientists, astronauts, and cosmonauts.

It is especially important to recognize that human spaceflight makes the universe both a destination and a destiny for the human race. Awareness of these exciting pathways should prove career inspiring to those students now in high school and college who will become the scientists, engineers, and astronauts of tomorrow.

Ever mindful of the impact of science and technology on society, *Human Spaceflight* examines the impact space travel has had on human development since the middle of the 20th century. This book also speculates about the expanded influence that human spaceflight can have on societal development for the remainder of this century and beyond.

The conquest of space by human explorers did not occur without technical problems, major financial commitments, and loss of life. Selected sidebars within the book address some of the most pressing contemporary issues associated with human spaceflight, including the biological consequences of extended exposure to microgravity and the persistent threat of the space radiation environment.

Human Spaceflight has been carefully crafted to help any student or teacher who has an interest in space travel discover what the physical and psychological conditions of human spaceflight are, where human-crewed spacecraft requirements and limitations arise, how life support systems work, and why crew training and psychological conditioning are so important. The back matter contains a chronology, glossary, and an array of historical and current sources for further research. These should prove especially helpful for readers who need additional information on specific terms, topics, and events associated with human beings traveling in outer space or visiting other worlds.

The Dream of Human Spaceflight

From the dawn of history, astronomical observations have played a major role in the evolution of human cultures. The field of archaeoastronomy unites astronomers, anthropologists, and archaeologists as they attempt to link contemporary knowledge of the heavens with the way humans' distant ancestors viewed the sky and interpreted the mysterious objects they saw. Throughout the world, most early peoples looked up at the sky and made up stories about what they saw but could not physically explain.

Prehistoric cave paintings (some up to 30 millennia old) provide a lasting testament that early peoples engaged in stargazing and incorporated such astronomical observations in their cultures. In some ancient societies, the leading holy men would carve special astronomical symbols in stones (petroglyphs) at ancient ceremonial locations. Modern archaeologists and astronomers now examine and attempt to interpret these objects, as well as other objects uncovered in ancient ruins that may have astronomical significance.

Many of the great monuments and ceremonial structures of ancient civilizations have alignments with astronomical significance. One of the oldest astronomical observatories is Stonehenge. During travel to Greece and Egypt in the early 1890s, the British physicist Sir Joseph Norman Lockyer (1836–1920) noticed how many ancient temples had their foundations aligned along an east-west axis—a consistent alignment that suggested to him some astronomical connection to the rising and setting Sun. To pursue this interesting hypothesis, Lockyer then visited Karnack, one of the great temples of ancient Egypt. He discussed the hypothesis in his 1894 book, *The Dawn of Astronomy.* This book is often regarded as the beginning of archaeoastronomy.

As part of his efforts, Lockyer studied Stonehenge, an ancient site located in south England. However, he could not accurately determine the site's construction date. As a result, he could not confidently project the solar calendar back to a sufficiently precise moment in history that would reveal how the curious circular ring of large, vertical stones topped by capstones might be connected to some astronomical practice of the ancient Britons. Lockyer's visionary work clearly anticipated the results of modern studies of Stonehenge—results that suggest the site could have served as an ancient astronomical calendar around 2,000 B.C.E.

The Egyptians and the Maya both used the alignment of structures to assist in astronomical observations and the construction of calendars. Modern astronomers have discovered that the Great Pyramid at Giza, Egypt, has a significant astronomical alignment, as do certain Mayan structures—such as those found at Uxmal in the Yucatán, Mexico. Mayan astronomers were particularly interested in times (called "zenial passages") when the Sun crossed over certain latitudes in Central America. The Maya were also greatly interested in the planet Venus and treated the planet with as much importance as the Sun. These Mesoamerican native peoples had a good knowledge of astronomy and calculated planetary movements and eclipses over millennia.

For many ancient peoples, the motion of the Moon, the Sun, and the planets and the appearance of certain constellations of stars served as natural calendars that helped regulate daily life. Since these celestial bodies were beyond physical reach or understanding, various mythologies emerged along with native astronomies. Within ancient cultures, the sky became the home of the gods, and the Moon and Sun were often deified.

While no anthropologist really knows what the earliest human beings thought about the sky, the culture of the Australian Aborigines—which has been passed down for more than 40,000 years through the use of legends, dances, and songs—gives collaborating anthropologists and astronomers a glimpse of how these early people interpreted the Sun, Moon, and stars. The Aboriginal culture is the world's oldest and most long-lived, and the Aboriginal view of the cosmos involves a close interrelationship between people, nature, and sky. Fundamental to their ancient culture is the concept of "the Dreaming"—a distant past when the spirit ancestors created the world. Aboriginal legends, dances, and songs express how in the distant past the spirit ancestors created the natural world and entwined people in a close relationship with the sky and with nature. Within the Aboriginal culture, the Sun is regarded as a woman. She awakes in her camp in the east each day and lights a torch that she then carries across the sky. In contrast, Aborigines consider the Moon as male, and, because of the coincidental association of the lunar cycle with the female menstrual cycle, they link the Moon with fertility and consequently give it a great magical status. These ancient peoples also regard a solar eclipse as the male Moon uniting with the female Sun.

For the ancient Egyptians, Ra (also called Re) was regarded as the all-powerful sun god who created the world and sailed across the sky each day. As a sign of his power, an Egyptian pharaoh would use the title "son of Ra." Within Greek mythology, Apollo was the god who pulled the Sun across the sky, riding in his golden chariot, and his twin sister Artemis (Diana in Roman mythology) was the Moon goddess.

But from the dawn of human history until the start of the scientific revolution in the early 17th century, the heavens were regarded as an essentially unreachable realm—the abode of deities and, for some civilizations and religions, the place where a good, just person (or at least their conscious spirit) would go after physical life on Earth. The legend of Hercules from Greco-Roman mythology is an example of a powerful and popular mortal hero, who upon death was allowed to join the gods in sky. The ancient Greeks even named a constellation of stars after him to emphasize Hercules's passage into the heavens. Even today, in the practice of many major religions, the religious devotees will often hold their hands to the heavens or raise their eyes to the sky in prayer.

Other legends reminded people how difficult it would be for mortal men to leave the ground and soar high above Earth. In Greek mythology, the brilliant engineer Daedalus was the grand architect of King Minos's labyrinth for the Minotaur on the island of Crete. But Daedalus also showed the Greek hero Theseus, who slew the Minotaur, how to escape from the labyrinth. An enraged King Minos imprisoned both Daedalus and his son Icarus. Undaunted, Daedalus fashioned two pairs of wings out of wax, wood, and leather. Before their aerial escape from a prison tower, Daedalus cautioned his son not to fly too high, so that the Sun would not melt the wax and cause the wings to disassemble. They made good their escape from King Minos's Crete, but while over the sea, Icarus, an impetuous teenager, ignored his father's warnings and soared high into the air. Daedalus (who reached Sicily safely) watched as his young son's wings collapsed, and the impetuous youth tumbled to his death in the sea below.

So what happened in the course of human events that changed the heavens from an unreachable realm to a place to be visited? In other words, what encouraged people to begin thinking about space travel?

The first major step in this transition took place in 1609, when the Italian scientist Galileo Galilei (1564–1642) learned about a new optical instrument (a magnifying tube) that had just been invented in Holland. Within six months, Galileo devised his own version of the instrument. Then, in 1610, he turned this improved telescope to the heavens and started the age of telescopic astronomy. With his crude instrument, he made a series of astounding discoveries, including the existence of mountains on the Moon, many new stars, and the four major moons of Jupiter—now called the *Galilean satellites* in his honor. Galileo published these important discoveries in the book *Sidereus Nuncius* (Starry messenger). The book stimulated both enthusiasm and anger. Galileo used the moons of Jupiter to prove that not all heavenly bodies revolve around Earth. This provided direct observational evidence for the Copernican model—a cosmological model that Galileo now began to endorse vigorously. And the mountains on the Moon and the dark regions, which Galileo thought were oceans and seas and mistakenly called mare, suddenly made the Moon a physical *place* just like Earth. If the Moon was indeed another world, and not some mysterious object in the sky, then inquisitive human beings would someday try to travel there. With the birth of optical astronomy in the early 17th century, not only was the scientific revolution accelerated, but the embryonic notion of space travel and visiting other worlds suddenly acquired a touch of physical reality.

But seeing other worlds in a telescope was just the first step. The next critical step that helped make the dream of space travel a reality was the development of a powerful machine that could not only lift objects off the surface of Earth but also operate in the vacuum of outer space. The modern rocket, as developed during World War II and vastly improved afterward in the cold-war era, became the enabling technology for human spaceflight—a pathway that would open many exciting future options for the human race.

But even with the modern rocket, there was one final step still needed to make human spaceflight a reality. One or more governments had to be willing to invest large quantities of money and engineering talent so people could travel beyond Earth's atmosphere. From a historic perspective, the fierce geopolitical competition of the cold-war era between the United States and the former Soviet Union provided the necessary social stimulus. In an effort to dominate world political opinion in the 1960s, both governments decided to make enormous resource investments in the superpower "race into space."

The remainder of this chapter shows how each of these steps: the vision, the enabling hardware, and the political will came together and made space travel a hallmark achievement of the human race in the latter portion of the 20th century. The apex of that technological achievement was the manned lunar landing missions of NASA's Apollo Project.

✦ Legend of Wan Hu

According to certain historical records, the Chinese were the first to use gunpowder rockets, which they called "fire arrows," in military applications. In the battle of Kai-fung-fu (1232 C.E.), for example, fire arrows helped the Chinese repel Mongol invaders.

Whatever the actual creative pathway of the rocket's discovery in ancient China, the Battle of Kai-fung-fu represents the first reported use of a gunpowder-fueled rocket in warfare. During this battle, Chinese troops used a barrage of rocket-propelled fire arrows to startle and defeat a band of invading Mongolian warriors. In an early attempt at passive guided missile control, Chinese rocketeers attached a long stick to the end of the fire arrow rocket. The long stick kept the center of pressure behind the rocket's center of mass during flight. Although the addition of this long stick helped somewhat, the flight of the rocket-propelled fire arrows still remained quite erratic and highly inaccurate. The heavy stick also reduced the range of these early gunpowder-fueled rockets.

Despite the limitations of the fire arrow, the invading Mongol warriors quickly learned from their unpleasant experience at the Battle of Kai-fung-fu and soon adopted the interesting new weapon for their own use. As a result, nomadic Mongol warriors spread rocket technology westward when they invaded portions of India, the Middle East, and Europe. It was in Europe at the end the 19th century that an obscure Russian schoolteacher and technical visionary named Konstantin Tsiolkovsky made the first credible scientific linkage between rocket propulsion and travel in space.

There is another interesting rocket story from China that deserves mention here. Rocketry legend suggests that around 1500, a lesser-known Chinese official named Wan Hu conceived of the idea of flying through the air in a rocket-propelled chair. He ordered the construction of a chair-kite structure to which were attached 47 fire arrow rockets. Then, serving as his own test pilot, Wan Hu bravely sat in the chair and ordered his servants to simultaneously light the fuses to all the rockets. Forty-seven servants, each carrying a small torch, rushed forward in response to their master's command. Dutifully, they lit the fuses and then dashed back to safety. Suddenly, there was a bright flash and a tremendous roar. The air was filled with billowing clouds of gray smoke. Unfortunately, Wan Hu and his rocket-propelled chair vanished in the explosion—perhaps reaching the heavens more suddenly than he intended.

While science historians regard this story as more legend than fact, it

According to early rocket lore, a Chinese official named Wan Hu (or Wan Hoo) attempted to use a rocket-propelled kite assembly to fly through the air in about 1500. This drawing shows a servant carefully lighting one of the kite's 47 gunpowder rockets, as Wan Hu awaits blastoff. Unfortunately, as the servants backed away, Wan Hu disappeared in a bright flash and explosion. *(NASA)*

TAIKONAUTS IN ORBIT

About 500 years after the legendary Chinese official named Wan Hu disappeared in a cloud of smoke—when he attempted to travel into the sky using rockets—several people called *taikonauts* successfully traveled into space from China. On October 15, 2003, the People's Republic of China became the third nation—following Russia (former Soviet Union) and the United States—to place a human being in orbit around Earth using a nationally developed launch vehicle. On that date, a Chinese Long March 2F rocket lifted off from the Jiuquan Satellite Launch Center and placed the *Shenzhou 5* spacecraft with taikonaut Yang Liwei on board into orbit around Earth. Within the international aerospace community, the word *taikonaut* is the suggested Chinese equivalent to *astronaut* and *cosmonaut*. *Taikong* is the Chinese word for space or cosmos, so the prefix "taiko-" assumes the same concept and significance as the use of "astro-" or "cosmo-" to form the words *astronaut* and *cosmonaut*. After 14 orbits around Earth, the spacecraft reentered the atmosphere on October 16, 2003, and Yang Liwei was safely recovered in the Chinese portion of Inner Mongolia.

About two years later, on October 12, 2005, the People's Republic of China successfully launched its second human-crewed mission from the Jiuquan Satellite Launch Center. A Long March 2F rocket blasted off with a pair of taikonauts on board the *Shenzhou 6* spacecraft, which has a general design similar to the Russian Soyuz spacecraft but with significant modifications. The *Shenzhou 6* had a reentry capsule, an orbital module, and a propulsion module. Taikonauts Fei Junlong and Nie Haisheng sat in the reentry module during takeoff and during the reentry/landing portion of the mission. During nearly five days (115.5 hours) in space, Junlong and Haisheng took turns entering the orbital module, which contained life support equipment and experiments. After 76 revolutions, their reentry capsule safely returned to Earth on October 16, making a soft, parachute-assisted landing in northern Inner Mongolia.

represents the first reported attempt (or at least suggestion) to use the rocket as a means of transportation. Previous applications of the gunpowder rocket were related to either warfare or fireworks for festivals. The first person to technically link human beings, space travel, and rockets was the German technical visionary Hermann Oberth. He made this important connection in the early part of the 20th century.

✧ Scientific Revolution

During the 16th and 17th centuries, Europe experienced a period of profound changes in intellectual thought—often referred to as the *scientific revolution*. Nicholaus Copernicus (1473–1543) began this process by causing a revolution in astronomy with his deathbed publication of *On the*

Revolutions of Celestial Spheres. The book flew in the face of almost two millennia of Aristotle's geocentric astronomy and endorsed a heliocentric model of the universe in which Earth, like the other known planets, revolved around the Sun.

In the early part of the 17th century, the telescopic observations of Galileo Galilei and the laws of planetary motion developed by Johannes Kepler (1571–1630) reinforced the Copernican revolution. Because of his meticulous experiments and careful attention to physical observations, Galileo is often regarded as the first modern scientist. Sir Isaac Newton (1642–1727) climaxed the scientific revolution in the late 17th century by developing and publishing his three laws of motion and the universal law of gravitation. These important scientific principles allowed scientists to explain in precise mathematical terms the motion of almost every object observed in the universe, from an apple falling to the ground, to the trajectory of projectiles fired from a cannon, to planets orbiting the Sun, to spacecraft carrying astronauts in orbit around Earth or going to the Moon.

Complementing Galileo Galilei's scientific accomplishments in the early 17th century, Kepler developed and presented the three laws of planetary motion—important physical principles that described the elliptical orbits of the planets around the Sun and provided the empirical basis for the acceptance of Nicholaus Copernicus's heliocentric hypothesis. Kepler's laws gave astronomy its modern, mathematical foundation.

Kepler's publication *De Stella Nova* (The new star) described the supernova in the constellation Ophiuchus that he first observed (with the naked eye) on October 9, 1604. The spectacular transient phenomenon of a supernova clearly refuted another long-held "astronomical" teaching of Aristotle—specifically that the heavens were immutable (unchanging).

Between 1618 and 1621, Kepler summarized all his planetary studies in the publication *Epitome Astronomica Copernicanae* (Epitome of Copernican astronomy). This work contained *Kepler's second law of planetary motion.* As a point of scientific history, Kepler actually based his second law (the law of equal areas) on a mistaken physical assumption that the Sun exerted a strong magnetic influence on all the planets. Later in the century, Sir Isaac Newton (through his universal law of gravitation) provided the "right physical explanation" (within the limits of classical physics) for the planetary motion correctly described by Kepler's second law. In addition to making these very important contributions to astronomy and orbital mechanics, Kepler was the first scientist to write about people traveling in space.

Before his death in 1630, Kepler wrote a very interesting novel called *Somnium* (The dream). It is a story about an Icelandic astronomer who travels to the Moon. While the tale contains demons and witches (who help get the hero to the Moon's surface in a dream state), Kepler's description of the lunar surface is quite accurate. Consequently, many historians treat this story, which was published after Kepler's death in 1634, as the first genuine piece of science fiction.

✦ Nineteenth-Century Visions of Human Spaceflight

Starting in the mid-19th century, the French writer and technical visionary Jules Verne (1828–1905) created modern science fiction and along with it the dream of space travel. Perhaps Verne's greatest influence on the development of space travel was his 1865 novel *De la terre à la lune* (*From the Earth to the Moon*).

In this fictional work, Verne gave his readers an apparently credible account of a human voyage to the Moon. In the visionary story, Verne's fictional travelers (Michel Ardan, Imply Barbicane, and Captain Nicholl) are blasted on a journey around the Moon in a special hollowed-out capsule that is fired from a very large cannon. The writer correctly located the cannon at a low-latitude site in Florida. Of course, scientists recognized that the acceleration of Verne's proposed capsule down the barrel of this huge cannon would have immediately crushed the three intrepid space travelers inside. If that were not bad enough, the capsule itself would have burned up traveling at escape velocity speed through Earth's atmosphere. Despite its obvious technical limitations, this tale made space travel by human beings appear possible for the first time in history.

Although Verne did not properly connect the rocket as the enabling technology for space travel, his famous story did correctly prophesize the use of small reaction rockets to control the attitude of the ballistic capsule during its flight through space. Verne was neither a scientist nor an engineer, but his literary skills served as an important source of inspiration for the scientists and engineers who actually responded to the challenge of interplanetary space travel. In particular, the three great pioneers of astronautics—Konstantin Tsiolkovsky, Robert Goddard, and Hermann Oberth—would soon independently make the important and necessary connection between powerful liquid-propellant rockets and space travel.

Each of these rocket pioneers also personally acknowledged the works of Jules Verne as a key childhood stimulus in developing their lifelong interest in space travel. The great French novelist, who died in Amiens, France, on March 24, 1905, not only wrote delightful stories that pleased millions of readers, but he also lit the flame of imagination for those who would actually create the modern rockets needed to free humankind from the bonds of Earth. Because of Jules Verne, rocket propulsion–based space travel became first the technical dream and then the technical reality of the 20th century.

Another very influential science fiction writer of the late 19th and early 20th century was Herbert George (H. G.) Wells (1866–1946). He inspired many future astronautical pioneers with his exciting fictional works that popularized the idea of space travel and life on other worlds. For example, in 1897, he wrote *War of the Worlds*—the classic tale about extraterrestrial invaders from Mars.

Wells was born on September 21, 1866, in Bromley, Kent, England. In 1874, a childhood accident forced him to recuperate with a broken leg. The prolonged convalescence encouraged him to become an ardent reader, and this period of intensive self-learning served him well. He went on to become an accomplished author of both science fiction and more traditional novels.

He settled in London in 1891 and began to write extensively on educational matters. His career as a science fiction writer started in 1895 with the publication of the incredibly popular book *The Time Machine*. At the turn of the century, he focused his attention on space travel and the consequences of alien contact. Between 1897 and 1898, *The War of the Worlds* appeared as a magazine serial and then a book. Wells followed this very popular space invasion story with *The First Men in the Moon*, which appeared in 1901. Like Jules Verne, Wells did not link the rocket to space travel, but his stories did excite the imagination. *The War of the Worlds* was the classic tale of an invasion of Earth from space. In his original story, hostile Martians land in 19th-century England and prove to be unstoppable, conquering villains until tiny terrestrial microorganisms destroy them.

In 1865, the French writer Jules Verne published the science fiction novel *From the Earth to the Moon*—the fantastic story about a human voyage around the Moon. This illustration comes from an early printing of Verne's work and shows his fictional characters (Michel Ardan, Imply Barbicane, and Captain Nicholl), as they experience weightlessness inside their bulletlike space capsule. Although Verne's gun-launch to space approach to space travel would have actually crushed his fictional characters before they left the giant gun barrel, the story itself made human spaceflight "appear" credible and thus was immensely inspirational for astronautical pioneers such as Konstantin Tsiolkovsky, Robert Goddard, and Hermann Oberth. *(NASA)*

In writing this story, Wells was probably influenced by the then popular (but incorrect) assumption that supposedly observed Martian "canals" were artifacts of a dying civilization on the Red Planet. This was a very fashionable hypothesis in late 19th-century astronomy. The "canal craze" started quite innocently in 1877, when the Italian astronomer Giovanni

Schiaparelli (1835–1910) reported linear features he observed on the surface of Mars as *canali*—the Italian word for channels. Schiaparelli's accurate astronomical observations became misinterpreted when translated as "canals" in English. Consequently, other notable astronomers such as the American Percival Lowell (1855–1916) began to search enthusiastically for and soon "discover" other surface features on the Red Planet that resembled signs of an intelligent Martian civilization.

H. G. Wells cleverly solved (or more accurately ignored) the technical aspects of space travel in his 1901 novel *The First Men in the Moon.* He did this by creating "cavorite"—a fictitious antigravity substance. His story inspired many young readers to think about space travel. However, Space Age missions to the Moon have now completely vanquished the delightful (though incorrect) products of this writer's fertile imagination, including giant moon caves, a variety of lunar vegetation, and even bipedal Selenites.

However, in many of his other fictional works, Wells was often able to correctly anticipate advances in technology. This earned him the status of a technical prophet. For example, he foresaw the military use of the airplane in his 1908 work *The War in the Air* and foretold of the splitting of the atom in his 1914 novel *The World Set Free.*

Following his period of successful fantasy and science fiction writing, Wells focused on social issues and the problems associated with emerging technologies. For example, in his 1933 novel *The Shape of Things to Come,* he warned about the problems facing Western civilization. In 1935, Alexander Korda produced a dramatic movie version of this futuristic tale. The movie closes with a memorable philosophical discussion on (technological) pathways for the human race. Sweeping an arm, as if to embrace the entire universe, one of the main characters asks his colleague: "Can it really be our destiny to conquer all this?" As the scene fades out, his companion replies: "The choice is simple. It is the whole universe or nothing. Which shall it be?"

The famous novelist and visionary died in London on August 13, 1946. He had lived through the horrors of two world wars and witnessed the emergence of many powerful new technologies, except space technology. His last book, *Mind at the End of Its Tether,* appeared in 1945. In this work, Wells expressed a growing pessimism about humanity's future prospects.

✧ The Birth of Astronautics

Astronautics is the science of spaceflight. The three great pioneers of astronautics—Konstantin Tsiolkovsky, Robert Goddard, and Hermann Oberth—independently made the important and necessary connection between powerful liquid-propellant rockets and space travel.

In chronological order, the first space-travel visionary was the Russian schoolteacher Konstantin E. Tsiolkovsky (1857–1935). He began writing a series of articles and books about the theory of rocketry and spaceflight at the end of the 19th century and continued his pioneering advocacy efforts for more than three decades. Among other things, his works suggested the necessity for liquid-propellant rockets—the very devices that the American physicist Robert H. Goddard would soon develop. Because of the geopolitical circumstances in czarist Russia, Goddard and many other scientists outside of Russia were unaware of Tsiolkovsky's work. Today Tsiolkovsky is regarded as the father of Russian rocketry, as well as one of the cofounders of astronautics.

The brilliant physicist Robert Hutchings Goddard (1882–1945) is regarded as the father of American rocketry and the developer of the practical modern rocket. In 1919, Goddard published the important technical paper "Method of Reaching Extreme Altitudes," in which he concluded that the rocket actually would work better in the vacuum of outer space than in Earth's atmosphere. At the time, Goddard's "radical" (but correct) suggestion cut sharply against the popular (but incorrect) belief that a rocket needed air to "push against." He also suggested that a multistage rocket could reach very high altitude and even attain sufficient velocity to "escape from Earth." This paper also included a final chapter that speculated about how scientists might use the rocket to send a modest payload to the Moon. Unfortunately, the press missed the true significance of his pioneering work and, instead, sensationalized his suggestion about reaching the Moon with a rocket. Goddard was given such unflattering nicknames as "Moony" and the "Moon Man." Offended by this negative publicity, Goddard chose to work in seclusion for the rest of his life. To avoid further public controversy, he published as little as possible. As a consequence, much of his leading-edge rocket research went unrecognized during his lifetime.

Despite his numerous technical accomplishments in rocketry, the U.S. government never really developed an interest in his important work. In fact, only during World War II did Goddard receive any government funding, and that was for him to design small rockets to help aircraft take off from navy carriers. By the time he died in 1945, Goddard held more than 200 patents in rocketry. Aerospace engineers and rocket scientists

Hermann J. Oberth was one of the cofounders of astronautics. Throughout his life, he vigorously promoted the concept of space travel. Unlike Robert Goddard and Konstantin Tsiolkovsky (the other founding fathers of astronautics), Oberth lived to see the arrival of the Space Age and human spaceflight, including the Apollo Project lunar landings. *(NASA)*

now find it essentially impossible to design, construct, or launch a modern liquid-propellant rocket without using some idea or device that originated from Goddard's pioneering work in rocketry.

The third cofounder of astronautics was Hermann Julius Oberth (1894–1989), whose writings and leadership promoted interest in rocketry in Germany following World War I. While Goddard worked essentially unnoticed in the United States, a parallel group of "rocketeers" thrived in Germany, centered originally within the German Rocket Society. In 1923, Hermann J. Oberth published a highly prophetic book entitled *The Rocket into Interplanetary Space*. This important work used mathematics to demonstrate that flight beyond the atmosphere was possible. One of the many readers inspired by this book was a brilliant young teenager named Wernher von Braun. In 1929, Oberth published another important book, *The Road to Space Travel*. Within this work, he proposed liquid-propellant rockets, multistage rockets, space navigation, and guided reentry systems.

✧ Pre–Space Age Visions of Stations in Space

A short time after the American Civil War, Edward Everett Hale published "The Brick Moon." This 1869 story was one of the earliest pieces of science fiction, describing a human-crewed space station. Several years later, both Tsiolkovsky and Oberth included space station concepts in their more technical, space travel–themed books.

Another person who helped develop the space station concept was Hermann Potocnik (1892–1929). Potocnik was an officer in the Imperial Austrian Army and an engineer who became attracted to space travel by the ideas and writings of Oberth. In 1928, writing under the pseudonym Hermann Noordung, Potocnik published *Das Problem der Befahrung des Weltraums* (The problem of space travel; The rocket motor)—a seminal work that concentrated on the engineering aspects of a space station. Decades ahead of his time, Potocnik (aka Noordung) addressed such important engineering and operational issues as the problem of weightlessness, communications between the crew and scientists on Earth, ways to maintain the habitability of the station, and extravehicular activity. Perhaps Potocnik's most important insight was the suggestion to rotate his wheel-shaped (about 100-foot- [30.5-m-] diameter) space station design in order to create artificial gravity in the living quarters and habitable work areas. The idea of Potocnik's (Noordung's) so-called *Wohnrad* (or living-wheel) diffused through the space-travel advocacy community in Europe and reappeared in the 1940s and 1950s with technical embellishments and engineering improvements.

Like Tsiolkovsky and Oberth before him, Potocnik made a bold stab at resolving the problem of providing a reliable source of electric power to

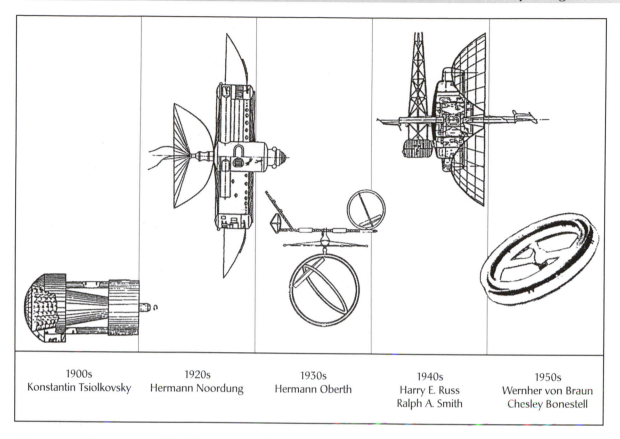

1900s	1920s	1930s	1940s	1950s
Konstantin Tsiolkovsky	Hermann Noordung	Hermann Oberth	Harry E. Russ Ralph A. Smith	Wernher von Braun Chesley Bonestell

Early space station concepts *(NASA)*

the space station. Since solar cells had not yet been invented, the resourceful Potocnik suggested using a large parabolic mirror to focus sunlight unto a conventional heat engine turbo-generator system, which used the basic principles of thermodynamics to generate electric power for the station. Today engineers refer to this engineering approach for electric power generation (on Earth or in space) as solar-thermal conversion. Potocnik also recommended using the space station for astronomical observations and suggested increased scientific value for the station by placing it in synchronous orbit around Earth.

A contemporary of Potocnik was Guido von Pirquet, an Austrian who wrote many technical papers on spaceflight, including the use of a space station as a refueling node for space tugs. In the late 1920s and early 1930s, von Pirquet also suggested the use of multiple space stations at different locations in cislunar space. After World War II, the German-American rocket scientist Wernher von Braun (with the assistance of the American space artist Chesley Bonestell) popularized the concept of a large, wheel-shaped space station.

Starting in the fall of 1952, von Braun also provided technical support for the production of a beautifully illustrated series of visionary space-travel articles, appearing in *Collier's* magazine. His detailed work represented the most comprehensive technical treatment of the space station concept to date. The series caught the eye of the American entertainment genius Walt Disney (1901–66). By the mid-1950s, von Braun had become a nationally recognized space-travel advocate through his frequent appearances on television. Along with Walt Disney, von Braun served as a host for an inspiring three-part television series on human spaceflight and space exploration. Thanks to von Braun's influence, when the Disneyland theme park opened in southern California in the summer of 1955, its "Tomorrowland" section featured a space station X-1 exhibit and a simulated rocket ride to the Moon.

In the early 1950s, Disney started planning an entirely new form of entertainment, a family-oriented amusement complex, which he called a "theme park." He built "Disneyland" in Anaheim, California, and the park had four major themes: Fantasyland, Adventureland, Frontierland, and Tomorrowland. Disney's previous cartoon and motion picture work had not ventured into the realm of future technology. Early in planning Disneyland, he recognized the power of television in promoting the new park. He also recognized the urgent need for a crowd-pleasing "future technology" to anchor Tomorrowland.

Disney turned to a longtime member of his staff, Ward Kimball (1914–2002), and asked for suggestions about Tomorrowland. Kimball mentioned that scientists were talking about the possibility of traveling in space. He showed Disney the *Collier's* magazine articles written by Wernher von Braun and other scientists that discussed space travel, space stations, missions to Mars, and the like. Disney's creative spirit recognized the opportunity. Space would become a major theme of Tomorrowland. He also developed a special television show to introduce the public to space travel—along with attractions at his new theme park.

Because he was busy developing Disneyland, Walt Disney gave Kimball a literal "blank check" to hire the best scientists and produce a space-travel

In the mid-1950s, Walt Disney (left) collaborated with rocket scientist Wernher von Braun (right) in the development of an animated three-part television series that popularized the dream of space travel for millions of Americans. *(NASA)*

television show that was both factual and entertaining. Disney frequently reviewed Kimball's progress and made creative suggestions but left Kimball in charge of the daily production activities.

On the evening of March 9, 1955, millions of television viewers across the United States tuned in to the popular Disneyland TV show. Suddenly, after the usual image of Sleeping Beauty's castle faded, Walt Disney himself appeared on the screen. He sat on the edge of his desk and held a futuristic model rocket. Unlike the format of previous shows, Disney now personally prepared his viewers for their trip into Tomorrowland. He began this special show with a powerful (but soft-spoken) introduction that described the important influence of science in daily living. He also mentioned how things that seemed impossible today could actually become realities tomorrow. Next, he described the concept of space travel as one of humanity's oldest dreams. He concluded the piece by suggesting that recent scientific discoveries had brought people to the threshold of a new frontier—namely, the frontier of interplanetary travel. Disney's show pleased and inspired millions of viewers. In a truly magic moment for the entertainment industry, Walt Disney, supported by the world's leading rocket scientists, spoke credibly to his audience about the possibility of the impossible—travel through interplanetary space.

Man in Space was the first of three extremely popular space-themed Disney television shows that energized the American population toward the possibility of space travel in the mid-1950s. Each show combined careful research and factual presentation with incredibly beautiful visual displays and a splash of Disney humor for good measure. After Disney's introduction, the first show continued with a history of rocketry (featuring German-American rocket historian Willy Ley), a discussion of the hazards of human spaceflight (featuring aerospace medicine expert Heinz Haber [1913–90]), and a detailed presentation by von Braun about a large, four-stage rocket that could carry six humans into space and safely return them to Earth. These space experts had previously provided technical support for the popular *Collier's* magazine series, and Disney (much to his credit) spared no expense in getting their expert opinions and participation for his own TV show.

Man in Space proved so popular with audiences that Disney rebroadcast the show on June 15 and again on September 7, 1955. One especially important person viewed the first show. President Dwight D. Eisenhower (1890–1969) liked the show so much that he personally called Disney and borrowed a copy of the show to use as a space education primer for the "brass" at the Pentagon. It is interesting to note that on June 30, 1955, Eisenhower announced that the United States would launch an Earth-orbiting artificial satellite as part of America's participation in the upcoming International Geophysical Year (1957). Coincidence? Perhaps. Or quite possibly, some of Disney's "visionary magic" worked like a much-needed catalyst in a sluggish federal bureaucracy that consistently failed to comprehend the emerging importance of space technology.

When Disneyland opened in the summer of 1955, the Tomorrowland section of the theme park had a large (82-foot- [25-m-] tall), needle-nosed rocket ship (designed by Ley and von Braun) to greet visitors to the Moon mission attraction.

Disney's second space-themed television show, *Man and the Moon,* aired on December 28, 1955. In this show, von Braun enthusiastically described his wheel-shaped, space station concept and how it could serve as the assembly platform for a human voyage around the Moon. Von Braun emerged from this show as the premier space-travel advocate in the United States. The enthusiastic public response to the first two Disney space-themed TV shows also attracted some journalistic skepticism. Certain reporters tried to "protect" their readers by cautioning them "not to get swept away by over-enthusiasm or arm-chair speculation." Despite the cautious warnings of these unimaginative skeptics, the Space Age arrived on October 4, 1957—when the former Soviet Union launched *Sputnik 1,* the first artificial Earth satellite.

At the dawn of the Space Age, Disney aired his third and final space-themed show, *Mars and Beyond,* on December 4, 1957. Von Braun appeared only briefly in this particular show because he was very busy trying to launch the first successful American satellite (*Explorer 1*). Through inputs from von Braun and his colleague, Ernst Stuhlinger, the highly animated show featured an armada of human-crewed, nuclear-powered interplanetary ships heading to Mars. The show also contained amusing, yet highly speculative, cartoon-assisted discussions about the possibility of life in the solar system. With the conclusion of this episode, Disney had reached millions of Americans and helped them recognize that space travel was real and no longer restricted to "Fantasyland." Through the use of television and his theme park, Disney vigorously promoted space technology in a truly imaginative, delightful way. Walt Disney died on December 15, 1966, in Burbank, California. Five years after his death, the world-famous Disney World vacation complex opened near Orlando, Florida—by interesting coincidence this world-famous entertainment complex lies just 61 miles (100 km) west of Cape Canaveral—America's spaceport.

✦ Riding in a Rocket Plane to the Threshold of Space

The North American X-15 research aircraft helped bridge the gap between human flight within the atmosphere and human flight in space. It was developed and flown in the 1960s to provide in-flight information and data on aerodynamics, structures, flight controls, and the physiological aspects of high-speed, high-altitude flight. For flight in the dense air of the lower ("aircraft-usable") portions of the atmosphere, the X-15 employed conventional aerodynamic controls. However, for flight in the thin, upper portions of Earth's atmospheric envelope, the X-15 used a ballistic control system. Eight hydrogen peroxide–fueled thruster rockets, located on the nose of the aircraft, provided pitch and yaw control.

The X-15 was a rocket-powered experimental aircraft 50 feet (15.24 m) long with a wingspan of 22 feet (6.71 m). It was a missile-shaped vehicle with an unusual, wedge-shaped, vertical tail; thin, stubby wings; and unique fairings that extended along the side of the fuselage. The X-15 had an empty mass of 13,950 pounds (6,340 kg) and a launch mass of 33,925 pounds (15,420 kg). The vehicle's pilot-controlled rocket engine was capable of developing 57,000 pounds-force (253,500 N) of thrust.

Because of its large fuel consumption, the X-15 was air-launched from a B-52 aircraft (i.e., a "mothership") at an altitude of about 44,950 feet (13,700 m) and an initial speed of about 500 miles per hour (805 km/h). Then, the pilot ignited the rocket engine, which provided thrust for the first 80 to 120 seconds of flight, depending on the type mission being flown. The remainder of the normal 10- to 11-minute duration flight was

NASA pilot Neil Armstrong is seen here next to the X-15 rocket plane after a successful hypersonic flight (1960). On July 20, 1969, as commander of the Apollo 11 lunar-landing mission, astronaut Armstrong became the first human being to walk on the surface of the Moon. (NASA)

powerless and ended with a 200 mph (322 km/h) glide landing at Edwards Air Force Base in California.

Generally, the X-15 pilot used one of two basic flight profiles: a high-altitude flight plan that called for the pilot to maintain a steep rate of climb or a speed profile that called for the pilot to push over and maintain a level altitude.

First flown in 1959, the three X-15 aircraft made a total of 199 flights. The X-15 flew more than six times the speed of sound and reached a maximum altitude of 67 miles (107.8 km) and a maximum speed of 4,520 mph (7,273 km/h). The final X-15 flight occurred on October 24, 1968. It is interesting to note that Apollo astronaut Neil Armstrong (the first human to walk on the Moon) was one of the pilots who flew the X-15 aircraft to the threshold of space.

✦ Early Space Race Accomplishments

In 1952, the International Council of Scientific Unions announced that the period 1957–58 would be an International Geophysical Year (IGY), with the primary scientific objective of exploring Earth and its atmosphere. The United States government responded with a pledge to launch an artificial Earth satellite as the culminating event of its participation in the international project. Officials from the former Soviet Union also declared that their government would launch a scientific satellite, but few observers in the West thought that the Soviets were technologically capable of doing so.

Against the advice of the German-American rocket expert Wernher von Braun, who vigorously recommended that the United States use a modified military missile as a launch vehicle, American officials made a political decision to develop a "civilian" rocket (called Vanguard) to put this scientific satellite into orbit. The government officials had reasoned that the use of a special civilian rocket would emphasize the peaceful uses of outer space and play down any public emphasis on military applications. The Soviet Union, on the other hand, simply kept their activities secret and planned to use a modified R-7 intercontinental ballistic missile (ICBM).

When it launched *Sputnik 1* (the world's first artificial satellite) on October 4, 1957, the former Soviet Union shattered the U.S. assumption of technological superiority. Less than one month later, the Soviets reinforced and confirmed their lead in the emerging "space race" with the launch of *Sputnik 2*—a much more massive spacecraft that carried a dog named Laika into orbit. Stunned by these early Soviet space achievements, the United States rushed the launch of the Vanguard rocket and its tiny mini-satellite (of the same name) on December 6, 1957. The widely publicized attempt ended in complete disaster. While the world watched, the Van-guard blew up after rising only a few inches (cm) from the launch pad at

Cape Canaveral. Its payload, a miniature spherical satellite, would end up hopelessly "beeping" at the edge of a raging palmetto-scrub inferno. Soviet premier Nikita Khrushchev sarcastically referred to the tiny three-pound (1.5-kg) test satellite as the "American grapefruit satellite." His taunting remarks, delivered on the world stage, heralded the start of a bitterly contested, decadelong space race between the two cold war–era superpowers.

Responding to the Vanguard disaster, American president Dwight D. Eisenhower assembled his advisers, who hastily mounted an emergency mission to save national prestige. The United States quickly formed a joint project involving Caltech's Jet Propulsion Laboratory (JPL) and the U.S. Army Ballistic Missile Agency (ABMA), with von Braun as the head rocket engineer. Von Braun's team supplied the Jupiter C launch vehicle (a modified intermediate-range ballistic missile [IRBM]), and JPL supplied the fourth-stage rocket, integrated with the *Explorer 1* satellite. Professor James A. Van Allen, Jr. (1914–2006), provided the satellite's scientific instrument package, which detected Earth's trapped radiation belts. When faced with this serious political crisis, American leaders quickly forgot the previously imposed distinction between military and civilian rockets. In the late evening on January 31, 1958 (local time), America's first satellite, *Explorer 1,* successfully achieved orbit after lifting off from Cape Canaveral, Florida. Having learned its lesson, the United States reverted to the use of modified military rockets for the early 1960s—including the use of Redstone, Atlas, and Titan ballistic missiles as launch vehicles in the nation's early human spaceflight projects.

The Soviet launch of *Sputnik 1* precipitated a race for technological supremacy in space that gave the early space exploration efforts of the 1960s a contest mentality. Throughout this period of the cold war, accomplishments in space technology and exploration served as globally recognized manifestations of national power. Superiority in space became emblematic of general technological superiority and, by simple extrapolation, the superiority of a nation's economic and political systems.

Determined to win the so-called space race, the United States began strengthening its civilian space program. During this period, the military space programs of both the United States and the Soviet Union remained cloaked in secrecy and, by official intention, drew essentially no public attention. The only space program that either superpower presented on the global stage was the civilian space program, involving both robotic spacecraft and human-crewed spacecraft.

On October 1, 1958, an act of the U.S. Congress and complementary action within the executive branch of the federal government transformed the National Advisory Committee for Aeronautics (NACA), which had been testing flights on the edge of space, into the National Aeronautics and Space Administration (NASA) and gave the new agency control over the nation's (civilian) space program. NASA was assigned the primary mission of the peaceful exploration of space for the benefit of all humankind. Within seven days of its birth, NASA officials announced the start of the Mercury Project, America's pioneering program to put human beings into orbit around Earth. The critical linkage of NASA's overall program with human spaceflight was forged. Two years later, von Braun and his team of rocket scientists at the Army Ballistic Missile Agency in Huntsville transferred to NASA and became the nucleus of the agency's space program. However, fearing budget imbalances, the Eisenhower administration was reluctant to commit the nation to a massive civilian space effort.

During the early years of the space race, the United States lagged behind the former Soviet Union. Sergei Korolev's large rockets helped the Soviet Union achieve many dramatic space technology "firsts." For example, the Soviet *Luna 1* spacecraft, launched on January 2, 1959, missed the Moon but became the first human-made object to escape the attractive force of Earth's gravity and orbit the Sun. The *Luna 2* spacecraft successfully impacted the Moon on September 14, 1959, and became the first space probe to crash-land on another world. Finally, the following month, *Luna 3* circumnavigated the Moon and took the first images of the lunar farside. In contrast, American attempts to send spacecraft to the Moon between 1958 and 1959 were unsuccessful, largely due to limitations of its launch vehicles.

On April 12, 1961, cosmonaut Yuri Gagarin lifted off in the *Vostok 1* spacecraft from the Baikonur Cosmodrome and successfully completed one orbit of Earth—making him the first human being to travel in space. This pioneering flight made Gagarin an international symbol for the Soviet space program and stimulated the superpower "Space Race" that dominated American and Soviet human spaceflight efforts in the 1960s. *(NASA)*

On April 12, 1961, the former Soviet Union achieved a dramatic space technology milestone by successfully launching the first human into space. Cosmonaut Yuri Gagarin (1934–1968) rode inside the *Vostok 1* spacecraft on top of one of Korolev's military rocket spacecraft and became the first person to observe Earth from an orbiting spacecraft. The United States responded by sending astronaut Alan B. Shepard, Jr., into space using a modified Redstone military rocket as the launch vehicle. As planned by NASA, Shepard achieved only a suborbital flight of approximately 15 minutes duration, because the Redstone rocket was simply not powerful enough to place the Mercury Project spacecraft into orbit. It was not until February 20, 1962, that American astronaut John H. Glenn, Jr., became the first American to orbit Earth. NASA used a modified version of the U.S. Air Force's more powerful Atlas ICBM to place Glenn in orbit around Earth. (See chapter 4 for additional details about the Mercury Project.)

FIRST AMERICAN TO TRAVEL IN SPACE

Selected as one of the original seven Mercury Project astronauts, Alan B. Shepard, Jr. (1923–98), became the first American to travel in outer space. The U.S. Navy officer and NASA astronaut accomplished this important space technology milestone on May 5, 1961, when he rode inside the *Freedom 7* space capsule as it was lifted off from Cape Canaveral Air Force Station, Florida, by a Redstone rocket. The suborbital *Mercury Redstone 3* mission hurled Shepard on a ballistic trajectory downrange from Cape Canaveral. After about 15 minutes, his tiny space capsule slashed down in the Atlantic Ocean some 280 miles (450 km) from the launch site. U.S. Navy recovery personnel plucked him and the *Freedom 7* space capsule from the ocean. Later in his astronaut career, Shepard made a second, much longer journey into space. In February 1971, he served as the commander of NASA's *Apollo 14* lunar landing mission. Together with astronaut Edgar Dean Mitchell, Shepard explored the Moon's Fra Mauro region.

This is a close-up picture of Mercury Project astronaut Alan Shepard, Jr., in his space suit seated inside the *Freedom 7* space capsule. On May 5, 1961, Shepard made a brief (15-minute) suborbital rocket flight from Cape Canaveral and became the first American astronaut to travel in outer space. *(NASA)*

✧ Sergei Korolev—The Man Who Started the Space Age

The Russian (Ukraine-born) rocket engineer Sergei Korolev (1907–66) was the driving technical force behind the initial intercontinental ballistic missile (ICBM) program and the early outer space exploration projects of the former Soviet Union. In 1954, he started work on the first Soviet ICBM, called the R-7. This powerful rocket system was capable of carrying a massive payload across continental distances. As part of cold-war politics, Soviet premier Nikita Khrushchev allowed Korolev to use this military rocket to place the first artificial satellite (named *Sputnik 1*) into orbit around Earth on October 4, 1957. This event is now generally regarded as the beginning of the Space Age. Korolev was also the technical expert responsible for the April 12, 1961, mission that placed the first human (Yuri Gagarin) in orbit around Earth in the *Vostok 1* spacecraft.

Korolev was trained in aeronautical engineering at the Kiev Polytechnic Institute and, after receiving a secondary education, cofounded the Moscow rocketry organization GIRD (Gruppa Isutcheniya Reaktivnovo Dvisheniya, Group for Investigation of Reactive Motion). Like the VfR (*Verein für Raumschiffahrt* [Society for Spaceship Travel]) in Germany and Robert H. Goddard in the United States, the Russian organizations were by the early 1930s testing liquid-fueled rockets of increasing size. In Russia, GIRD lasted only two years before the military, seeing the potential of rockets, replaced it with the RNII (Reaction Propulsion Scientific Research Institute). RNII developed a series of rocket-propelled missiles and gliders during the 1930s, culminating in Korolev's RP-318, Russia's first rocket-propelled aircraft. Before the aircraft could make a rocket-propelled flight, however, Korolev and other aerospace engineers were thrown into the Soviet prison system in 1937–38, during the peak of Joseph Stalin's political purges.

Korolev at first spent months in transit on the Trans-Siberian railway and on a prison vessel at Magadan. This was followed by a year in the Kolyma gold mines, the most dreaded part of the Gulag. However, Stalin recognized the importance of aeronautical engineers in preparing for the impending war with Hitler and retrieved Korolev and other technical personnel from incarceration. He reasoned that these prisoners could help the Red Army by developing new weapons. Consequently, a system of sharashkas (prison design bureaus) was set up to exploit the jailed talent. Korolev was saved by the intervention of senior aircraft designer Sergei Tupolev, himself a prisoner, who personally requested Korolev's services in the TsKB-39 sharashka.

Following World War II, Korolev was released from prison and appointed chief constructor for development of a long-range ballistic missile. By April 1, 1953, as Korolev was preparing for the first launch of the R-11 rocket, he received approval from the Council of Ministers for development of the world's first ICBM, the R-7. To concentrate on development of the R-7, Korolev's other projects were spun off to a new design bureau in Dnipropetrovs'k headed by Korolev's assistant, Mikhail Kuzmich Yangel. This was the first of several design bureaus that would spin off from Korolev's work. It was Korolev's R-7 ICBM that launched *Sputnik 1* on October 4, 1957. This historic launch also served to galvanize American concern about the capability of the Soviet Union to attack the United States with nuclear weapons using ballistic missiles. During the early 1960s, Korolev campaigned to send a Soviet cosmonaut to the Moon.

Following the initial reconnaissance of the Moon by the *Luna 1, 2, and 3* spacecraft, Korolev established three largely independent efforts aimed at achieving a Soviet lunar landing before the Americans. The first objective, met by *Vostok* and *Voskhod* spacecraft, was to prove that human spaceflight was possible. The second objective was to develop lunar vehicles, which would soft-land on the Moon's surface to ensure that a cosmonaut would not sink into the dust accumulated by 4 billion years of meteorite impacts. The third objective, and the most difficult to achieve, was to develop a huge booster to send cosmonauts to the Moon. Beginning in 1962, his design bureau began work on the N-1 launch vehicle, a counterpart to the American Saturn V. This giant rocket was to be capable of launching a maximum of 110,000 pounds (50,000 kg) into low-Earth orbit. Although the project continued until 1971 before cancellation, the N-1 never made a successful flight.

On January 14, 1966, Korolev died during a botched routine surgery at a hospital in Moscow. He was only 58 years old. Some of Korolev's contributions to space technology include the powerful, legendary R-7 rocket (1956); the first artificial satellite (1957); pioneering lunar spacecraft missions (1959); the first human spaceflight (1961); a spacecraft to Mars (1962); and the first space walk (1965). Even after his death, the Soviet government chose to hide Korolev's identity by publicly referring to him only as the "Chief Designer of Carrier Rockets and Spacecraft." Despite this official anonymity, Korolev is now properly recognized as the brilliant rocket engineer who ushered in the Space Age.

✧ John F. Kennedy and the Race to the Moon

In the midst of numerous Soviet technical triumphs in space during the cold war, President John Fitzgerald Kennedy (1917–63) boldly proposed to a joint session of the U.S. Congress on May 25, 1961, that NASA send astronauts to the Moon to demonstrate American space technology superiority over the Soviet Union. Shot by an assassin on November 22, 1963, Kennedy did not live to see the triumphant Apollo Project lunar landings (1969–72)—a magnificent technical accomplishment that his vision and leadership set in motion almost a decade earlier.

Kennedy was born on May 29, 1917, in Brookline, Massachusetts. He graduated from Harvard University in 1940 and then served in the United States Navy as a commissioned officer during World War II. Following the war, Kennedy became the Democratic representative from the 11th Massachusetts Congressional District and served his district in the House of Representatives from 1946 to 1952. He ran for the U.S. Senate in 1952 and defeated the Republican incumbent, Henry Cabot Lodge, Jr. In the 1960 presidential election, Kennedy narrowly defeated his Republican opponent, Richard M. Nixon, and became the 35th president of the United States.

During his brief term in office (January 20, 1961 to November 22, 1963), President Kennedy had to continuously deal with conflicts involving the former Soviet Union, led by an aggressive premier, Nikita Khrushchev. Kennedy's challenges included clashes over Cuba and Berlin, as well as a

THE VOSTOK, VOSKHOD, AND SOYUZ SPACECRAFT

The Vostok (meaning "East") spacecraft was the first Russian- (Soviet-) manned spacecraft. This spacecraft was occupied by a single cosmonaut and consisted of a spherical cabin (about 7.5 feet [2.3 m] in diameter) that was attached to a biconical instrument module. *Vostok 1* was launched on April 12, 1961, carrying cosmonaut Yuri Gagarin, the first human to fly in space. Gagarin's flight made one orbit of Earth and lasted about 108 minutes. Since the *Vostok 1* spacecraft did not have a retrorocket system to support a soft landing, as the capsule performed its parachute-assisted descent following reentry, Gagarin ejected from the spacecraft when it was at an altitude of about 23,000 feet (7,000 m) above Earth and completed the remainder of his historic journey by parachute.

Years later, when Russian officials admitted that Gagarin had ejected from the *Vostok 1* spacecraft during descent and did not land in the same craft in which he started his journey, some FAI officials raised a technicality that questioned the official status of his "first-in-spaceflight" record. Founded in 1905, the Fédération Aéronautique Internationale (FAI) is the world organization responsible for setting standards and keeping records within the fields of aeronautics and astronautics. In 1961, the FAI rules required that a pilot (that is, an astronaut or cosmonaut) must land with the spacecraft to be considered as having achieved an official spaceflight worthy of entry into the FAI book of records. At the time of Gagarin's mission, Soviet officials insisted

that the cosmonaut had landed with the *Vostok 1* spacecraft and so the FAI certified the flight. Despite the subsequent FAI record challenge based on a technicality within the rules, Gagarin's mission and marvelous accomplishment is still almost universally recognized as the first human spaceflight.

In August 1961, Russian cosmonaut Gherman S. Titov (1935–2000) became the second person to travel in orbit around Earth. His *Vostok 2* spacecraft made 17 orbits of Earth, during which he became the first of many space travelers to experience space sickness. Cosmonaut Valentina Tereshkova (b. 1937) holds the honor of being the first woman to travel in outer space. She accomplished this feat on June 16, 1963, by riding the *Vostok 6* spacecraft into orbit. During her historic mission, she completed 48 orbits of Earth. Upon her return, Tereshkova received the Order of Lenin and was made a hero of the Soviet Union by Premier Khrushchev.

The Voskhod (meaning "sunrise") spacecraft was an early Russian three-person spacecraft that evolved from the Vostok spacecraft. *Voskhod 1* was launched on October 12, 1964, and carried the first three-person crew into space. The cosmonauts, Vladimir Komarov, Konstantin Feoktistov, and Boris Yegorov, flew on a one-day Earth orbital mission. *Voskhod 2* was launched on March 18, 1965, and carried a crew of two cosmonauts, including Alexei Leonov, who performed the world's first space walk (about 10 minutes in duration) during the orbital mission.

growing world community perception that the United States had lost its technical superiority to the Soviet Union. During Kennedy's presidency, the Soviet premier constantly flaunted his nation's space technology accomplishments as an illustration of the superiority of Soviet communism over Western capitalism. President Kennedy worked hard to maintain a balance between American and Soviet spheres of influence in global politics. While not a space technology enthusiast per se, Kennedy recognized that civilian

The Soyuz (meaning "union") spacecraft is an evolutionary family of crewed spacecraft that have been used by the Soviet Union and later the Russian Federation on a wide variety of space missions. The first Soyuz spacecraft, called *Soyuz 1*, was launched in April 1967. Unfortunately, upon reentry, a parachute failed to open properly, and the spacecraft was destroyed on impact; its occupant, cosmonaut Vladimir M. Komarov (1927–67), was killed. Cosmonaut Komarov was an air force officer and the first person to make two trips into space. As a result of his fatal landing accident on April 24, 1967, he also became the first person to die while engaged in space travel. During the final stage reentry over the Kazakh Republic, the recovery parachute became entangled, causing his *Soyuz 1* spacecraft to impact the ground at high speed. He died instantly and was given a hero's state funeral.

The second Russian space tragedy occurred at the end of the *Soyuz 11* mission (June 1971), when a valve malfunctioned as the spacecraft was separating from the *Salyut 1* space station, allowing all the air to escape from the crew compartment. This particular early version of the Soyuz spacecraft did not have sufficient room for the crew to wear their pressure suits during reentry; consequently, the three cosmonauts, Georgi Dobrovolsky, Victor Patseyev, and Vladislav Volkov, suffocated during the reentry operation. They were found dead by the Russian recovery team after touchdown.

In July 1975, the *Soyuz 19* spacecraft was used successfully by cosmonauts Alexei Leonov and Valeri Kubasov in the Apollo–Soyuz Test Project—an international rendezvous and docking mission. The next major variant of this versatile spacecraft, called the Soyuz-T (with the "T" standing for transport), was first flown in December 1979. The Soyuz-TM is a modernized version of the Soyuz-T. It was flown in May 1986 and has been used to ferry crew and supplies to an orbiting station, such as the *Mir* and later the *International Space Station* (*ISS*).

The *Soyuz TMA-1* is a Russian automatic passenger spacecraft designed for launch by a Soyuz launch vehicle from the Baikonur Cosmodrome. Following launch, the spacecraft proceeds in an automated fashion to rendezvous and dock with the *ISS*. The *Soyuz TMA-1* is a larger craft with a more comfortable interior than the previous Soyuz TM models. After docking, the spacecraft remains parked at the *ISS*, serving as an emergency escape spacecraft until it is relieved by the arrival of another Soyuz spacecraft.

For example, in late October 2002, a *Soyuz TMA-1* was launched from the Baikonur Cosmodrome and successfully carried three cosmonauts (two Russian and one Belgian) to the *ISS*. The *Soyuz TMA-1* automatically docked with the *ISS*. After 10 days of microgravity research, the three visiting cosmonauts departed from the *ISS* using the previously parked *Soyuz TM-34* spacecraft. The *Soyuz TMA-1* spacecraft that carried them into space remained behind as a lifeboat for the permanent crew of the *ISS*.

space technology achievements were giving the former Soviet Union greater influence in global politics. Driven by political circumstances early in his presidency, Kennedy took steps to respond to this challenge.

In the spring of 1961, Kennedy needed something special to restore America's global image as leader of the Free World. Space technology was the new, highly visible arena for Soviet-American superpower competition. On April 12, 1961, the Soviet Union launched the first human into orbit around Earth (cosmonaut Yuri Gagarin). The relatively mild American response was the suborbital mission of Mercury Project astronaut Alan B. Shepard, Jr., on May 5, 1961.

During mid-May 1961, Kennedy consulted with his advisers and reviewed many space achievement options with his Vice President, Lyndon B. Johnson (1908–73), who headed the National Aeronautics and Space Council. After much thought, Kennedy selected the Moon-landing project. Kennedy did so not to promote space science or to satisfy a personal, long-term space exploration vision but because this mission was a truly daring project that would symbolize American strength and technical superiority

President John F. Kennedy during his historic May 25, 1961, message to a joint session of the U.S. Congress in which he declared: "I believe this nation should commit itself to achieving the goal, before the decade is out, of landing a man on the Moon and returning him safely to Earth." Shown in the background are (left) Vice President Lyndon B. Johnson and (right) Speaker of the House Sam T. Rayburn. *(NASA)*

This inspirational view of the "rising" Earth greeted the *Apollo 8* astronauts (Frank Borman; James A. Lovell, Jr.; and William Anders) as they came from behind the Moon after performing the lunar-orbit insertion burn (December 1968). *(NASA)*

in head-to-head cold-war competition with the Soviet Union. During his special message to the U.S. Congress, Kennedy announced the Moon landing mission with these immortal words: "I believe that this nation should commit itself to achieving the goal, before this decade is out, of landing a man on the Moon and returning him safely to Earth. No single space project in this period will be more impressive to mankind, or more important for the long-range exploration of space; and none will be so difficult or expensive to accomplish. . . ."

This speech gave NASA the mandate to expand and accelerate its Mercury Project activities and configure itself to accomplish the "impossible" through the Apollo Project. When Kennedy made his decision, the United States had not yet successfully placed a human being in orbit around Earth. Kennedy's mandate galvanized the American space program and marshaled incredible levels of technical and fiscal resources. Science historians often compare NASA's Apollo Project to the Manhattan Project (World War II atomic bomb program) or the construction of the Panama Canal in extent, complexity, and national expense.

As a result of Kennedy's bold vision, three American astronauts (Frank Borman, James A. Lovell, Jr., and William Anders) became the first human beings to escape from the grasp of Earth's gravity and travel around the Moon. Their historic lunar circumnavigation journey in December 1968 bore a remarkable similarity to the around-the-Moon journey prophesied by Jules Verne about a century earlier. Sent on their amazing journey by one of von Braun's powerful Saturn V rockets, the *Apollo 8* astronauts gazed back at Earth in a way never before experienced by human beings. The *Apollo 8* mission gave the United States a much-needed, widely publicized victory in the superpower space race. But this triumphant mission to another world was just the first lap in the overall race to send humans to the Moon. The checkered flag of total victory was just a few months away.

An entire world watched as the three astronauts of NASA's *Apollo 11* mission left for the Moon on July 16, 1969. On July 20, 1969, two *Apollo 11* astronauts, Neil A. Armstrong and Edwin E. "Buzz" Aldrin, stepped on the

The crew of NASA's Apollo 11 lunar-landing mission. From left to right, they are Neil A. Armstrong (commander), Michael Collins (command module pilot), and Edwin E. (Buzz) Aldrin, Jr. (lunar module pilot). On July 20, 1969, Armstrong became the first human being to walk on the Moon. Minutes later, Aldrin joined him on the lunar surface, while Collins orbited overhead in the Apollo command and service module, named *Columbia. (NASA)*

lunar surface and successfully fulfilled Kennedy's bold initiative. (Chapter 6 discusses the Apollo Project in detail.)

Sadly, the young President who launched the most daring space exploration project of the cold war did not personally witness its triumphant conclusion. An assassin's bullet had taken his life in Dallas, Texas, on November 22, 1963. NASA's Kennedy Space Center—site of Launch Complex 39 from which humans left Earth to explore the Moon—bears his name.

Often forgotten in the glare of the Moon-landing announcement are several other important space technology initiatives that President Kennedy called for in his historic speech on May 25, 1961. Kennedy accelerated development of the Rover nuclear rocket program as a means of preparing for more ambitious space exploration missions beyond the Moon. (Due to a dramatic change in space program priorities, the Nixon administration canceled this program in 1972.)

Kennedy also requested additional funding to accelerate the use of communication satellites to expand worldwide communications and the use of satellites for worldwide weather observation. Both of these initiatives quickly evolved into major areas of space technology that now serve the global community.

President John F. Kennedy, in responding to the Soviet space technology challenge, satisfied a major dream of space pioneers down through the ages. Through his bold and decisive leadership and with the steadfast support of his successor, President Lyndon Baines Johnson, human beings traveled through interplanetary space and walked on another world for the first time in history.

The United States won the race to the Moon and demonstrated a clear superiority in space technology. National pride soared. But in the mercurial world of government budgets and nonvisionary politics, great technical success does not necessarily support new opportunities. While millions cheered the triumphant lunar astronauts, the Nixon administration was already making major cutbacks in the space program. The Apollo Project represents the farthest distance human beings have thus far ventured into the cosmos. As discussed in the later chapters of this book, one of the problems that has haunted NASA officials since the triumph of Apollo is how best to successfully respond to the difficult question: "Where does the American space program go after it has sent men to the Moon?"

The U.S. government issued this commemorative airmail stamp on September 9, 1969, as part of the worldwide celebration that followed the successful landing of the *Apollo 11* astronauts on the Moon on July 20. Hundreds of millions of people around the world witnessed this major milestone in technology, and most of these people generally regarded the event as a brilliant triumph for all humankind. *(Author)*

Living in Space

Because the inertial trajectory of a spacecraft compensates for the force of Earth's gravity, an orbiting spacecraft and all its contents approach a state of free fall. In this state of free fall, all objects inside the spacecraft appear "weightless."

Sir Isaac Newton's law of gravitation states that any two objects have a gravitational attraction for each other that is proportional to their masses and inversely proportional to the square of the distance between their centers of mass. It is also interesting to recognize that a spacecraft orbiting Earth at an altitude of 249 miles (400 km) is only 6 percent farther away from the center of Earth than it would be if it were on Earth's surface. Using Newton's law, physicists find that the gravitational attraction at this particular altitude is only 12 percent less than the attraction of gravity at the surface of Earth. In other words, an Earth-orbiting spacecraft and all its contents are very much under the influence of Earth's gravity. The phenomenon of weightlessness occurs because the orbiting spacecraft and its contents are in a continual state of free fall.

Albert Einstein's principle of equivalence states that the physical behavior inside a system in free fall is identical to that inside a system far removed from other matter that could exert a gravitational influence. Therefore, the term *zero gravity* (also called zero g) is often used to describe a free-falling system (and its contents) in orbit around Earth (or other primary celestial body).

So what is the difference between mass and weight? Why do people say, for example, "weightlessness" and not "masslessness"? According to the laws of classical physics, *mass* is the physical substance of an object—it has the same value everywhere in the universe. *Weight*, on the other hand, is the product of an object's mass and the local acceleration of gravity—in accordance with Newton's second law of motion, namely $F = ma$. For example, an Apollo Project astronaut would weigh about

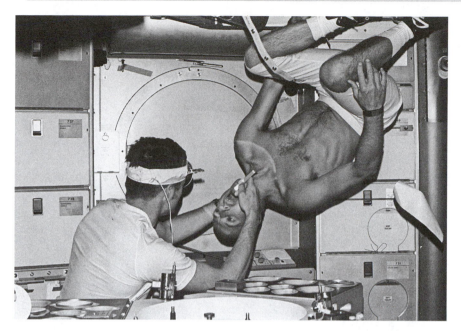

Making an extraterrestrial house call, *Skylab 2* astronaut and medical officer Joseph P. Kerwin, M.D., gives astronaut Charles P. (Pete) Conrad a dental examination (June 1973). In the absence of an examination chair, Conrad took advantage of microgravity conditions inside the *Skylab* space station and rotated his body to an upside-down position to facilitate the procedure. As shown, providing routine, as well as emergency, medical care in an orbiting spacecraft not only presents special challenges but also offers unusual opportunities. *(NASA)*

one-sixth as much on the Moon as on Earth, but his mass remains the same in both places.

A zero-gravity environment is really an ideal situation that can never be totally achieved in an orbiting spacecraft. The venting of gases from the space vehicle, the minute drag exerted by a very thin, residual, terrestrial atmosphere at low-altitude orbits, and even crew motions create nearly imperceptible forces on people and objects alike. These tiny forces are collectively called "microgravity." In a microgravity environment, astronauts and their equipment are almost, but not entirely, weightless.

✧ Living in a Continual State of Microgravity

Microgravity represents an intriguing experience for space travelers. However, life in microgravity is not necessarily easier than life on Earth. For example, the caloric (food-intake) requirements for people living in microgravity are the same as those on Earth. Living in microgravity also calls for special design technology. A beverage in an open container, for instance will cling to the inner or outer walls and, if shaken, will leave the container as free-floating droplets or fluid globs. Such free-floating droplets are not merely an inconvenience. The droplets can annoy crewmembers, and they represent a definite hazard to certain onboard equipment, especially sensitive electronic devices and computers.

Therefore, water usually is served in microgravity through a specially designed dispenser unit that can be turned on or off by squeezing and releasing a trigger. Other beverages, such as orange juice, typically are served in sealed containers through which a plastic straw can be inserted. When the beverage is not being sipped, the straw is simply clamped shut.

Microgravity living also calls for special considerations in handling solid foods. Crumbly foods are provided only in bite-size pieces to avoid having crumbs floating around the space cabin. Gravies, sauces, and

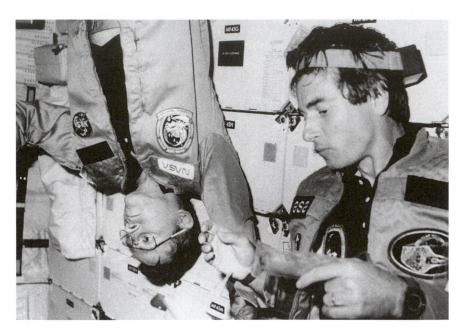

In this picture, American astronaut John W. Young and German physicist Ulf Merbold are having a meal in the mid-deck of the space shuttle *Columbia* during the STS-9 mission (December 1983). Young served as the commander of this shuttle mission, and Merbold was a payload specialist, representing the European Space Agency on the first orbital flight of the European-built Spacelab, which was carried in the shuttle's cargo bay. Since there is really no up or down under microgravity conditions, astronauts can enjoy a meal while anchored in a place and at an orientation that is comfortable. Merbold's "headband" was part of a test to monitor the astronaut during his waking hours throughout the 10-day orbital mission. As a result of this inaugural shuttle/Spacelab flight, Merbold became the first non-U.S. citizen to fly on an American spacecraft. *(NASA)*

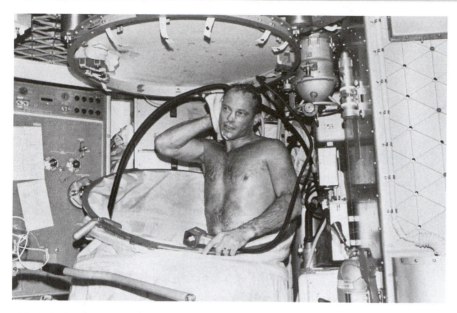

This is a close-up view of *Skylab 3* astronaut Jack R. Lousma drying off after taking a hot shower in the crew quarters of the *Skylab* space station (July 1973). Bathing and controlling droplets of water is tricky under microgravity conditions. To use this particular shower facility, the *Skylab* astronauts would first pull the curtain from the floor to the ceiling. Water for bathing came through a push-button showerhead (shown in Lousma's left hand), which was attached to a flexible hose. Used bathwater within this enclosure was then drawn off by a vacuum system. *(NASA)*

dressings have a viscosity (thickness) that generally prevents them from simply lifting off food trays and floating away. Typical space food trays are equipped with magnets, clamps, and double-adhesive tape to hold metal, plastic, and other utensils. Astronauts are provided with forks and spoons. However, they must learn to eat without sudden starts and stops if they expect the solid food to stay on their eating utensils.

Personal hygiene is also a bit challenging in microgravity. Because of volume limitations inside the crew cabin, space shuttle astronauts have to take sponge baths rather than showers or regular baths. For longer duration flights, such as on the former American *Skylab* space station, special microgravity bathing facilities were made available for *Skylab* astronauts. Current *International Space Station* (*ISS*) crewmembers must use specially designed bathing/shower facilities to maintain personal hygiene. (Because water adheres to the skin in microgravity, perspiration can be annoying, especially during strenuous activities.) Waste elimination in microgravity represents another challenging design problem. Special toilet facilities help keep an astronaut in place (that is, prevent drifting). The waste products themselves are flushed away by a flow of air and a mechanical "chopper-type" device.

Sleeping in microgravity is another interesting experience. For example, shuttle and space station astronauts can sleep either horizontally or vertically while in orbit. Their fireproof sleeping bags attach to rigid padded boards for support. But the astronauts themselves quite literally sleep "floating in air." On the space shuttle, astronauts can sleep in the commander's seat, the pilot's seat, or in bunk beds. There are only four bunk beds in the crew cabin. Consequently, on shuttle missions with five or more astronauts, the other crewmembers have to sleep in a sleeping bag attached to their seat or to a convenient cabin wall.

On the *International Space Station,* there are two small crew cabins. Each cabin is only big enough for one person and contains a sleeping bag and a large window to look out into space. What happens when the space station has another crewmember for months at a time? Where does the extra astronaut or cosmonaut sleep? If the ISS expedition commander approves, the third crewmember can sleep pretty much anywhere inside the station he or she likes, so long as they attach their sleeping bag and themselves to something. Otherwise, instead of sleepwalking they might get caught sleep-floating around the station. For example, on ISS Expedition Two American astronaut Susan Helms slept in the huge Destiny laboratory module, while her crewmates (cosmonaut Yury Usachev and astronaut James Voss) slept on the opposite side of the 171 feet (52 m) long (at the time) Zvezda service module.

Generally, astronauts are scheduled for eight hours of sleep at the end of each mission day. Just like people on Earth, however, they may wake up in the middle of the night (that is, during their assigned sleep period) to use the toilet, or they may simply be restless and decide to stay up late so they can look out the window. During their sleep period, astronauts have reported having dreams and nightmares. A few have even reported snoring in space.

The excitement of space travel or a case of space sickness (space adaptation syndrome) can easily disrupt an astronaut's sleep pattern. In addition, sleeping in close quarters can be quite disruptive, since crewmembers can easily hear each other, the whirring of onboard machinery, and the occasional thud of a tiny micrometeorite slamming into the outer structure of

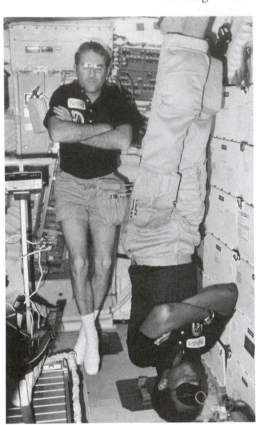

STS-8 mission commander Richard "Dick" Truly and mission specialist Guion Bluford demonstrate how astronauts can sleep in space. In this image, they are resting while floating peacefully in front of the forward lockers and port side wall of the space shuttle *Challenger's* mid-deck (September 1983). Truly sleeps with his head at the ceiling and his feet to the floor. Bluford, wearing a sleep mask (blindfold), is oriented with the top of his head at the floor and his feet on the ceiling. The microgravity conditions found inside an Earth-orbiting spacecraft allow such unusual sleeping arrangements. *(NASA)*

Astronaut Susan J. Helms (left) pauses from her work, while cosmonaut Yury V. Usachev (right) speaks into a microphone on board the U.S. laboratory module (Destiny)—a component of the *International Space Station.* This image was recorded with a digital still camera during Expedition 2 on April 5, 2001. *(NASA/JSC)*

the station. Sleeping on the shuttle's upper deck (cockpit area of the crew cabin) can also be quite difficult since the Sun "rises" every 90 minutes or so during a mission. The sunlight and warmth entering the cockpit window might easily disturb a sleeper who is not wearing a sleep mask.

When it is time for the crew to wake up, NASA's mission control center (MCC) in Houston, Texas, sends wake-up music to the shuttle crew. Typically, personnel at mission control will select a song for a different crewmember each day. Sometimes a family member makes a request for an astronaut's favorite song, and this provides a special surprise for the particular loved one in orbit. The selection of wake-up music depends on the crew. Selections have included classical, rock and roll, country and western, light contemporary music, and even Russian music when a cosmonaut is riding aboard the shuttle. However, only the shuttle crew receives wake-up music each day in orbit; the space station crew uses an alarm clock.

Working in microgravity requires the use of special tools (e.g., torqueless wrenches), handholds, and foot restraints. These devices are needed to balance or neutralize reaction forces. If these devices were not available, an astronaut might find him-/herself helplessly rotating around a "work piece." When astronauts work inside the space shuttle cabin or in

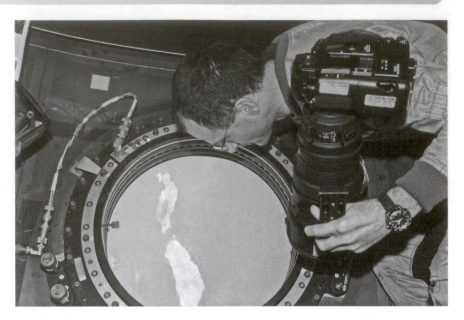

NASA astronaut Donald R. Pettit, *ISS* Expedition Six science officer, holds a still camera as he looks through the nadir window in the Destiny laboratory module on the *International Space Station* (January 2003). The islands of Lanzarote and Fuerteventura of the Canary Island chain in the Atlantic Ocean off the coast of Morocco are visible. Looking back at Earth from orbit is a popular activity of astronauts and cosmonauts, providing an opportunity to perform Earth observation science as well as giving each space traveler "quiet time" for relaxation and reflection. *(NASA)*

the pressurized, habitable modules of the space station, they usually wear comfortable, everyday type clothing, and, in general, they are free to select the colors. But when astronauts and cosmonauts work outside the space-craft or space station, they are performing an extravehicular activity and must wear a space suit.

Exposure to microgravity also causes a variety of physiological (bodily) changes. For example, space travelers appear to have smaller eyes because their faces have become puffy. They also get rosy cheeks and distended veins in their foreheads and necks. They may even be a little bit taller than they are on Earth, because their body masses no longer "weigh down" their spines. Leg muscles shrink, and anthropometric (measurable postural) changes also occur. Astronauts tend to move with a slight crouch, with head and arms forward.

Some space travelers suffer from a temporary condition resembling motion sickness. This condition is called space sickness or space adaptation syndrome. In addition, sinuses become congested, leading to a condition similar to a cold.

EXTRAVEHICULAR ACTIVITY

Extravehicular activity, or EVA, is defined as the activities conducted in space by an astronaut or cosmonaut outside the protective environment of his/her spacecraft, aerospace vehicle, or space station. In the U.S. space program, Astronaut Edward H. White II performed the first EVA on June 3, 1965, when he left the protective environment of his *Gemini 4* space capsule and ventured into space (while constrained by an umbilical tether). Since that historic demonstration, EVA has been used successfully during a variety of American and Russian space missions to make critical repairs, perform inspections, help capture and refurbish failed satellites, clean optical surfaces, deploy equipment, and retrieve experiments. The term *EVA* (as applied to the space shuttle and

International Space Station) includes all activities for which crewmembers don their space suits and life support systems and then exit the orbiter vehicle's crew cabin to perform operations internal or external to the cargo bay.

American astronaut Jerry L. Ross, STS-88 mission specialist, is pictured during one of three space walks that were conducted on the 12-day space shuttle *Endeavour* mission in December 1998, a mission that supported the initial orbital assembly of the *International Space Station* (ISS). Astronaut James H. Newman, mission specialist, recorded this image while perched on the end of *Endeavour*'s remote manipulator system arm. Newman can be seen reflected in Ross's helmet visor. The solar array panel for the Russian-built Zarya module for the *ISS* appears along the right edge of the picture. *(NASA)*

Many of these microgravity-induced physiological effects appear to be caused by fluid shifts from the lower to the upper portions of the body. So much fluid goes to the head that the brain may be fooled into thinking that the body has too much water. This can result in an increased production of urine.

Extended stays in microgravity tend to shrink the heart, decrease production of red blood cells, and increase production of white blood cells. A

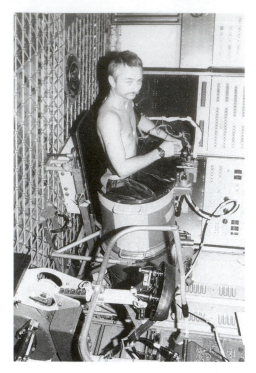

Skylab 3 scientist-astronaut Owen K. Garriott testing out the lower body negative pressure device (LBNPD) in the work and experiments area of the American Skylab space station (August 1973). The LBNPD provided data on the rate of cardiovascular adaptation of an astronaut to microgravity during the course of a long-duration spaceflight. The device also provided in-flight data useful for projecting the degree of orthostatic intolerance and impairment of physical capacity that might be expected when an astronaut returned to normal gravity on Earth. The bicycle ergometer, used by Skylab astronauts as part of their exercise regimen in orbit, appears in the right foreground of the picture. (NASA)

process called resorption occurs. This is the leaching of vital minerals and other chemicals (e.g., calcium, phosphorous, potassium, and nitrogen) from the bones and muscles into the body fluids that are then expelled as urine. Such mineral and chemical losses can have adverse physiological and psychological effects. In addition, prolonged exposure to a microgravity environment might cause bone loss and a reduced rate of bone-tissue formation.

While a relatively brief stay (say from seven to 70 days) in microgravity may prove a nondetrimental experience for most space travelers, long-duration (i.e., one to several years) missions, such as a human expedition to Mars, could require the use of artificial gravity (created through the slow rotation of the living modules of the spacecraft) to avoid any serious health effects that might arise from such prolonged exposure to a microgravity environment. While cruising to Mars, this artificial gravity environment also would help condition the astronauts for activities on the Martian surface, where they will once again experience the "tug" of a planet's gravity.

Being cooped up inside a relative small volume of living space with several other human beings for extended periods of time can produce adverse psychological affects on the crewmembers. Some of the factors that will produce such psychological stress include excessively heavy workloads and inflexible schedules, physical separation from family and friends, food menus that provide limited variety, noise and constant vibration throughout the spacecraft, and sleep disruption or deprivation. All of these factors can cause an astronaut to make a simple but potentially lethal mistake in an unforgiving environment and may trigger anxiety, depression, mental breakdowns, or other behavioral problems.

✦ Space Suit

Outer space is a very hostile environment. If astronauts and cosmonauts are to survive there, they must take part of Earth's environment with them. Air to breathe, acceptable ambient pressures, and moderate temperatures have to be contained in a shell surrounding the space traveler. This can be accomplished by providing a very large enclosed structure or habitat or, on an individual basis, by encasing the astronaut in a protective flexible capsule called the space suit.

Space suits used on previous NASA missions from the Mercury Project up through the Apollo-Soyuz Test Project have provided effective protection for American astronauts. However, certain design problems have handicapped the suits. These suits were custom-fitted garments. In some suit models, more than 70 different measurements had to be taken of the astronaut in order to manufacture the space suit to the proper fit. As a result, the space suit could be worn by only one astronaut on only one mission. These early space suits were stiff, and even simple motions such as grasping objects quickly drained an astronaut's strength. Donning the suit was an exhausting process that at times lasted more than an hour and required the help of an assistant.

For example, the Mercury Project space suit was a modified version of a U.S. Navy high-altitude jet aircraft pressure suit. It consisted of an inner layer of Neoprene-coated nylon fabric and an outer layer of aluminized nylon. Joint mobility at the elbows and knees was provided by simple break lines sewn into the suit; but even with these break lines, it was difficult for the wearer to bend arms or legs against the force of a pressurized suit. As an elbow or knee joint was bent, the suit joints folded in on themselves, reducing suit internal volume and increasing pressure. The Mercury space suit was worn "soft," or unpressurized, and served only as backup for possible spacecraft cabin pressure loss—an event that never happened.

NASA space suit designers then followed the U.S. Air Force approach toward greater suit mobility when they developed the space suit for the two-man Gemini Project spacecraft. Instead of fabric-type joints used in the Mercury suit, the Gemini space suit had a combination of a pressure bladder and a link-net restraint layer that made the whole suit flexible when pressurized.

The gas-tight, human-shaped pressure bladder was made of Neoprene-coated nylon and covered by load-bearing link-net woven from Dacron and Teflon cords. The net layer, being slightly smaller than the pressure bladder, reduced the stiffness of the suit when pressurized and served as a type of structural shell. Improved arm and shoulder mobility resulted from the multilayer design of the Gemini suit.

Walking on the Moon's surface presented a new set of problems to space suit designers. Not only did the space suits for the "Moonwalkers" have to offer protection from jagged rocks and the intense heat of the lunar day, but the suits also had to be flexible enough to permit stooping and bending as the Apollo astronauts gathered samples from the Moon and used the lunar rover vehicle for transportation over the surface of the Moon.

The additional hazard of micrometeoroids that constantly pelt the lunar surface from deep space was met with an outer protective layer on the Apollo space suit. A backpack portable life support system provided oxygen for breathing, suit pressurization, and ventilation for moonwalks lasting up to seven hours.

Apollo space suit mobility was improved over earlier suits by use of bellowslike molded rubber joints at the shoulders, elbows, hips, and knees. Modifications to the suit waist for the *Apollo 15* through *17* missions provided flexibility and made it easier for astronauts to sit on the lunar rover vehicle.

From the skin out, the Apollo A7LB space suit began with an astronaut-worn liquid-cooling garment, similar to a pair of longjohns with a network of spaghetti-like tubing sewn onto the fabric. Cool water, circulating through the tubing, transferred metabolic heat from the astronaut's body to the backpack, where it was then radiated away to space. Next came a comfort and donning improvement layer of lightweight nylon, followed by a gas-tight pressure bladder of Neoprene-coated nylon or bellowslike molded joints components, a nylon restraint layer to prevent the bladder from ballooning, a lightweight thermal superinsulation of alternating layers of thin Kapton and glass-fiber cloth, several layers of Mylar and spacer material, and, finally, protective outer layers of Teflon-coated, glass-fiber Beta cloth.

Apollo space helmets were formed from high-strength polycarbonate and were attached to the space suit by a pressure-sealing neck ring. Unlike Mercury Project and Gemini Project helmets, which were closely fitted and moved with the astronaut's head, the Apollo Project helmet was fixed, and the astronaut's head was free to move within it. While walking on the Moon, the Apollo crew wore an outer visor assembly over the polycarbonate helmet to shield against eye-damaging ultraviolet radiation and to maintain head and face thermal comfort.

Lunar gloves and boots completed the Apollo space suit. Both were designed for the rigors of exploring; the gloves also could adjust sensitive instruments. The lunar gloves consisted of integral structural restraint and pressure bladders, molded from casts of the crewperson's hands, and covered by multilayered superinsulation for thermal and abrasion protection. Thumb- and fingertips were molded of silicone rubber to permit a degree of sensitivity and "feel." Pressure-sealing disconnects, similar to the helmet-to-suit connection, attached the gloves to the space suit arms.

The lunar boot was actually an overshoe that the Apollo astronaut slipped on over the integral pressure boot of the space suit. The outer layer of the lunar boot was made from metal-woven fabric, except for the ribbed silicone rubber sole; the tongue area was made from Teflon-coated glass-fiber cloth. The boots' inner layers were made from Teflon-coated glass-fiber cloth followed by 25 alternating layers of Kapton film and glass-fiber cloth to form an efficient, lightweight thermal insulation.

Modified versions of the Apollo space suit were used also during the Skylab Program (1973–74) and the Apollo-Soyuz Test Project (1975).

A new space suit was developed for shuttle-era astronauts that provided many improvements in comfort, convenience, and mobility over previous models. This suit, which is worn outside the orbiter during extravehicular activity (EVA), is modular and features many interchangeable parts. The torso, pants, arms, and gloves come in several different sizes and can be assembled for each mission in the proper combination to suit individual male and female astronauts. The design approach is cost-effective because the suits are reusable and not custom-fitted.

The shuttle space suit is called the extravehicular mobility unit (EMU) and consists of three main parts: liner, pressure vessel, and primary life support system (PLSS). These components are supplemented by a drink bag, communications set, and helmet and visor assembly.

Containment of body wastes is a significant problem in space suit design. In the shuttle-era EMU, the PLSS handles odors, carbon dioxide, and the containment of gases in the suit's atmosphere. The PLSS is a two-part system consisting of a backpack unit and a control and display unit located on the suit chest. A separate unit is required for urine relief. Two different urine-relief systems have been designed to accommodate both male and female astronauts. Because of the short-time durations for extravehicular activities, fecal containment is considered unnecessary.

The manned maneuvering unit (MMU) is a one-person, nitrogen-propelled backpack that latches to the EMU space suit's PLSS. Using rotational and translational hand controllers, the astronaut can fly with precision in or around the orbiter's cargo bay or to nearby free-flying payloads or structures and can reach many otherwise inaccessible areas outside the orbiter vehicle. Astronauts wearing MMUs have deployed, serviced, repaired, and retrieved satellite payloads.

The MMU has been called "the world's smallest reusable spacecraft." The MMU propellant (non-contaminating gaseous nitrogen stored under pressure) can be recharged from the orbiter vehicle. The reliability of the unit is guaranteed with a dual-parallel system rather than a backup redundant system. In the event of a failure in one parallel system, the system would be shut down, and the remaining system would be used to return the MMU to the orbiter's cargo bay. The MMU includes a 35-mm still camera that is operated by the astronaut while working in space.

Shuttle-era space suits are pressurized at 4.3 psi (29.6 kilopascals), while the shuttle cabin pressure is maintained at 14.6 psi (101 kilopascals). Because the gas in the suit is 100 percent oxygen (instead of 20 percent oxygen as is found in Earth's atmosphere), the person in the space suit actually has more oxygen to breathe than is available at an altitude of 9,840 feet (3,000 m) or even at sea level without the space suit. However, prior to leaving the orbiter to perform tasks in space, an astronaut has to spend several hours breathing pure oxygen. This procedure (called "prebreathing") is necessary to remove nitrogen dissolved in body fluids and thereby prevent its release as gas bubbles when pressure is reduced, a condition commonly referred to as "the bends."

In addition to new space-walking tools and philosophies for astronaut-assisted assembly of the *International Space Station* (*ISS*), American astronaut space walkers have an enhanced space suit. The shuttle space suit (the extravehicular mobility unit) was originally designed for sizing and maintenance between flights by skilled specialists on Earth. Such maintenance and refurbishment activities would prove difficult, if not impossible, for astronauts aboard the station.

The shuttle space suit has been improved for use on the *ISS*. It can now be stored in orbit and is certified for up to 25 extravehicular activities before it must be returned to Earth for refurbishment. The space suit can be adjusted in flight to fit different astronauts and can be easily cleaned and refurbished between EVAs on board the station. The modified space suit has easily replaceable internal parts, reusable carbon dioxide removal cartridges, metal sizing rings that accommodate in-flight suit adjustments to fit different crewmembers, new gloves with enhanced dexterity, and a new radio with more channels to allow up to five people to communicate with one another simultaneously.

Due to orbital motion–induced periods of darkness and component-caused shadowing, assembly work on the space station is frequently being performed at much colder temperatures than those encountered during most of the space shuttle mission EVAs. Unlike the shuttle, the *ISS* cannot be turned to provide an optimum amount of sunlight to moderate temperatures during an extravehicular activity, so a variety of other enhancements now make the shuttle space suit more compatible for use aboard the space station. Warmth enhancements include fingertip heaters and the ability to shut off the space suit's cooling system. To assist assembly work in shadowed environments, the space suit has new helmet-mounted floodlights and spotlights. There is also a jet-pack "life jacket" to allow an accidentally untethered astronaut to fly back to the station in an emergency. In 1994, as part of the STS-64 mission, astronaut Mark Lee performed an EVA during which he tested a new mobility system called the Simplified Aid for EVA Rescue (SAFER). This system is similar to, but smaller and simpler than, the MMU.

Before the arrival of the joint airlock module (called Quest), as part of the STS-104 mission to the *ISS* (in July 2001), space walks conducted from the space station could use only Russian space suits unless the space shuttle was present. The facilities of the Zvezda service module limited space station–based EVAs to only those with Russian Orlan space suits. The Quest module, now attached to the *ISS*, gives the station's occupants the capability to conduct EVAs with either Russian- or American-designed space suits. Prebreathing protocols and space suit design differences no longer limit EVA activities by astronauts and cosmonauts on the space station.

Astronauts wear white space suits for a number of practical reasons. Perhaps the most important technical reason is that white reflects incident solar radiation (heat) to keep the astronaut from getting too warm. While wearing a space suit during an EVA, astronauts can also sometimes get too cold, but this condition is usually limited to their hands and fingers. NASA engineers have placed heaters in the gloves to prevent any frigid finger phenomenon from ruining an important EVA work session. Another reason American space walkers wear white space suits is because the color white is readily visible against the jet black background of outer space. A white space suit makes it easier for other astronauts to see the person who is space walking. Many times, astronauts will work in pairs while performing an EVA. In that case, one of the white space suits will have red stripes in four distinct places to make it easier for the other astronauts to tell one space walker from the other.

✧ Hazards to Space Travelers and Workers

Current experience with human performance in space is mostly for individuals operating in low Earth orbit (LEO). However, the establishment of permanent lunar bases, three-year duration and longer expeditions to Mars, and the construction of large space settlements will require extensive human activities in space beyond the protection offered travelers in LEO by Earth's magnetosphere. The maximum continuous time spent by any single human being in space is now a few hundred days, and the people who have experienced extended periods of space travel generally represent a very small number of highly trained and highly motivated individuals.

Expeditions to the *International Space Station* (*ISS*) are expanding human spaceflight experience, but the space shuttle *Columbia* accident (February 2003) has caused a significant curtailment in the crew size, crew rotation, and scientific and biophysical experiment schedules.

The available technical database, although limited to essentially low-Earth-orbit spaceflight, now suggests that with suitable protection, people can live and work in space safely for extended periods of time and then enjoy good health after returning to Earth. Data from the three crewed *Skylab* missions, numerous long-duration *Mir* space station missions, and the first 14 expeditions to the *ISS* are especially pertinent to answering the important question of whether people (in small, isolated groups) can live and work together effectively in space for more than a year in a confined and relatively isolated habitat. Human factor data collected from international crew performance during extended *ISS* expeditions will provide an idea of how best to prepare human beings for one-year or more remote duty in establishing a lunar surface base or three-year or more remote duty on the first human expedition to Mars.

One recurring issue is the overall hazard of space travel. Three major accidents in the American space program—the *Apollo One* mission fire (January 27, 1967), the *Challenger* accident (January 28, 1986), and the *Columbia* accident (February 1, 2003)—have impressed upon the national consciousness that space travel is and will remain for a significant time a hazardous undertaking. Some of the major cause-effect factors related to space traveler health and safety are shown below. Many of these factors require "scaling up" from current medical, safety, and occupational analyses to achieve the space technologies necessary to accommodate large groups of space travelers and permanent inhabitants.

Some of these health and safety issues include preventing launch-abort, spaceflight, and space-based assembly and construction accidents; preventing failures of life support systems; protecting space vehicles and habitats from collisions with space debris and meteoroids; protecting the crew from the ionizing radiation hazards encountered in outer space; and providing habitats and quality living conditions that minimize psychological stress.

The biomedical effects of the substantial acceleration and deceleration forces when leaving and returning to Earth, living and working in a weightless (microgravity) environment for long periods of time, and chronic exposure to space radiation are three main factors that must be dealt with if people are to live in cislunar space (space between Earth and the Moon) and eventually populate heliocentric space.

Astronauts and cosmonauts have adapted to microgravity conditions for extended periods of time in space and have experienced maximum acceleration forces up to an equivalent of six times Earth's gravity (that is, 6 g). No acute operational problems, permanent physiological deficits, or adverse health effects on the cardiovascular or musculoskeletal systems have been observed from these experiences. However, short-term physical difficulties, such as space adaptation syndrome, or "space sickness," as well as occasional psychological problems (such as feelings of isolation and stress) and varying post-flight recovery periods after long-duration missions have been encountered. One special concern is that chronic exposure to microgravity decreases the need of a person's muscles. After a long interplanetary flight, a significant loss of muscle strength and endurance may so weaken space explorers that they cannot live and work on the surface of the destination planet, such as Mars.

Some physiological deviations have been observed in American astronauts and Russian cosmonauts during and following extended space missions, including *Skylab, Mir,* and the *ISS*. Most of these observed effects appear to be related to the adaption to microgravity conditions, with the

This photograph dramatically depicts the fatal *Challenger* launch ascent accident on January 28, 1986—an accident that destroyed the orbiter and claimed the lives of all seven astronauts on board the vehicle. Hurtling out of the conflagration at 78 seconds after liftoff during the STS 51–L mission are *Challenger*'s left wing, main engines (still burning residual propellant), and forward fuselage (crew cabin). *(NASA)*

affected physiological parameters returning to normal ranges either during the missions or shortly thereafter. No apparent persistent adverse consequences have been observed or reported to date. Nevertheless, some of these deviations could become chronic and might have important health consequences, if they were to be experienced during extended missions in space—such as an approximately three-year expedition to Mars, repeated long-term tours in a space assembly facility at Lagrangian point four or five in cislunar space, or at a permanent lunar surface base with its significantly reduced (one-sixth g) gravitational environment. The physiological deviations experienced by space travelers due to microgravity have usually returned to normal within a few days or weeks after returning to Earth. However, bone calcium loss appears to require an extended period of recovery after a long-duration space mission.

Strategies are now being developed to overcome these physiological effects of weightlessness. An exercise regimen can be applied, and body fluid shifts can be limited by applying lower-body negative pressure. Anti-motion medication is also useful for preventing temporary motion

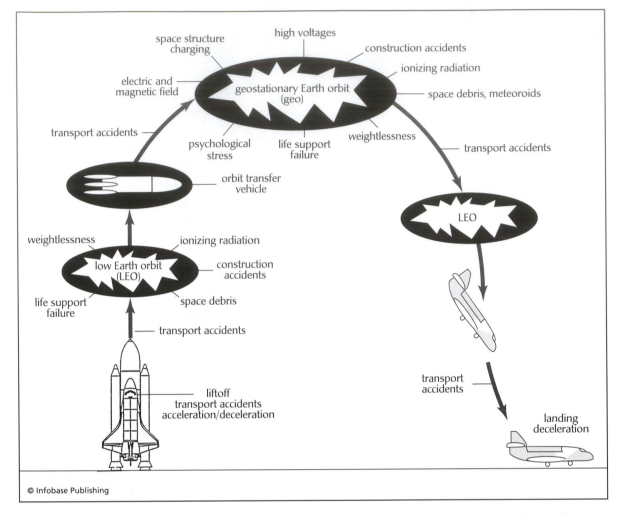

The illustration describes some of the major factors related to the health and safety of space travelers and workers who might be involved in construction operations in low Earth orbit or in geostationary orbit. *(NASA and the U.S. Department of Energy)*

sickness. Proper nutrition, with mineral supplements, and regular exercise also appear to limit other observed effects. One way around this problem in the long term, of course, is to provide acceptable levels of artificial gravity in larger space bases and orbiting space settlements. In fact, very large space settlements will most likely offer the inhabitants a wide variety of gravity levels, ranging from microgravity up to a normal terrestrial gravity level (that is, one g). (See chapter 12.) This multiple-gravity-level option will not only make space settlement lifestyles more diverse than on Earth but will also prepare planetary settlers for life on their new worlds or help other space travelers gradually adjust to the "gravitational rigors" of returning to Earth.

The ionizing-radiation environment encountered by workers and travelers in space is characterized primarily by fluxes of electrons, protons, and energetic atomic nuclei. In low Earth orbit, electrons and protons are trapped by Earth's magnetic fields, forming the Van Allen belts. The amount of ionizing radiation in LEO varies with solar activity. The trapped radiation belts are of concern when space-worker crews transfer from low Earth orbit to geostationary Earth orbit (GEO) or to lunar surface bases. In geostationary locations, solar-particle events (SPEs) represent a major ionizing radiation threat to space workers. Throughout cislunar space and interplanetary space (beyond the protection of the Earth's magnetosphere), space travelers are also bombarded by galactic cosmic rays. These are very energetic atomic particles, consisting of protons, helium nuclei, and heavy nuclei (that is, nuclei with an atomic number greater than two). Shielding, solar flare–warning systems, and excellent radiation dosimetry equipment should help prevent space travelers from experiencing ionizing radiation doses in excess of the standards established for various space missions and occupations. Lunar surface bases will require extensive radiation shielding. Because of the serious but unpredictable hazard posed by an anomalously large solar-particle event (ALSPE), when lunar surface workers venture an appreciable distance away from their well-shielded base, they will need ready access to a temporary radiation "storm cell." Similarly, any human-crewed Mars expedition vehicle will also require significant amounts of radiation shielding to protect against chronic exposure to galactic cosmic rays during the long interplanetary journey. The vehicle will also have to provide additional emergency shielding provisions (possibly some type of radiation-safe room) where the crew can scramble and hide for a few hours to avoid acute radiation doses from an ALSPE.

Mars expedition personnel and lunar surface base workers might also experience a variety of psychological disorders, including the solipsism syndrome and the shimanagashi syndrome. The solipsism syndrome is a state of mind in which a person feels that everything is a dream and is not real. It might easily be caused in an environment (such as a small space base or confined expedition vehicle) where everything is artificial or human-made. The shimanagashi syndrome is a feeling of isolation in which individuals begin to feel left out, even though life may be physically comfortable. (See also chapter 12.) Careful design of living quarters and good communication with Earth should relieve or prevent such psychological disorders. Techniques used to maintain the psychological well-being and happiness of *ISS* crewmembers provide some valuable human factor guidelines for ensuring lunar base worker or Mars expedition crew-member mental fitness.

Living and working in space in this century will present some interesting challenges as well as some ever-present dangers and hazards. However, the rewards of an extraterrestrial lifestyle, for certain pioneering individuals, will more than outweigh any such personal risks.

✧ Space Food

Eating is a basic survival need that an astronaut or cosmonaut must accomplish in order to make space a suitable place to work and accomplish mission objectives. The space food (meals) he or she consumes must be nutritious, safe, lightweight, easily prepared, convenient to use, require little storage space, need no refrigeration, and be psychologically acceptable (especially for crews on long-duration missions).

Since the beginning of human spaceflight in the early 1960s, eating in space has become more natural and "Earth-like," while better meeting the other criteria. Spaceflight feeding has progressed from squeezing paste-like foods from "toothpaste" tubes to eating a "sit-down" dinner complete with normal utensils, except for the addition of scissors, which today's astronauts use to cut open packages.

In general, there are five basic approaches to preparing food for use in space: rehydratable food, intermediate-moisture food, thermostabilized food, irradiated food, and natural-form food.

Rehydratable food has been dehydrated by a technique such as freeze-drying. In the space shuttle program, for example, foods are dehydrated to meet launch vehicle mass and volume restrictions. They are rehydrated later in orbit when they are ready to be eaten. Water used for rehydration comes from the orbiter vehicle's fuel cells, which produce electricity by combining hydrogen and oxygen; water is the resultant by-product. More than 100 different food items, such as scrambled eggs and strawberries, go through this dehydration/rehydration process. For example, when a strawberry is freeze-dried, it remains full size in outline, with its color, texture, and quality intact. The astronaut can then rehydrate the strawberry with either saliva (mouth moisture) as it is chewed or by adding water to the package.

Twenty varieties of drinks, including tea and coffee, also are dehydrated for use in space travel. But pure orange juice and whole milk cannot be included. If water is added to dehydrated orange juice, orange "rocks" form in water; they do not rehydrate. Dehydrated whole milk does not dissolve properly upon rehydration. It floats around in lumps and has a disagreeable taste. So skim milk must be used. Back in the 1960s, General Foods Corporation developed a synthetic orange juice product (called Tang) that could be used in place of orange juice.

Intermediate-moisture food is partially dehydrated food, such as dried apricots, dried pears, or dried peaches. Thermostabilized food is cooked at moderate temperatures to destroy bacteria and then sealed in cans or aluminum pouches. This type of space food includes tuna, canned fruit in heavy syrup, and ground beef. Irradiated food is preserved by exposure to ionizing radiation. Various types of meat and bread are processed in this manner. Finally, natural-form food is low in moisture and taken into space in much the same form as found on Earth. Peanut butter, nuts, graham crackers, gum, and hard candies are examples. Salt and pepper are packaged in liquid form because crystals would float around the crew cabin and could cause eye irritation or contaminate equipment.

All food in space must be packaged in individual serving portions that allow easy manipulation in the microgravity environment of an orbiting spacecraft. These packages can be off-the-shelf thermostabilized cans, flexible pouches, or semirigid containers.

In the space shuttle program, the variety of food carried into orbit is sufficiently broad that crewmembers can enjoy a six-day menu cycle. A typical dinner might consist of a shrimp cocktail, steak, broccoli, rice, fruit cocktail, chocolate pudding, and grape drink.

When the *International Space Station* becomes fully operational (sometime beyond 2008 because of the *Columbia* accident), the station's permanent crew of three could eventually increase in number to a maximum of seven persons. Their food and other supplies must be replenished at regular intervals. *ISS* residents are now using an extension of the joint U.S.-Russian food system that was developed during the initial shuttle-*Mir* phase of the *ISS* program. *ISS* crewmembers have a menu cycle of eight days, meaning the menu repeats every eight days. Half of the food system is American and half is Russian. However, there are plans to include the foods of other *ISS* partner countries, including Europe, Japan, and Canada. The packaging system for the daily menu food is based on single-service, disposable containers. Single-service containers eliminate the need for a dishwasher, and the disposal approach is quite literally out of this world.

Since electrical power for the *ISS* is generated by solar panels rather than fuel cells (as on the space shuttle), there is no extra water generated on board the station. Water is recycled from cabin air, but the amount recovered is not enough for use in the food system. Consequently, the percentage of shuttle-era rehydratable foods is being decreased and the percentage of thermostabilized foods increased over time.

Generally, the American portion of the current *ISS* food system is similar to the shuttle food system. It uses the same basic types of foods—thermostabilized, rehydratable, natural form, and irradiated—and the same packaging methods and materials. As on the shuttle, beverages on the *ISS* are in powdered form. The water temperature is different on the station; unlike the shuttle, there is no chilled water. Crewmembers have only ambient, warm, and hot water available to them.

Space station crewmembers usually eat breakfast and dinner together. They use the food preparation area in the Russian *Zvezda* service module to prepare their meals. The module has a fold-down table designed to accommodate three astronauts or cosmonauts eating together under microgravity conditions. Used food-packaging materials are bagged and placed along with other trash in a Progress supply vehicle. Then the robot spacecraft assumes a secondary mission as an extraterrestrial garbage truck. It is jettisoned from the *ISS* and burns up upon reentry into Earth's atmosphere. Garbage management is a major problem, especially when regularly scheduled resupply missions are interrupted and delayed.

Long-duration human missions beyond Earth orbit (where resupply is difficult or impossible) will require food supplies capable of extended storage. The menu cycle will have to be greatly expanded to support crew morale and nutritional well-being. A permanent lunar base could have a highly automated greenhouse to provide fresh vegetables and fruits. (Cosmonauts and astronauts have performed several modest greenhouse experiments on the *ISS*.) A rotating "space greenhouse" (to provide an appropriate level of artificial gravity for plant development) might accompany human expeditions to Mars and beyond. A permanent Martian surface base most likely would include an "agricultural facility" as part of its closed environment life support system (CELSS).

✧ Space Radiation Environment

One of the major concerns associated with the development of a permanent human presence in outer space is the ionizing radiation environment, both natural and human-made. The natural portion of the space radiation environment consists primarily of Earth's trapped radiation belts (also called the Van Allen belts), solar-particle events (SPEs), and galactic cosmic rays (GCRs). Ionizing radiation sources associated with human activities can include space nuclear power systems (fission reactors and radioisotope), the detonation of nuclear explosives in the upper portion of Earth's atmosphere or in outer space (activities currently banned by international treaty), space-based particle accelerators, and radioisotopes used for calibration and scientific activity.

Earth's trapped radiation environment is most intense at altitudes ranging from 622 miles (1,000 km) to 18,645 miles (30,000 km). Peak intensities occur at about 2,485 miles (4,000 km) and 13,670 miles (22,000 km). Below approximately 6,215 miles (10,000 km) altitude, most trapped particles are relatively low-energy electrons (typically, a few million electron volts [MeV]) and protons. In fact, below about 310 miles (500 km) altitude, only the trapped protons and their secondary nuclear interaction products represent a chronic ionizing radiation hazard.

CLOSED ECOLOGICAL LIFE SUPPORT SYSTEM

A closed ecological life support system (CELSS) is a system that can provide for the maintenance of life in an isolated living chamber or facility through complete reuse of the materials available within the chamber or facility. This is accomplished, in particular, by means of a cycle in which exhaled carbon dioxide, urine, and other waste matter are converted chemically or by photosynthesis into oxygen, water, and food. On a grand (macroscopic) scale, the planet Earth itself is a closed ecological system; on a more modest scale, a "self-sufficient" space base or space settlement also would represent another example of a closed ecological system. In this case, however, the degree of "closure" for the space base or space settlement would be determined by the amount of makeup materials that had to be supplied from Earth.

Material recycling in a life support system can be based on physical and chemical processes, can be biological in nature, or can be a combination of both. Chemical and physical systems are designed more easily than biological systems but provide little flexibility or adaptability to changing needs. A life support system based solely on physical and chemical methods

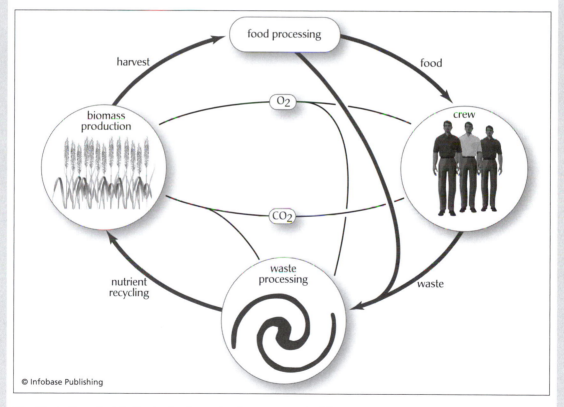

The illustration depicts the basic elements of a regenerative life-support system. *(NASA)*

(continues)

(continued)

also would be limited because it would still require resupply of food and some means of waste disposal. A *bioregenerative* life support system incorporates biological components in the creation, purification, and renewal of life support elements. Plants and algae are used in food production, water purification, and oxygen release. While the interactions of the biomass with the environment are very complex and dynamic, creating a fully closed ecological system—one that needs no resupply of materials (although energy can cross its boundaries, as does sunlight into Earth's biosphere)—such a system appears possible and even essential for future, permanently inhabited human bases and settlements within the solar system. This type of closed ecological system is sometimes called a controlled ecological life support system.

Trapped electrons collide with atoms in the outer skin of a spacecraft, creating penetrating X-rays and gamma rays (called secondary radiations) that can cause tissue damage. Trapped energetic protons can penetrate several grams of material (typically 0.4-inch- [1-cm-] to 0.8-inch- [2-cm-] thick shields of aluminum are required to stop them), causing ionization of atoms as they terminate their passage in a series of nuclear collisions. Most human-occupied spacecraft missions in low Earth orbit (LEO) are restricted to altitudes below 310 miles and inclinations below about 60 degrees to avoid prolonged (chronic) exposure to this type of radiation.

For orbits below 310 miles (500 km) altitude and inclinations less than about 60 degrees, the predominant part of an astronaut's overall radiation exposure will be due to trapped protons from the South Atlantic Anomaly (SAA). The SAA is a region of Earth's inner radiation belts that dips close to the planet over the southern Atlantic Ocean southeast of the Brazilian coast. Passage through the SAA generally represents the most significant source of chronic natural space radiation for space travelers in LEO. Earth's geomagnetic field generally protects astronauts and spacecraft in LEO from cosmic-ray and solar-flare particles.

However, spacecraft in highly elliptical orbits around Earth will pass through the Van Allen belts each day. Furthermore, those spacecraft with a high-apogee altitude (say greater than 18,645 miles [30,000 km]) will also experience long exposures to galactic cosmic rays and solar-flare environments. Similarly, astronauts traveling through interplanetary space to the Moon or Mars will be exposed to both a continuous galactic cosmic-ray environment and a potential solar-flare environment (that is, a solar-particle event).

A solar flare is a bright eruption from the Sun's chromosphere that may appear within minutes and then fade within an hour. Solar flares cover a wide range of intensity and size. They eject high-energy protons that represent a serious hazard to astronauts traveling beyond LEO. The SPEs associated with solar flares can last one to two days. Anomalously large solar-particle events (ALSPEs), the most intense variety of solar-particle event, can deliver potentially lethal doses of energetic particles—even behind modest spacecraft shielding (e.g., 0.23 ounce per square inch [1 g/cm^2] to 0.45 ounce per square inch [2 g/cm^2] of aluminum). The majority of SPE particles are energetic protons, but heavier nuclei also are present.

Galactic cosmic rays originate outside the solar system. GCR particles are the most energetic of the three general types of natural ionizing radiation in space and contain all elements from atomic number 1 to 92. Specifically, galactic cosmic rays have the following general composition: protons (82–85 percent), alpha particles (12–14 percent), and highly ionizing heavy nuclei (1–2 percent), such as carbon, oxygen, neon, magnesium, silicon, and iron. The ions that are heavier than helium have been given the name HZE particles, meaning high atomic number (Z) and high energy (E). Iron (Fe) ions appear to contribute substantially to the overall HZE population. Galactic cosmic rays range in energy from 10s of MeV to 100s of GeV (a GeV is 1 billion [10^9] electron volts) and are very difficult to shield against. In particular, HZE particles produce high-dose ionization tracks and kill living cells as they travel through tissue.

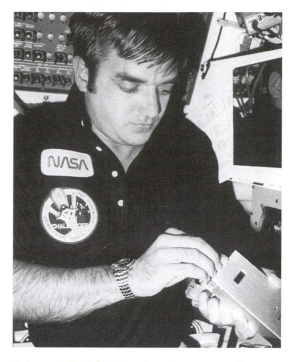

Astronaut Daniel C. Brandenstein operates the radiation monitoring equipment experiment in the crew cabin of the space shuttle *Challenger* during the STS-8 mission (September 1983). *(NASA)*

An effective space radiation protection program for astronauts and cosmonauts on extended missions in interplanetary space or on the lunar or Martian surface should include sufficient permanent shielding (of the spacecraft or surface base habitat modules), adequate active dosimetry, the availability of "solar storm shelters" (zones of increased shielding protection) on crewed spacecraft or on the planetary surface, and an effective solar-particle event warning system that monitors the Sun for ALSPEs. The ionizing radiation environment found in space also can harm sensitive electronic equipment (e.g., a single-event upset) and spacecraft materials. Design precautions, operational planning, localized shielding, device redundancy, and computer-memory "voting" procedures are techniques used by aerospace engineers to overcome or offset space radiation–induced problems that can occur in a spacecraft, especially one operating beyond LEO.

Astrochimps, Canine Cosmonauts, and Other Space-Traveling Animals

The word *astrochimp* is the nickname commonly given to the nonhuman primates (especially chimpanzees) used during the early U.S. space program. Prior to and during NASA's Mercury Project, astrochimps were used to test space capsule and launch vehicle–system hardware before their commitment to human flight. In a similar fashion, the former Soviet Union used animals, especially dogs, to help test and demonstrate the space travel hardware used by cosmonauts. Life scientists in both countries also sent a variety of biological specimens and animals into space while conducting experiments designed to study the impact of rocket launches and orbital flight on living things. Experiments with living specimens has continued in the space shuttle era, but under more stringent animal-use regulations and more well-defined biophysical testing protocols.

The concept of using animals, especially higher-order mammals, in scientific research is a very controversial issue. On one side of this controversy, proponents (such as researchers) point out that the use of animals as biological pioneers and surrogates for human beings accelerates scientific progress and ultimately saves human lives. For these people, the use of animals in research is considered justified, especially if the animals are treated in an ethical way—that is, with pain and suffering (including fatalities) being kept to an absolute minimum. Opponents of using animals in research (such as animal rights advocates) say it is unethical and unnecessary to subject higher-order animals, especially more developed living creatures like dogs and chimpanzees, to pain and suffering in the name of science. Animal rights advocates suggest that such testing is often done under cruel, shoddy circumstances with little or no regard for the well-being of the animals. They also point out that the animals are really involuntary subjects in such experiments, regardless of how noble stated scientific goals are.

With respect to life sciences experiments related to spaceflight, astronauts and cosmonauts, who are often used as test specimens in biophysical research, understand and voluntarily accept the risks involved in venturing into outer space. But, as the animal rights advocates claim, the American astrochimps and the Soviet space dogs blindly trusted their human masters and had no choice in surrendering their bodies—and in many cases their lives—in the name of scientific progress.

While there is truth on both sides of this inflammatory debate, it is beyond the intent or scope of this chapter to resolve the controversy or to take sides. The objective here is to describe as accurately as possible the higher-order animal experiments that did take place as part of the early space programs of the United States and the former Soviet Union. Why were such tests performed? The paramount purpose of the vast majority of these early biological experiments was to provide life sciences data that directly supported upcoming human spaceflight missions by astronauts and cosmonauts at the dawn of the Space Age.

This chapter summarizes the technical circumstances and results for the most notable animal missions in space. However, the chapter makes no attempt to condone or condemn the actions taken by American or Russian scientists and space mission managers in the early days of the Space Age. From the perspective of history—that is, from a vantage point of nearly five decades—it is quite difficult to reconstruct fully or properly appreciate how and why certain choices and pathways were chosen. Without question, there were serious concerns within the scientific community that the human body might not be able to survive the rigors of a rocket launch or exposure to extended periods of microgravity (so-called weightlessness) during orbital flight. What may now appear as controversial or unwise decisions were actually choices made by scientists and mission managers who were seeking expedient solutions to the most pressing uncertainties regarding the anticipated perils of human spaceflight.

In this fast-paced, politically driven environment, the use of animal-in-space experiments to pave the way for human spaceflight appeared reasonable, expedient, and justified. Modern ethics continues to assert that the end, no matter how noble, should never justify the means. However, the political expediencies of the early cold-war era appear to have diluted or obscured any rigorous examination of potentially questionable experimental means in pursuit of technical ends, which were generally deemed of high national security significance. As a point of historic reference, during the same period (namely the 1950s) thousands of American and Russian military personnel were involuntarily exposed to hazardous levels of ionizing radiations by their respective governments as a result of certain atmospheric nuclear weapons tests. Today these so-called atomic veterans are considered unintentional "casualties" of the cold war nuclear arms race.

Undoubtedly, some of these early animal-in-space experiments would be performed quite differently if they were accomplished in the context of modern technology and the contemporary sociopolitical environment. Today it appears quite reasonable to ask: Why did the United States not use a well-instrumented humanoid robot to test the Mercury space capsule instead of chimpanzees? One response is that the appropriate level robotic system technology (including miniaturized electronics and computers) was simply not available in the late 1950s to do an adequate job. This simple answer might satisfy most people, but there will be others who remain unmoved.

✧ Research with Animals in Space

Immediately after World War II, scientists began to wonder whether a human being, or any other living creature, could survive the anticipated perils of space travel. Not only were there great concerns about a person's ability to survive the high acceleration levels (g-forces) associated with a rocket ride into space, but also there were equally grave predictions that the human body would not function properly under the extended periods of microgravity within a spacecraft orbiting around Earth. There were also many other biophysical concerns and unanswered questions regarding the perils of spaceflight—such as the potentially lethal effects of the space radiation environment.

No one could provide valid technical answers to most of these pressing questions. So, starting in the late 1940s, American and Russian scientists began sending small animals into space in the nose cones (payload compartments) of captured and refurbished German V-2 rockets. Other cold war–era rocket vehicles, such as the U.S. Army's Redstone and Jupiter ballistic missiles, also supported American space life sciences research by carrying animal payloads into space. Unfortunately, many of these animal-in-space experiments suffered some level of equipment failure, causing the tiny space travelers to perish in the act of providing important technical data that supported human space travel. In 1957, Russian scientists launched *Sputnik 2* and its living passenger, a mixed breed female dog named Laika. The pioneering orbital flight showed that a living creature could survive in microgravity (weightlessness). Following this mission, animal-in-space research entered an important new realm—namely, providing assistance to engineers in demonstrating the safety of life support systems intended for human-rated space vehicles.

NASA scientists still send animals into space to perform special research—but only on rare occasions and when absolutely necessary. Life scientists interested in studying how the human body reacts to extended exposure to microgravity now prefer to conduct their research with computer models or by directly involving astronauts or cosmonauts on the *International Space Station* (*ISS*). People aboard the *ISS* can perform many interesting life sciences experiments. However, there are still some experiments that people cannot accomplish, because the research protocols would seriously interfere with the crew's other duties. For example, many interesting life sciences experiments require the test animals (including astronauts and cosmonauts) to have a very closely controlled and carefully monitored diet. But, spaceflight experience has shown that astronauts and cosmonauts are generally not very willing to eat exactly the same type and amount of food each day at some precisely assigned time. This type of controlled biological experiment is very burdensome to the astronauts. Therefore, researchers must perform such experiments with test animals such as mice, which can more easily have monitored feedings. Life scientists will then attempt to transfer any data collected during carefully controlled animal-in-space experiments to human physiological behavior models.

Due to ethical constraints and the practicalities of housekeeping in space, researchers will generally use the lowest form of life considered scientifically appropriate for the experiment. Many times results involving snails or fish can be applied to human conditions. Some interesting genetic studies, for example, have been conducted with fish. This approach avoids the use of higher-evolved mammals. While there is not always a one-to-one transfer of the resulting data, the similarities are often sufficient to gain some useful knowledge about the consequences of long-duration spaceflight on living systems. But there remain some space travel–related biomedical issues that still require experiments involving small mammals.

Taking animals into space is no easy task. If laboratory mice were to fly aboard the *ISS,* for example, traditional aquarium-style cages would not provide enough traction for the mice to "move" around. Instead, the space mice require special wire mesh cages so their tiny toes could grip a rougher surface. Of course, wood chips could not be used for their bedding, since the chips would float around in microgravity. The traditional gravity-feed water bottles, the kind used for laboratory mice on Earth, would not work either. So aerospace engineers have designed special, pressurized water containers. Traditional food pellets are not practical because they, too, would float around the cage. NASA scientists prefer to use compressed food bars instead. And of course there is the problem that plagues all animal maintenance workers both on Earth and in space, waste management. Collecting feces and urine in a microgravity environment is definitely a challenging task. Fortunately, NASA life scientists have designed special waste-containment systems for small animals, such as mice and monkeys.

Do space mice like living in microgravity? Does floating around instead of scurrying about really confuse them? According to the results of some NASA experiments, within about five minutes of exposure to microgravity, the mice float around in their engineered living spaces, eating and grooming themselves, in ways they would on Earth.

Of course not all animals tested on orbit adapted that quickly to microgravity. Inside an orbiting spacecraft, there is no up or down to orient them, so space-faring fish and tadpoles ended up swimming in loops rather than straight lines, as they do on Earth. Baby mammals also had a difficult time in space. For one thing, they normally huddle together for warmth here on Earth. But, as they floated about in all directions in their special spaceflight cages, the baby mammals (mice and rats) found it nearly impossible to huddle. Researchers also observed that it is difficult for the baby mammals to nurse in microgravity, because they cannot easily locate their mother's nipple.

Today, when NASA life scientists want to send an animal into space, that animal's welfare becomes a key concern. NASA officials must ensure that any animal traveling in space is cared for ethically and humanely. Because people can give their consent, current government rules regulating the use of animals in research are more stringent than those governing the use of people in research. Animals cannot give their consent, so government regulations exist to intervene on their behalf. Research animals sent into space are protected by the United States Department of Agriculture under the provisions of the Animal Welfare Act, as well as by animal care and use committees within NASA.

✧ Nonhuman Primates—The First American Space Travel Pioneers

On June 11, 1948, a captured and refurbished German V-2 rocket (designated number 37) roared off its launch pad at the U.S. Army's White Sands Missile Range in southern New Mexico and ascended to an altitude of 38.7 miles (62 km). This launch was conducted at White Sands under a post–World War II effort known as the Hermes Project—the first United States ballistic missile program. The U.S. Army headquartered the Hermes Project team at Fort Bliss, Texas, where Wernher von Braun and other rocket scientists were relocated from Germany so they could work together with American engineers in reassembling, refurbishing, and then launching captured German V-2 rockets.

The fledgling U.S. Air Force's Air Research and Development Command (ARDC) sponsored this flight under Project Blossom—an upper atmosphere research project. What was special about Project Blossom is that it involved the parachute recovery of a payload canister designed to carrying biological specimens, such as seeds and fruit flies. At the time, scientists wanted to expose these living organisms to cosmic rays during the high-altitude flight. The rocket, sometimes called a V-2 Blossom rocket, also carried the first American space monkey, an anesthetized rhesus monkey (*Macaca mulatta*) named Albert. The small primate (also known as a rhesus macaque) was restrained in an extended position, by means of nylon netting, on a specially designed couch padded with foam rubber. Researchers had positioned a thermocouple in the rubber mask, which was held in place over the monkey's face. Scientists used this instrument to monitor the monkey's respiration. The U.S. Air Force biomedical researchers had also placed electrodes in the monkey's chest and in one leg, so they could record electrocardiograms during the rocket flight. Unfortunately, the rocket flight proved fatal for Albert, who was the apparent victim of suffocation due to an equipment malfunction in scientific payload compartment. The capsule was not recovered because the recovery parachute failed to deploy during descent.

The next V-2 Blossom rocket was launched from White Sands on June 14, 1949. Since the refurbished V-2 rocket (number 47) reached a maximum altitude of 83 miles (134 km), the primate passenger—another anesthetized rhesus macaque monkey, named Albert II—became the first monkey in space. (By an arbitrary rule within the American space program, when a human being reaches an altitude of 50 miles (80 km) that person qualifies for spaceflight status.) Although monitored biophysical data indicated that Albert II survived the high-altitude rocket flight into outer space, he died on impact, when the canister recovery system again failed to provide a survivable soft-landing.

Space travel proved equally perilous for Albert III, a rhesus monkey provided by the U.S. Air Force Aero-Medical Laboratory (AML) at Wright-Patterson Air Force Base, Ohio. This Project Blossom monkey died on September 16, 1949, when the V-2 rocket (number 32) malfunctioned and exploded over the White Sands Missile Range at an altitude of just 2.6 miles (4.2 km). The last so-called V-2 monkey flight under Project Blossom took place on December 8, 1949, when a V-2 rocket (number 31) blasted off from White Sands and reached a maximum altitude of 81 miles (130 km). Its instrumented astrochimp passenger was Albert IV—another AML rhesus monkey. Data from the biophysical monitoring instruments indicated that Albert IV had a very successful flight and was showing no ill effects—that is, until the primate experienced a fatal, hard-impact landing caused by another recovery canister failure. As a historic note, there appears to be a slight conflict within Department of Defense and NASA historic records concerning what type of monkeys Albert III and IV were. While most available records state that Albert III and IV were rhesus monkeys, another otherwise credible NASA life sciences reference suggests that Albert III and Albert IV were cynomolgus monkeys (also known as crab-eating macaques). Since neither monkey was recovered to support post-flight biomedical evaluations, the significance of the conflict in historic records is considered minimal with respect to the overall purpose of this chapter.

Between 1951 and 1952, biomedical researchers from the Holloman Aero-Medical Research Laboratory conducted several Aerobee sounding

U.S. AIR FORCE CHIMPANZEES

Why was the newly established U.S. Air Force so interested in monkeys in the late 1940s? One reason is that Aero-Medical Laboratory (AML) researchers at Wright-Patterson AFB, Ohio, were concerned about the physiological and psychological consequences of high-speed, high-altitude flight. No one was sure what would happen to human pilots if jet aircraft flew faster and higher. Nonhuman primates, such as the West African chimpanzee (*Pan troglodytes*), were of considerable interest as surrogates for human beings in hazardous military aviation experiments. The field of aviation medicine broadened in the mid-1950s to include the anticipation of human spaceflight. As the newly integrated field of aerospace medicine developed, some researchers saw the chimpanzee as a convenient live test subject for extremely hazardous experiments involving high-speed ejection equipment and (later on) space capsule life-support systems.

To meet these biomedical research challenges, the U.S. Air Force established a new field test facility at Holloman Air Force Base, New Mexico. At this facility, which soon became known as the Holloman Aero-Medical (HAM) Research Laboratory, a variety of pioneering biodynamic studies were performed using human *volunteers* such as John Paul Stapp, who was a medical doctor and U.S. Air Force officer. Many of these tests involved a horizontal rocket sled demonstration of the biophysical consequences of high-speed accelerations and very sudden decelerations. Other experiments centered involved the body's tolerance to

total pressure change (from sea level to near vacuum), as experienced during very-high-altitude balloon flights into the stratosphere. For some of the military aviation experiments at the Holloman Aero-Medical Research Laboratory, the researchers used large animals, including hogs, bears, and chimpanzees.

In the early 1950s, the U.S. Air Force obtained 65 young and infant chimpanzees that had been captured in Africa. Complemented by a captive-breeding program, this colony of chimpanzees was to provide a continuous supply of live specimens (that is, human surrogates) for the more hazardous tests within the military flight research program at Holloman AFB. From the mid-1950s, these chimpanzees were used as living specimens to test the biophysical consequences of acceleration and deceleration forces, high-speed movement—including those conditions that might imperil or prevent travel in space by human beings. Once the Mercury Project formed, several chimpanzees from this air force–run colony, most notably Ham and Enos, were specially trained and conditioned to help NASA qualify the Mercury space capsule before astronauts flew it.

Once the United States conducted its first successful human orbital flight (John Glenn, February 1962), all serious interest in using astrochimps to test spaceflight equipment ended. While millions of jubilant Americans celebrated the success of the human astronauts, who traveled in space under NASA's Mercury Project, the important role of the nonhuman primate space pioneers was rocket launches at the White Sands Missile Range, using monkeys and mice as passengers. The objective of these sounding rocket tests was to study the effects of cosmic radiation and rocket flight–induced changes in animal cardiovascular systems. On September 20, 1951, an anesthetized and instrumented rhesus monkey (named Yorick), along with

In this picture (taken at Cape Canaveral in early January 1961), Ham is seated comfortably in his specially designed bio–pack launch couch during preflight testing. The chimpanzee was born wild in West Africa, captured as an infant, and then sent to the chimpanzee colony at Holloman Air Force Base, New Mexico. His name derives from the U.S. Air Force's Holloman Aerospace Medical (HAM) Laboratory. On loan to NASA, the astrochimp was blasted into space by a U.S. Army Redstone rocket on January 31, 1961. The success of Ham's suborbital mission raised engineering confidence that the Mercury Project spacecraft would be suitable for use by human astronauts. (NASA)

either totally ignored or quickly forgotten. No longer needed for spaceflight research, officials at the Holloman Aero-Medical Research Laboratory once again used the remaining population of wild-caught chimpanzees and their descendants as live specimens in hazardous military aviation-related tests, such as helmet and restraint performance experiments that simulated aircrew ejections at very high speeds. By the mid-1970s, the U.S. Air Force abandoned the use of chimpanzees as test specimens and, then, in a cost-saving move, elected to lease the majority of the surviving colony members to other biomedical research organizations. Unfortunately, once leased, little official scrutiny took

place regarding the treatment of the chimpanzees at recipient research facilities. As a result, several of these nonmilitary facilities were eventually cited by the U.S. Department of Agriculture for violations of the Animal Welfare Act (AWA).

In June 1997, the U.S. Air Force announced that all of its remaining chimpanzees would be given away by means of a public divestiture process as authorized and approved by the U.S. Congress. Under the terms of this divestiture, any surviving chimpanzees (including astrochimps) and their descendants would either be given to a research laboratory or else retired to a sanctuary. Among other things, this action resulted in the U.S. Air Force getting completely out of the chimpanzees-for-research business and stimulated several lengthy legal challenges on behalf of the chimpanzees. One of these legal challenges led to the establishment of a permanent astrochimp sanctuary in Florida run by the Save the Chimps organization.

11 unanesthetized mice, rode as the biological experiment payload onboard an Aerobee rocket that reached an altitude of about 45 miles (72 km). Electrocardiogram, respiration, and blood pressure measurements were performed on Yorick during the flight. Both Yorick and the mice survived the sounding rocket flight. However, the rhesus monkey died several hours after landing due to the effects of heat exposure—primarily because the recovery team was delayed in reaching the rocket's payload capsule, which had properly soft-landed in the desolate desert regions of southern New Mexico. Despite the post-landing fatality, Yorick was the first American space monkey to survive a rocket flight into space and be recovered alive. A bit more precisely, Yorick reached a region of the upper atmosphere called "near space," since the rocket's maximum altitude was just under 50 miles [80 km]—the nominal threshold for spaceflight in the American space program.

On May 22, 1952, the U.S. Air Force launched another Aerobee sounding rocket with a biological research payload from White Sands. The nose cone of this rocket carried a pair of anesthetized Philippine monkeys, named Patricia and Mike. The Philippine monkey (*Macaca philippinensis*) is also known as the long-tailed macaque (*Macaca fascicularis*). To examine any biophysical differences in rocket flight acceleration, scientists placed Patricia in a seated position (to receive exposure to head-to-tail acceleration) and Mike in a supine (lying on back) position (to receive chest-to-back acceleration). As the rocket flew to its maximum altitude of about 36 miles (58 km), researchers observed signals from the payload compartment that described the biophysical behavior of the monkeys and two companion mice during acceleration, deceleration, and weightlessness. The animals survived the flight and the subsequent parachute-assisted soft-landing of the rocket's nose cone. According to one historic account from NASA, Patricia and Mike were then allowed to retire at the National Zoological Park in Washington, D.C. Patricia died there two years later, while Mike survived at the zoo until 1967.

The successful launch of the Russian space dog, Laika, in early November 1957 and the dog's survival (for a few days) as a living passenger aboard *Sputnik 2* triggered a renewed interest by American biomedical researchers in flying additional monkeys into space, as precursors to anticipated travel by humans. The first effort involved a cooperative effort between the U.S. Army and Navy in which one or more monkeys rode inside the Jupiter missile, as it was being test-fired from Cape Canaveral. The Jupiter was a U.S. Army–developed intermediate range ballistic missile (IRBM)—the brainchild of von Braun and his V-2 rocket team. The transplanted team of rocket scientists was now working at the Army Ballistic Missile Agency (ABMA) in Huntsville, Alabama.

The Jupiter ballistic missile was a liquid-propellant rocket that burned liquid oxygen and RP-1 (essentially a refined kerosene mixture). With a single engine, the Jupiter developed a sea-level thrust of 150,000 pounds-force (667,000 N) and a maximum range of about 1,500 miles (2,415 km). In November 1956, there was a decision within the hierarchy of the Department of Defense to transfer the Jupiter IRBM program from the U.S. Army to the U.S. Air Force. But this programmatic transfer did not lessen the U.S. Army's participation in the two Jupiter missile biomedical launches, known as Jupiter AM-13 and Jupiter AM-18.

The two Jupiter biomedical launches from Cape Canaveral carried monkeys into space in late 1958 and in mid-1959 on high-altitude sub-orbital trajectories. The Jupiter AM-13 launch took place on December 13, 1958. The U.S. Army provided the Jupiter missile and the U.S. Navy's School of Aviation Medicine (located in Pensacola, Florida) provided Gordo, a trained South American squirrel monkey (*Saimiri sciureus*). Gordo (also called Old Reliable) was allowed to fly on a noninterference basis as part of the missile test mission. But, because of missile test instrumentation, there was only a small volume available within the Jupiter's nose cone. So, Gordo's tiny life support enclosure was just 750 cubic inches (12,290 cm^3) in volume. The enclosure contained instruments to provide researchers an indication of the cabin temperature, pressure, and radiation level. U.S. Navy doctors and veterinarians also instrumented Gordo and restrained the monkey on a specially designed launch couch. The instrumentation allowed researchers to receive biomedical telemetry, such as heart rate, heart sounds, and body temperature, from the unanesthetized monkey during the flight.

Blasting off from Cape Canaveral, the flight of the Jupiter rocket went well and carried the space-traveling primate to a maximum altitude of 300 miles (480 km), while traveling downrange. The preliminary objective of biomedical experiment on the Jupiter AM-13 flight was to demonstrate that, if an adequate life support system was provided, a test animal could survive a high-speed, high-acceleration ballistic missile flight unharmed. Scientists viewed this as an important step in determining whether human beings could participate in space travel. The secondary objective was to design, build, test, and demonstrate a life support system suitable for rocket flight and by extrapolation—eventually suitable for human space travel. Other secondary objectives included the demonstration of the ability to recover a living test specimen at sea following a rocket flight into space and a practical demonstration of the principle of biomedical telemetry by which the physiological characteristics and behavior status of the test animal are observed by researchers throughout the mission.

On this flight, the life support system functioned well, and the biomedical telemetry indicated that Gordo survived the flight with no

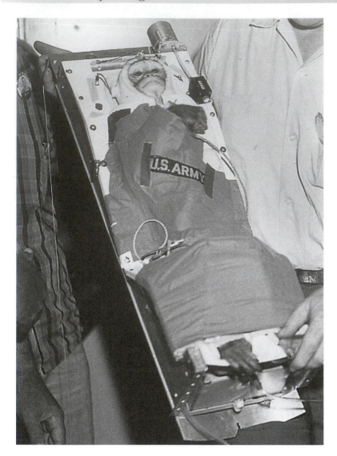

Launch support personnel prepare to place Able, a female rhesus monkey, into the special capsule within the nose cone of the U.S. Army's Jupiter AM-18 rocket, as part of preflight testing at Cape Canaveral. As shown in the photograph (taken on May 18, 1959), the American-bred monkey is restrained on a special launch couch and extensively instrumented with implanted medical electrodes. *(NASA)*

obvious adverse effects. Unfortunately, the monkey could not be recovered and was lost at sea, because the nose cone's floatation device failed. Gordo perished as the rocket's nose cone sank beneath the waves before U.S. Navy recovery craft could arrive at the splashdown location.

The Jupiter AM-18 flight took place on May 28, 1959. The biomedical payload included two monkeys: an American-born, female rhesus monkey named Able and a female South American squirrel monkey named Baker. Medical research teams from the U.S. Army provided Able, while the Naval Aviation Medical Center in Pensacola provided Baker. Both monkeys were instrumented to provide similar biomedical measurements as collected from Gordo during the Jupiter AM-13 flight. However, placing two monkeys and their life support equipment into the nose cone of the Jupiter missile proved quite a challenge for the launch support crew. Able was installed into the nose cone on her biopack couch three days before launch and then fed intraperitoneally, while any waste products were allowed to accumulate in her diapers.

The Jupiter IRBM successfully lifted off its pad at Cape Canaveral and traveled about 1,500 miles downrange from Cape Canaveral, reaching a maximum altitude of 300 miles and a maximum speed in excess of 10,000 miles per hour (16,000 km/h). Both monkeys were capable of withstanding g-forces in excess of 38 times the normal force of gravity and successfully endured about nine minutes of weightlessness. Following splashdown of the nose cone, Able and Baker were recovered alive and unharmed from the surface of the ocean. However, Able died several days later (on June 1) from the effects of anesthesia as veterinarians were operating on her to remove an infected, implanted biomedical electrode. Baker proved to be a much hardier space traveler. After surviving the high-g-force, high-speed suborbital rocket flight, Baker was retired from space monkey duty and placed on exhibit in a zoological environment. Baker lived to the age of 27 and died on November 2,

1984, of kidney failure at the U.S. (formerly Alabama) Space and Rocket Center in Huntsville, Alabama.

✧ Mercury Project Monkey Missions

On October 7, 1958, the newly formed civilian space agency NASA announced the start of an overall project to demonstrate that human beings could travel in space. Despite the survival of several American and Russian animals during suborbital flights (such as the American monkeys Able and Baker) or orbital flights (such as the Russian dog Laika), aerospace medical experts in the United States still expressed serious concerns. Little was known about the combined biophysical effects of space travel–induced high stresses that an astronaut would experience in the proposed Mercury Project missions. So many of these medical experts pressed for additional monkey-in-space flights as part of the Mercury Project and prior to the first attempt by an American astronaut to travel in space.

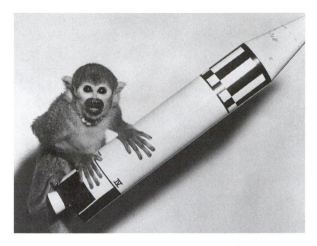

A photograph of Baker (a female South American squirrel monkey) alongside a small-scale model of the Jupiter AM-18 launch vehicle. On May 28, 1959, the Jupiter rocket (a U.S. Army ballistic missile converted to space launch duties) sent this astrochimp and her primate companion (a female rhesus monkey, named Able) on a high-altitude parabolic trajectory into space from Cape Canaveral. Riding in a special compartment inside the rocket's nose cone, both monkeys were recovered unharmed in the Atlantic Ocean. (Date of photograph: May 29, 1959.) *(NASA)*

As a result of these debates and lingering concerns, NASA officials agreed to form an animal program within the Mercury Project. The Mercury animal program consisted of two phases: Phase one involved flights of small primates (rhesus monkeys) in the Little Joe rocket vehicle out of Wallops Island, Virginia; phase two involved flights of medium-size primates in the Mercury space capsule launched by first the Redstone rocket on a suborbital mission and then by the Atlas rocket (sometimes called Big Joe) on an orbital mission. The decision to use chimpanzees rather than other primates for phase two of the Mercury Project animal program was aimed at providing the highest level of performance short of human.

THE RHESUS MONKEYS: SAM AND MISS SAM

The Little Joe launch vehicle was a relatively inexpensive solid-propellant rocket designed by a team of engineers at NASA's Langley Research Center in Virginia specifically to test the Mercury Project spacecraft abort system in a series of suborbital flights. The rocket's interesting name derives from the casino game craps and the throw of a double deuce (pair of twos) on the dice. The original Little Joe concept called for four solid rocket motors fired two at a time—that is, a pair of twos. The stubby Little Joe rocket's four large stabilizing fins equally spaced around the aft airframe also helped to perpetuate the name.

To support phase one of the Mercury Project animal program, space was made available in the boilerplate space capsule used in the Little Joe rocket tests. The U.S. Air Force School of Aviation Medicine at Brooks AFB, Texas, was given the responsibility for preparing and installing the biological packages (bio-packs) for the two American-born rhesus monkeys used in the tests. A bio-pack is a container for housing a living organism (for example, an insect, fish, or small mammal) in a habitable environment and for recording biological activities during the organism's spaceflight. (The term *boilerplate space capsule* means a nonfunctioning volume and mass replica, as opposed to a much more elaborate and expensive flight-qualified space capsule.) Although not considered as critical as the upcoming chimpanzee spaceflights, the rhesus monkey suborbital flights on the Little Joe rocket would provide biomedical evaluation of the accelerations expected during the abort of a Mercury Project flight at liftoff and shortly after liftoff. The Mercury space capsule had a rocket-propelled escape tower that was designed to quickly pull the capsule away from an exploding launch vehicle in the event of an abort during the booster's ascent to space.

On December 4, 1959, NASA rocket test personnel at Wallops Flight Facility in Virginia launched the Little Joe 2 (LJ-2) mission to check the high-altitude performance of the Mercury capsule's launch-escape system. For this suborbital flight, NASA officials had also decided to allow U.S. Air Force aeromedical specialists to run all the experiments they wanted. Therefore, personnel from the School of Aerospace Medicine at Brooks AFB provided the primate passenger with a bio-pack. Sam was a small male rhesus monkey whose name was the acronym for his training alma mater. During his brief (about 11-minute) flight, Sam flew inside the boilerplate version of the Mercury space capsule and was given life support protection by the bio-pak from the School of Aerospace Medicine. Just before noon, the Little Joe rocket ripped through the air under full power and burned out at an altitude of 100,000 feet (30,500 m). As planned, the escape tower and Mercury capsule separated, and the tower's escape rocket provided an additional boost, throwing the space capsule and its primate passenger into a suborbital coasting trajectory that reached its zenith just short of 280,000 feet (85,365 m), or 53 miles (85 km). This peak altitude was actually about 100,000 feet lower than expected because of a serious wind effect miscalculation. As a result of this altitude error, Sam experienced only three minutes of weightlessness instead of the anticipated four minutes. Nevertheless, the monkey survived the reentry, the not-so-mild splashdown and ocean impact, and about six hours of confinement in his bio-pak within the bobbing space capsule before being recovered from the Atlantic Ocean by the destroyer USS *Borie*. Postflight examinations indicated that Sam apparently suffered no ill effects from his brief journey into space and was thus returned to the colony in which he trained. He died in November 1982.

The Little Joe 1-B launch took place on January 21, 1960. During this test, NASA officials wanted to examine the performance of the Mercury capsule launch-escape system at the point of maximum dynamic pressure (max q). The primate passenger on this Little Joe test flight was Miss Sam, a female rhesus monkey from the U.S. Air Force School of Aviation Medicine. The Little Joe rocket blasted off at Wallops Island, rose to an altitude of slightly less than nine miles (15 km), and attained a maximum velocity of just over 2,000 miles per hour (3,220 km/h). Then, as planned, the escape rocket ignited and rapidly pulled the boilerplate Mercury space capsule along with its primate passenger away from the point of the simulated booster abort. Miss Sam, secure in her bio-pack, survived the 8.5-minute flight in good condition and splashed down in the Atlantic Ocean about 11 miles (18 km) downrange from the launch site. The monkey was recovered from the ocean's surface almost immediately by the crew of a U.S. Marine Corps helicopter and returned by air to the Wallops Flight Facility.

For about 30 seconds after the escape rocket fired, Miss Sam was apparently badly shaken up during the test flight and did not respond to test stimuli, as she was trained to do. Otherwise the small rhesus monkey acted the role of a perfectly trained primate automaton. In reviewing Miss Sam's biotelemetry, medical experts suspected the primate suffered from nystagmus (a spasmodic, involuntary motion of the eyeball) right after the escape rocket fired and immediately after splashdown. These conditions caused some concern about a similarly impaired astronaut's effectiveness in activating certain backup systems. Miss Sam's brief flight produced some modification in the backup procedures used by the Mercury astronauts under similar conditions of physical impairment. Following the rocket flight, Miss Sam was returned to the monkey colony where she had been trained.

ASTROCHIMP HAM

Phase two of the animal program involved the use of chimpanzees to flight-demonstrate the safety of the Mercury Project space capsule, before its commitment to use by American astronauts. Following the successful completion of the primate flights on the Little Joe rockets, NASA officials met with aviation biomedical experts from the U.S. Air Force, Army and Navy to plan the chimpanzee spaceflight program. The primary objective was to conduct these primate spaceflights in such a way as to verify the feasibility of human spaceflight. The aviation medicine experts also wanted to collect as much data as possible on the level of mental and physical activity that could be expected during space travel. Since the chimpanzee is a higher-order primate that closely resembles humans, it was reasoned that one or more could be trained to perform simple tasks that mimicked how an astronaut might perform under the physical rigors of both suborbital and orbital flight.

Personnel from the U.S. Air Force's Holloman Aero-Medical (HAM) Research Laboratory became responsible for providing and training the chimpanzees for flight. A NASA representative served as a coordinator to assist in the integration of the animal flights into the total Mercury Project flight program. The decision to use chimpanzees rather than other primates for phase two of the Mercury animal program was aimed at providing the highest level of performance short of human. Restraint of the chimpanzees would be minimal during the flights to make it possible for them to perform simple psychomotor tests. The electrocardiograms, body temperature, and respiratory movement would be recorded, using the same techniques planned for use with human astronauts.

Although the HAM Research Laboratory possessed the chimpanzees, veterinarians, and aerospace physiologists, the complex lacked facilities to obtain behavioral measurements of the animals. Accordingly, arrangements were made to train several chimpanzees under contract with the Wenner-Gren Aeronautical Research Laboratory at the University of Kentucky. With the additional assistance of training specialists from the Walter Reed Army Institute of Research, the air force team at Holloman AFB began the training of eight chimpanzees using standard operator condition equipment and special restraint chairs. As part of the program, this group of chimpanzees was trained to do simple tasks like pushing a lever within a certain time period in response to a certain electric light signal. Failure to respond properly to a task would result in the application of negative reinforcement (or undesirable feedback), typically in the form of a discomforting electric shock administered on the soles of the chimpanzee's feet.

As the training of the eight chimpanzees progressed, the Veterinary Services Branch at the HAM Research Laboratory was also collecting normal baseline data on the entire colony of immature chimpanzees. Personnel were also designing and fabricating methods of restraint. In order to complete the design of the chimpanzee's launch couch system, the staff also conducted a series of simulated flights (using a centrifuge) to determine the effects of acceleration and vibration on a chimpanzee.

On January 2, 1961, a colony of six chimpanzees (four females and two males), along with 20 medical specialists and animal handlers from Holloman AFB, moved into facilities at Cape Canaveral, Florida. The chimpanzees were divided into two separate compounds in order to prevent the spread of any contagion throughout the entire colony. In preparation for the first chimpanzee rocket flight, the handlers led the chimpanzees through daily exercises and psychomotor performance tasks. To condition each of the astrochimp candidates to respond properly, banana pellets were used as rewards and mild electric shocks as punishments. A mock-up of the Mercury space capsule was also provided in each training compound. By January 29, each of the six astrochimp candidates was bored but well trained in lever-pulling. All that remained was to pick one primary and one backup for the actual flight on top of the U.S. Army's Redstone rocket.

The competition to become the first American chimpanzee in space was fierce. The honor went to Ham, a 37-pound (17-kg) chimpanzee who was named after the Holloman Aero-Medical Research Laboratory. According to the veterinarian on the final selection committee, Ham was "exceptionally frisky and in good humor." Ham had been born free in West Africa, captured as an infant in July 1957 soon after his birth, and brought to the Aero-Medical Research Laboratory at Holloman AFB in July 1959. The selection team chose a female chimpanzee as Ham's alternate. At 19 hours before launch, both Ham and his female backup were each put on low-residue diets, fitted with biomedical sensors, and then checked out in the specially designed launch couches—bio-packs that functioned as pressurized life support cabins. Ninety minutes before launch, Ham, still frisky and enthusiastic though encased in his bio-pack, rode the launch site gantry elevator for placement in the Mercury space capsule that was connected to the top of the Redstone missile.

Although some pesky launch vehicle problems delayed Ham's flight, officially designated by NASA as the Mercury-Redstone 2 (MR-2) mission, at 16:55 (UTC) on January 31, 1961, the Redstone rocket roared off its pad and carried astrochimp Ham on his brief but very eventful suborbital flight downrange from Cape Canaveral. Ham had coped with the launch delays quite well, and at liftoff, he was dutifully working his levers accurately enough to avoid the punishment (an electric shock to the bottom of his feet) that would result from performance errors or inattention.

Behavioral scientists had placed a dashboard-like device at the waist level of the chimpanzee's launch couch. This dashboard contained two lights and two levers that required some level of effort to depress. As a result of his preflight training program, Ham understood how to stay comfortable by avoiding a series of electric shocks. Cued by a white warning light, each successful operation (depression) of the right-hand lever postponed the next scheduled shock for 15 seconds. At the same time, Ham had to push the left-hand lever within five seconds after a blue light flashed in order to avoid another series of electric shocks to the soles of his feet. The blue light flashed about once every two minutes or so.

During the MR-2 flight, the rocket's thrust controller ran above nominal performance and depleted the propellant one-half a second before deactivation of the abort pressure sensor. The launch vehicle's abort system detected the early shutdown of the rocket engine and aborted the Mercury spacecraft, as if Ham's rocket was experiencing a serious problem. The higher-than-normal cutoff velocity, combined with the added thrust of the escape tower rocket, hurled the primate-carrying space capsule well beyond the planned recovery area. As designed, the rocket's abort action also sent a distress signal to the recovery forces, which immediately responded and headed for a computed impact point farther downrange.

Because of the over-performance of the launch vehicle and the added thrust of the escape tower rocket, Ham's Mercury capsule achieved a maximum speed of 5,840 miles per hour (9,400 km/h), as opposed to an intended maximum speed of 4,350 miles per hour (7,000 km/h). The errant space capsule also reached a maximum height of 155 miles (250 km) versus the planned altitude of 115 miles (185 km). The astrochimp's flight lasted a total of 16.7 minutes, some 2.4 minutes longer than intended, and the capsule flew 416 miles (670 km) downrange from the launch site, about 130 miles (209 km) farther than planned.

Astrochimp Ham reaches for an apple after his suborbital flight on January 31, 1961. This photograph was taken on board the USS *Donner* a few minutes after the chimpanzee and Mercury space capsule were recovered from the Atlantic Ocean. Ham performed well during his 16.5-minute suborbital flight, and his pioneering mission paved the way for astronaut Alan B. Shepard, Jr.'s historic suborbital flight on May 5, 1961. Most people interpret Ham's expression as a postflight "grin," but modern primatologists (like Jane Goodall) suggest that the chimpanzee's expression is actually one of great fear. *(NASA)*

Not only did Ham's flight experience an overzealous rocket engine thrust but also, just prior to the abort sequence (about 2.3 minutes after liftoff), the cabin pressure in the space capsule suddenly dropped from 5.5 pounds per square inch (psi) (37.9 kilopascals) to 1 psi (6.9 kilopascals). Fortunately, Ham's space suit and enclosed biopack prevented him from being harmed or killed by this rapid depressurization of the space capsule. Postflight investigations showed the problem was a faulty air-inlet snorkel valve. While the unintentionally opened valve did not prove a fatal mishap for the chimpanzee, after splashdown it allowed seawater to rush into the capsule, almost drowning the simian space traveler. Despite all the mechanical mishaps, Ham performed very well during the flight and dutifully pulled levers when signaled as he had been trained.

Some 27 minutes after the chimpanzee splashed down in the Atlantic Ocean, a U.S. Navy search plane spotted the astrochimp's upright space capsule bobbing in the water. Because all the primary recovery ships were too far away, the U.S Navy decided to dispatch rescue helicopters from the USS *Donner* to the impact site. When the helicopters arrived on the scene, the crew saw that the space capsule was now floating on its side and taking on more water. The helicopter crew plucked Ham's spacecraft out of the water. Ham appeared in good shape and readily accepted an apple after his harrowing suborbital.

Because of the launch vehicle's over-performance, Ham traveled about 422 miles (679 km) downrange, reached a peak altitude of 157 miles (253 km), experienced about six minutes of weightlessness, and endured a peak reentry deceleration of 14 g (almost 3 g greater than planned). How did the astrochimp react to his historic rocket flight? Well, four strong men could not bring Ham back to the Mercury capsule to allow photographers to snap several postflight pictures. Ham wanted nothing else to do with his Mercury spacecraft.

After the Mercury Redstone 2 mission, Ham enjoyed a bit of celebrity—even gracing the cover of the February 18, 1961, issue of *Life* magazine. His pioneering flight paved the way for astronaut Alan B. Shepard, Jr., to make the first American suborbital flight during the Mercury Redstone 3 (MR-3) mission on May 5, 1961. (See chapter 4 for a discussion of the manned Mercury Project missions.) Both of these important milestones in spaceflight history were somewhat overshadowed on the world stage by the orbital flight of cosmonaut Yuri Gagarin, who became the first human to orbit around Earth in a spacecraft on April 12, 1961.

After his rocket flight into space, Ham was retired from astrochimp duty and sent to live at the National Zoo in Washington, D.C. For all his fame and contributions to the American space program, he remained a lonely celebrity for the next 17 years. Then, in 1981, at the urging of primatologists, Ham was given a new home—this time at a zoological park in North Carolina, where he could enjoy the company of other chimpanzees. He died there on January 19, 1983. As a tribute to the primate's great contributions to space travel, Ham's body is buried in a prominently marked gravesite at the International Space Hall of Fame in Alamogordo, New Mexico.

ASTROCHIMP ENOS

Astrochimp Enos was another male chimpanzee from the Holloman AFB chimp colony. This chimpanzee was born in West Africa, captured as an infant in August 1956, and eventually purchased from the Miami Rare Bird Farm in April 1960 by the Holloman Aero-Medical Research Laboratory. Animal handlers and trainers at Holloman AFB came up with the primate's name from the Hebrew word *Enosh*, meaning "man." Enos and several other chimpanzees completed more than 1,250 hours of behavioral training at Holloman AFB and at the University of Kentucky in preparation for an orbital flight test of the Mercury Project space capsule. Enos received more intense training than Ham because his orbital mission would be more complicated and include longer exposure to weightlessness.

Mercury Atlas 5 (MA-5) was the second and final orbital qualification flight of the Mercury Project space capsule prior to the system's release for use in orbital flight by American astronaut John H. Glenn, Jr. In late 1961, there was a great deal of debate within NASA concerning the necessity of conducting this astrochimp mission. Earlier that year, the suborbital flights of the astrochimp Ham (January 31) and of the astronauts Alan B. Shepard, Jr. (May 5), and Virgil "Gus" I. Grissom (July 21) had already demonstrated that the Mercury space capsule worked—at least on very short missions into space. Furthermore, orbital flights by the Russian cosmonauts Yuri A. Gagarin (April 12) and Ghermann S. Titov (August) clearly demonstrated that, with a properly functioning life support system, a human being could indeed survive travel in space. In the end, the more conservative engineering actions within NASA's Mercury Project prevailed, and astrochimp Enos was given the pioneering task of demonstrating that the Mercury space capsule would function properly during an orbital mission.

On November 29, 1961, an Atlas rocket blasted off from Cape Canaveral and placed the nonhuman primate–carrying Mercury space capsule in a 98-mile (158-km) perigee by 147-mile (237-km) apogee orbit around Earth. Enos's spacecraft traveled around Earth at an inclination of 32.5 degrees and with a period of 88.5 minutes. During his flight, the 37-pound (17-kg) chimpanzee continued to perform various psychomotor activities—just as he had been trained and despite the fact that certain equipment malfunctions resulted in him receiving negative feedback (that is, electric shocks) even though he performed the lever-pushing tasks correctly.

Other capsule hardware problems plagued the mission and caused NASA's engineers to end the flight on the second orbit, rather than the planned third orbit. A metal clip in a fuel supply line caused a problem with the space capsule's attitude control system, leading to a failure of a roll reaction jet. This equipment malfunction resulted in Enos's spacecraft drifting 30 degrees from the nominal attitude, at which point the automatic controls would bring the spacecraft back to the nominal attitude. The entire malfunction and automatic correction process caused the Mercury capsule to consume an extra pound (0.45 kg) of precious attitude control propellant on each orbit. In addition to excessive propellant consumption, a problem also developed in an inverter to the spacecraft's electrical system, causing an elevated temperature condition inside the capsule. Both of these in-flight anomalies could have been corrected by a human pilot on board the spacecraft. Since Enos was not trained to perform such corrective actions, NASA mission controllers decided to bring the capsule down after the second orbit around Earth.

With a total mission elapsed time of approximately three hours and 21 minutes, the capsule splashed down in the Atlantic Ocean about 255 miles (410 km) southeast of Bermuda. A U.S. Navy search plane had spotted the descending space capsule when it was at an altitude of 5,000 feet (1,525 m), so officials quickly dispatched recovery ships to the impact area. About 75 minutes after the space capsule landed in the water, the crew of USS *Stormes* pulled alongside and plucked the spacecraft out of the water. Once the recovery crew opened the capsule's hatch, they could see and hear that Enos had survived the flight. As he danced around the deck of the ship, he appeared in good health despite experiencing the g-forces of launch and reentry, as well as approximately three hours of weightlessness. Reports from the recovery crew also indicated that the happy chimpanzee ran around shaking the hands of those who had freed him from the space capsule. The success of Enos's flight resolved all remaining issues within the NASA management hierarchy and paved the way for astronaut John H. Glenn's historic orbital mission in the *Friendship 7* spacecraft on February 20, 1962. (Chapter 4 describes the details of Glenn's flight.)

Despite his pioneering efforts, Enos quickly became an essentially forgotten chapter in the Mercury Project story. Without much fanfare, he was transferred back to the chimp colony at Holloman AFB. There, almost a year after his historic spaceflight, he died on November 4, 1962, due to an antibiotic resistant case of shigella dysentery. For two months prior to his death, the biomedical team at the Holloman Aero-Medical Research Laboratory had kept Enos under careful observation and care. In their postmortem evaluation, the facility's animal pathologists ruled that the astrochimp's fatal illness was not related to his orbital flight. No ceremony acknowledged his passing nor was the brave astrochimp given a marked gravesite.

✧ Russian Space Dogs

Although the theme of this chapter focuses on the use of nonhuman primates in the American space program, no treatment of the pathway to human spaceflight is complete without an acknowledgment of the pioneering role that dogs played in the space program of the former Soviet Union. Like primates in the American space program, dogs served as the biological precursors in the Soviet human spaceflight program, which culminated in cosmonaut Yuri Gagarin's historic orbital flight in the *Vostok 1* spacecraft on April 12, 1961.

After World War II, Russian rocket engineer Sergei Korolev (1907–66) received permission from the Soviet government to manufacture the R-1 rocket, which was basically a copy of the German V-2 rocket, produced in postwar Soviet factories. To assist in the manufacture of this rocket, the Soviets used parts and documents seized when the Red Army captured

Peenemünde (the German rocket test range on the Baltic Sea) and various V-2 manufacturing facilities in the closing days of the war. Korolev also made significant improvements in the R-1 rocket and by 1949 was able to launch test flights of this improved version, which he designated the R-2 rocket.

Mindful of the work being performed by the Americans at White Sands with captured V-2 rockets under the Hermes Project and Project Blossom, Korolev and his own team of biomedical experts began using the R-1 rocket to carry small animals, such as mice, rats, and rabbits, on one-way flights into the upper atmosphere. These rocket flights, essentially a mirror image of the American Project Blossom, attempted to collect fundamental biophysical data about whether living creatures could travel in space. As the Soviet program matured and more powerful rockets (such as the R-2) became available, Korolev's team of biomedical researchers shifted their living specimen interest and began conducting rocket flights with small dogs.

According to somewhat fragmented historic records from the former Soviet Union, between 1951 and 1952 the biomedical R-1 rocket flight series conducted at Kapustin Yar involved a total of nine dogs, with three dogs actually flying twice. Kapustin Yar is a minor launch complex located on the banks of the Volga River near the city of Volgograd. This complex was originally constructed to support the early Soviet ballistic missile test program, including the R-1 and R-2 rockets. The first missile launch at the complex took place in 1947. Starting in 1951, R-1 biomedical rocket flights would carry a pair of dogs in a pressurized (hermetically sealed) nose-cone compartment that was then recovered by parachute.

Like their American counterparts, Russian biomedical researchers and aerospace engineers were looking ahead at the possibility of human space-flight, and they needed data from living specimens to help in the design of a human-rated space capsule and its associated life support subsystem. Once the small-mammal rocket flight program passed beyond mice and rats, the Russian researchers chose small dogs over monkeys because they felt that dogs would be less difficult to experiment with than monkeys. They decided to use two animals on each rocket flight in order to obtain more accurate results. (The Russian scientists also preferred to use female dogs because they felt it was easier to control their waste products.)

The first of the dog-carrying R-1 rocket flights took place in August 1951, and the two canine cosmonauts, Dezik and Tsygan (Gypsy), were recovered successfully. The second R-1 launch took place in early September 1951. This time Dezik and her companion, Lisa, were lost, but the flight's biomedical data recorder survived. Reports suggest that the loss of the two dogs greatly disturbed Korolev, who cherished his own pet dog. The Soviets soon launched Smelaya (Bold) and Malyshka (Little One).

The third rocket flight was successful, and the dogs were recovered. The fourth launch in the R-1 series failed, resulting in two more space dog fatalities. The fifth launch proved successful, and the two space dogs were recovered. The sixth launch in the series took place on September 15, 1951, and the rocket reached an altitude of 62 miles (100 km). Both space dogs returned to Earth successfully.

Unfortunately, the available Soviet records for these early dogs-in-space launches are generally incomplete with respect to precisely what dogs flew on what rocket flights. Furthermore, the test protocols imposed during these early biomedical experiments were probably not as stringent as those imposed during later orbital flights for which more reliable data, including the canine cosmonaut names, are available. For example, one of the dogs reportedly used on the sixth R-1 flight had the name ZIB—the Russian language acronym for the words *substitute for missing dog Bobik*. Apparently Bobik was one of the two dogs originally scheduled for this rocket flight, but he escaped. So, a replacement dog was hastily found. The replacement canine was a small stray dog the scientists found scrounging for food near a canteen at the launch site. Soviet researchers gave this frisky mutt the unusual name ZIB. By escaping into the wilderness around the Kapustin Yar complex, the missing dog Bobik lost the chance to contribute to spaceflight history.

On October 4, 1957, the former Soviet Union was able to shatter the prevailing global assumption about the overwhelming technical superiority of the United States by launching the world's first Earth-orbiting satellite, *Sputnik 1*—a simple 184-pound- (83.5-kg-) mass, hollow, steel sphere, containing batteries and a radio transmitter. Less than a month later, on November 3, 1957, the Soviets reinforced this techno-surprise by launching a 1,118-pound (508-kg) satellite, called *Sputnik 2*. The new, more massive satellite was the second spacecraft ever placed into orbit around Earth. As he had done before launching *Sputnik 1*, Korolev obtained permission from the Soviet leadership to use a modified R-7 ICBM to place *Sputnik 2* into orbit. Following its successful launch from the Baikonur Cosmodrome, *Sputnik 2* went into a 132-mile (212-km) by 1,031-mile (1,660-km) altitude orbit with an inclination of 65.3 degrees and a period of 103.7 minutes. In addition to being very massive (for the time), *Sputnik 2* was the world's first biological spacecraft, or biosatellite.

The spacecraft contained several compartments for radio transmitters, a telemetry system, a programming unit, a regeneration and temperature control system for the cabin, and scientific instruments. *Sputnik 2* also had a separate sealed cabin, containing Laika, a mixed-breed female dog. Laika ("barker" in Russian) was the first live animal to orbit Earth in a spacecraft.

At launch, the part-Samoyed terrier had a mass of about 13 pounds (6 kg). *Sputnik 2*'s pressurized cabin was padded and provided enough room for the small dog to lie down or stand. An air regeneration system provided oxygen, while food and water were dispensed in a gelatinized form. Laika was fitted with a harness, a bag to collect waste, and electrodes to monitor the animal's vital signs during orbital flight. According to one report, the early biotelemetry indicated that Laika was agitated but nevertheless ate her food. Because *Sputnik 2* was constructed early in the Soviet space program, the satellite had no capability for safely returning Laika back to Earth. Soviet scientists had originally estimated that Laika's oxygen supply would run out after about 10 days in orbit. However, this pioneering canine cosmonaut most likely died after just one or two days in orbit because of the thermal control (cabin heating) problems experienced by the satellite. Laika became an internationally recognized "space pioneer," and the little dog's mission provided Soviet scientists with the world's first biophysical data on the behavior of a living organism in the microgravity environment of an orbiting spacecraft. On April 14, 1958—after 162 days in orbit—*Sputnik 2* decayed and reentered Earth's atmosphere.

The Soviet scientists followed Laika's trailblazing mission with a series of other dogs-in-space Sputnik launches. Each canine mission provided additional biomedical data and improved confidence in the spacecraft's design. These demonstration flights ultimately allowed cosmonaut Yuri Gagarin to become the first human being to travel in space and safely return to Earth.

On July 28, 1960, two dogs Bars (Lynx) and Lisichka (Little Fox) died on board a satellite called the *Korabl Sputnik* when the booster rocket exploded on ascent to space. Reports indicate that the spacecraft was an engineering prototype of the Vostok spacecraft later used in the initial Soviet manned space missions. The next space dog launch took place on August 19, 1960. This time, the Soviets successfully placed a 10,120-pound (4,600-kg) satellite, called *Sputnik 5,* into an approximately 190-mile- (306-km-) altitude orbit at an inclination of 65 degrees and with an orbital period of 90.7 minutes. *Sputnik 5* (also referred to as *Korabl Sputnik 2* [*Korabl* is Russian for "ship"]) carried two dogs: Strelka (Little Arrow) and Belka (Squirrel), along with a television system and other scientific instrumentation. After one day in orbit, the spacecraft's recovery capsule reentered Earth's atmosphere, functioned properly, and the dogs safely landed. Strelka and Belka became the first two living creatures to orbit Earth and then return safely to Earth. Some time after her space mission, Strelka gave birth to a litter of puppies. In an unusual gesture of goodwill during the cold war, Soviet premier Nikita S. Khrushchev (1894–1971) presented one of Strelka's puppies to President John F. Kennedy (1917–63), as a gift for his young daughter Caroline.

Sputnik 6 (also called *Korabl Sputnik 3*) was the next Soviet mission involving dogs. A modified SS-6 ballistic missile lifted off from the Baikonur Cosmodrome on December 1, 1960, and successfully placed the 10,030-pound (4,560-kg) satellite into a 103-mile (166-km) perigee by 144-mile (232-km) apogee orbit at an inclination of 65 degrees with a period of 88.5 minutes. On board *Sputnik 6* were two dogs: Pchelka (Little Bee) and Mushka (Little Fly). The flight lasted one day, and things went well on orbit. Then, during an attempted capsule recovery, the two dogs were killed when their life support cabin burned up during atmospheric reentry.

On December 22, 1960, Soviet researchers attempted to launch another Korabl Sputnik spacecraft with two dogs aboard: Damka (Little Lady) and Krasavka (Beauty). During the rocket ride into space from the Baikonur Cosmodrome, the upper-stage rocket failed, causing an abort in the overall launch attempt. However, both dogs were safely recovered and were without any apparent injuries from their brief, unplanned suborbital flight.

The next Russian space dog mission was *Sputnik 9*, which was launched on March 9, 1961. The 10,340-pound (4,700-kg) satellite, also known as *Korabl Sputnik 4,* carried a dog named Chernushka (Blackie) and a "dummy" cosmonaut—that is, a wooden mannequin that replicated the approximate size, mass, and volume of a human being. *Sputnik 9* had an orbit around Earth characterized by a 108-mile (173-km) perigee, a 149-mile (239-km) apogee, an inclination of 65 degrees, and a period of 88.6 minutes. The flight lasted for just a single orbit, followed by the successful recovery of the capsule, its canine passenger, and wooden cosmonaut mannequin.

The final space dog Sputnik mission took place on March 25, 1961. The 10,250-pound (4,695-kg) *Sputnik 10* satellite carried the dog Zvezdochka (Little Star) into orbit. This little dog also had a dummy cosmonaut as its companion. A modified SS-6 ICBM launch vehicle placed the satellite, also known as *Korabl Sputnik 5,* into a 102-mile (164-km) perigee by 143-mile (230-km) apogee orbit at an inclination of 65 degrees and with a period of 88.4 minutes. After one orbit, the capsule containing Zvezdochka and "Ivan Ivanovich" (the nickname for the cosmonaut mannequin) returned safely to Earth. This was the final test demonstration that the Soviet rocket engineers needed to commit the Vostok spacecraft to Gagarin's historic orbital flight. Thanks to the pioneering orbital missions of all these canine cosmonauts, the age of human spaceflight dawned on April 12, 1961.

✧ Animal–in–Space Missions after 1961

Once human beings successfully traveled in space in 1961, aerospace medical experts turned their research focus on the biophysical consequences to living creatures of long-term spaceflight and extensive exposure to microgravity. While cosmonauts and astronauts served as test subjects for most of these biomedical investigations, there were still some experimental circumstances in which other living creatures, including small mammals, flew in space in the name of life sciences research.

The American Biosatellite program was the first major effort by NASA to exploit Earth-orbital missions to conduct scientific investigations involving basic biological processes in space. From the perspective of the scientific method, test specimen controls during previous sounding rocket and orbital flights were usually inadequate. As a result, scientists often found it difficult to relate causes to the effects observed in biological specimens after travel in space. With the advent of the NASA's Biosatellite program, missions devoted exclusively to carefully controlled experiments in biology could be performed in an unmanned Earth-orbiting spacecraft.

Biosatellite 3 was the last mission in this American program. The 3,400-pound (1,545-kg) spacecraft was launched from Cape Canaveral by a Delta rocket and placed into a 137-mile (221-kg) perigee by 149-mile (240-km) apogee orbit around Earth. The satellite's orbit had an inclination of 33.5 degrees and a period of 92 minutes. The major objective of the *Biosatellite 3* mission was to study the behavior of a 13-pound (6-kg), male pigtailed monkey (*Macaca nemestrina*) named Bonnie in orbit around Earth for 30 days. Researchers used four similarly instrumented and restrained rhesus monkeys as ground experiment controls. However, after only 8.8 days in orbit, the NASA life scientists terminated the mission because of the monkey's deteriorating health. The monkey died eight hours after return capsule recovery, ostensibly from a massive heart attack brought on by dehydration. At the time of death, the monkey had a body mass of 9.7 pounds (4.4 kg). Some researchers suggested that the monkey's weight loss was due to the marginally palatable food pellets, which were used to accommodate experimental requirements. Mission scientists also speculated that the animal's demise was likely due to chronic restraint and over-instrumentation, rather than the consequences of microgravity. This hypothesis appears plausible, since subsequent joint Russian and American experiments with rhesus monkeys in the Cosmos (or Bion) biosatellite program had test specimens that survived five to 14-day orbital missions. In addition, shortly after the termination of the orbital flight phase of *Biosatellite 3,* two of the similarly instrumented and restrained monkeys used as ground control specimens also died.

Starting in 1966 with the launch of *Cosmos 110,* Soviet scientists conducted a series of biosatellite missions that combined studies in numerous areas within the overall field of space life sciences. This ambitious and successful program allowed scientists, including invited specialists from a number of European countries and the United States, to investigate many organisms from very different taxonomic orders.

Cosmos 110 was a 12,540-pound (5,700-kg) Earth-orbiting satellite launched from the Baikonur Cosmodrome on February 22, 1966, by a modified SS-6 ICBM. The biosatellite traveled around Earth in a 118-mile (190-km) perigee by 548-mile (882-km) apogee orbit that had an inclination of 51.9 degrees and a period of 95.3 minutes. The major biological specimens of this mission were two dogs, Veterok (Breeze) and Ugolyok (Small Piece of Coal). After traveling in space for 22 days, the return canister carried both dogs and the other test specimens safely back to Earth.

Beginning with *Cosmos 605,* the Soviet biosatellite program included small mammals, such as rats. *Cosmos 605* (also called *Bion 1*) was the first in an extensive series of biosatellite launched from the Plesetsk Cosmodrome, situated south of Archangel in the northwest corner of Russia. On October 31, 1973, a modified SS-6 ICBM placed this satellite into a 137-mile perigee by 264-mile (424-km) apogee orbit around Earth at an inclination of 62.8 degrees and with a period of 90.7 minutes. The *Cosmos 605* (*Bion 1*) spacecraft was engineered to conduct long-term experiments in space and to return the biological subjects to Earth. This biosatellite's landing module (a modified Vostok design) was a complex, autonomous compartment capable of housing biological specimens and test subjects.

Through the mid-1990s, the United States participated in eight Cosmos-designated biosatellite missions—namely, *Cosmos 782, 936, 1129, 1514, 1667, 1887, 2044,* and *2229. Cosmos 782* (also called *Bion 3*) was launched on November 25, 1975. This biosatellite was the first joint United States–Soviet biomedical research flight, and subject animals included white rats and tortoises.

As part of a joint American-Russian biological research program, Rhesus monkeys were carried into Earth orbit in pairs on the following Cosmos biosatellite missions: *Cosmos 1514, 1667, 1887, 2044,* and *2229. Cosmos 1514* (*Bion 6*) was launched on December 14, 1983, out of the Plesetsk Cosmodrome by a Soyuz rocket and placed into low-altitude (approximately 160-mile [257-km]), high-inclination (82.3 degree) orbit around Earth. This spacecraft (based on the Russian Zenith reconnaissance satellite) carried two rhesus monkeys (named Abrek and Bion) and 18 pregnant white rats as part of its biological specimen payload. Scientists used the white rats to study the combined effects of microgravity and radiation on small mammals during pregnancy. The *Cosmos 1514* (*Bion 6*) mission ended after five days. The space-traveling rats produced normal litters.

The *Cosmos 1667* (*Bion 7*) biosatellite was launched on July 10, 1985, and carried two rhesus monkeys, Verny (Faithful) and Gordy (Proud), who were recovered seven days later. The *Cosmos 1887* (*Bion 8*) biosatellite was launched on September 29, 1987, from the Plesetsk Cosmodrome. The spacecraft's biological payload included two rhesus monkeys, Yerosha (Drowsy) and Dryoma (Shaggy), who traveled in orbit for 13 days. During the flight, Yerosha partially freed himself from his restraints and, while weightless, tried to explore his orbital cage. The *Cosmos 2044*

(*Bion 9*) biosatellite carried two rhesus monkeys, Zhakonya and Zabiyaka (Troublemaker), into orbit. They were safely recovered after 14 days in orbit, although other biological specimens on the same flight died due to a failure in the spacecraft's thermal control system, which resulted in elevated temperatures.

The 13,200-pound (6,000-kg) *Cosmos 2229* (*Bion 10*) satellite was launched on December 29, 1992, from the Plesetsk Cosmodrome by a Russian Soyuz rocket. As an interesting point of comparison, *Sputnik 2* (the world's first biological satellite) had a mass of 1,118 pounds (508 kg). The *Cosmos 2229* (*Bion 10*) spacecraft carried two Rhesus monkeys, Krosh (Tiny) and Ivasha, as well as other living specimens. It orbited Earth on a planned 12-day flight. As on previous Cosmos biosatellite missions, the monkeys were trained to activate food and juice dispensers. For this mission, they were trained to operate a foot pedal so that researchers could

SAVING THE CHIMPS—SOCIAL MOVEMENTS, LEGAL CHALLENGES, AND ETHICAL EFFORTS

The concern for the welfare of captive, nonhuman primates rose considerably in the 1980s in civilized societies around the globe. In 1985, the U.S. Congress amended the Animal Welfare Act (AWA) to include, among other things, providing for the psychological well-being of nonhuman primates held captive in the United States, whether in zoological parks, amusement complexes, or research laboratories.

Over time, the concept of primate psychological well-being has become synonymous with the terms *environmental enrichment* and *environmental enhancement*. The congressional delegates responsible for the 1985 AWA amendments intended to provide captive, nonhuman primates more exercise, play, and compatible social interactions. In 1989, the Animal and Plant Health Inspection Service (APHIS) of the United States Department of Agriculture (USDA) responded to the new AWA amendments by drafting environmental enrichment regulations based on advice received from a group of primate experts. The proposed regula-

tions contained requirements for social housing of nonhuman primates, inanimate enrichment items, and exercise. With some modifications, the APHIS proposed regulations that became part of the Code of Federal regulations (namely, 9 CFR Section 3.81) in 1991. To assist APHIS personnel in the promulgation and enforcement of these regulations, the USDA's Animal Welfare Information Center prepared a seminal document entitled *Environmental Enrichment for Nonhuman Primates Resource Guide*—the latest version of which was released in March 2006.

In June 1997, the U.S. Air Force decided to divest itself of all its research chimpanzees. Since the 1950s, these higher-order primates had been used as surrogates for human beings in a variety of hazardous tests supporting military aviation or space travel–related research at U.S. Air Force biomedical facilities such as the Holloman Aero-Medical Research Laboratory in New Mexico. Under the public divestiture as envisioned and authorized by the U.S. Congress, the U.S. Air Force study muscle responses in flight. To accommodate in-flight neurovestibular testing, the monkeys were also trained to make hand and head movements in response to visual stimuli. The monkeys and other biological specimens were recovered two days earlier than scheduled because thermal control problems with the spacecraft resulted in unacceptably high onboard temperatures. After landing, both monkeys received treatment for dehydration and subsequently recovered. *Cosmos 2229* was the last mission in the Cosmos biosatellite series. The Russian Space Agency decided to identify subsequent biosatellites as simply numbered Bion satellites.

The Russian Space Agency launched the *Bion 11* biosatellite from the Plesetsk Cosmodrome on December 24, 1996, using a Soyuz rocket. Included in *Bion 11*'s complement of biological specimens were two rhesus monkeys, Lapik and Multik (Cartoon). The orbital flight lasted 14 days. Upon return to Earth, Multik died just one day after capsule recovery,

would either give the chimpanzees away to a research laboratory, or else allow the primates to retire in an animal sanctuary.

Despite the well-expressed congressional concern for the welfare of captive, nonhuman primates, U.S. Air Force officials decided to award most of the chimpanzees to research laboratories, including facilities that had already been cited by the USDA for serious violations of the Animal Welfare Act (such as negligent chimpanzee deaths). Animal rights organizations protested the decision, and lengthy legal battles began.

Save the Chimps (STC), a nonhuman primate advocacy and protection organization, was founded in 1997 in direct response to the U.S. Air Force announcement that it was divesting itself of research chimpanzees. After most of the chimpanzees were awarded to the Coulston Foundation (a biomedical research laboratory in Alamogordo, New Mexico) in the divestiture process, the Save the Chimps organization took legal action in the federal court system to contest the U.S Air Force's decision. In October 1999, after a year of legal wrangling, Save the Chimps and the Coulston Foundation entered into an agreement, which gave STC permanent legal custody of 21 chimpanzees who were the survivors or descendants of the chimpanzees originally captured in Africa in the 1950s and then trained by the U.S. Air Force in support of NASA's astrochimp effort within the Mercury Project. Today these 21 chimpanzees are retired at STC's specially designed primate sanctuary in Fort Pierce, Florida.

There is an interesting epilogue to the U.S. Air Force chimpanzee story. In September 2002, the Save the Chimp organization gained possession of the Coulston Foundation's facility in New Mexico and all 266 of the chimpanzees located there. Since then, STC personnel have modified living conditions for the chimpanzees in Alamogordo and encouraged the primates to form social groups. This initial effort is part of an overall STC strategic plan that includes enlargement of the organization's original astrochimp sanctuary in Florida. Over the next decade, all 266 chimpanzees in New Mexico will be relocated to an expanded primate sanctuary in Florida, where they will have a tranquil, environmentally enriched permanent home. In the insightful words of British playwright William Shakespeare, "All's well, that ends well."

during his post-landing medical checkup and operation (to remove implanted electrodes). In the modern information age, with news traveling quickly across international boundaries, Multik's death raised renewed ethics questions concerning the continued use of nonhuman primates in space life sciences research.

NASA scientists have conducted life sciences experiments using the space shuttle orbiter and, on certain shuttle flights, *Spacelab*—the European Space Agency's space research module. Flown in a variety of configurations, *Spacelab* rode into orbit nestled within the shuttle's massive cargo bay. As part of the *Spacelab 3* (*SL-3*) mission, two adult male squirrel monkeys (*Saimiri sciureus*) were carried into space by the space shuttle *Challenger* on April 29, 1985. During the STS-51B shuttle mission, each monkey traveled unrestrained in orbital flight in his own specially designed primate cage. The cages had a removable solid window through which astronauts could view the animal. A perforated window beneath the solid window provided physical access to the animal. Each cage had a temporary restraint system that could be activated to secure the monkey in-flight in the event of an emergency. A flow of air through the cage directed urine and feces to absorbent, removable trays beneath the cage's grid floor. Scientists used two infrared lights and two activity sensors located at the opposite sides of the cage to monitor each primate's movements. Researchers also made periodic video recordings of the monkeys to evaluate how they were responding to spaceflight. NASA life scientists also used the *Spacelab 3* monkey experiment to evaluate how the research animal holding facility would perform in space in comparison to vivarium housing on Earth.

The STS 51-B/SL-3 mission ended successfully on May 6, 1985. After seven days in space, both monkeys were returned to Earth in good condition. Each soon adapted to normal gravity following their weeklong exposure to microgravity. Postflight reports noted that both monkeys ate less food and were less active in-flight than on the ground. One monkey adapted quickly to microgravity, while the other monkey exhibited symptoms characteristic of space adaptation syndrome. The sick monkey consumed no food and drank little water during the first four days of flight. Then, on the fifth day of flight, after being hand-fed banana pellets by members of the crew, the monkey's behavior became more comparable to that of the other squirrel monkey. Purposefully, NASA officials did not give either *Spacelab 3* monkey a name. The primates were officially identified only as test specimens Number 3165 and Number 384-80.

Tiny Space Capsules and the Mercury Project

Established on October 7, 1958, only a year and three days after the former Soviet Union launched the world's first unmanned satellite (*Sputnik 1*), NASA's Mercury Project was the pioneering American effort to put a human being into orbit. (Sometimes an equivalent term "Project Mercury" is encountered in the aerospace literature.) The Mercury Project involved a series of six flights—two suborbital and four orbital. The overall project was designed to demonstrate that human beings could withstand the high acceleration of a rocket launching, a prolonged period of weightlessness, and then a period of high deceleration during reentry.

NASA's original seven astronauts were chosen for the Mercury Project in April 1959 after a nationwide call for jet pilot volunteers. President Dwight David Eisenhower (1890–1969) greatly facilitated the overall selection process by allowing the military to provide the highly qualified pilots NASA needed. The men chosen to be America's first seven astronauts were (in alphabetical order): Malcolm Scott Carpenter (b. 1925–   ); Leroy Gordon Cooper, Jr. (1927–2004); John Herschel Glenn, Jr. (b. 1921–   ); Virgil Ivan "Gus" Grissom (1926–67); Walter M. Schirra, Jr. (b. 1923–   ); Alan B. Shepard, Jr. (1923–98); and Donald "Deke" Slayton (1924–93). Each of these individuals had to pass a rigorous physical screening process as well as a stringent battery of psychological examinations. They were all superior test pilots with extensive flight experience in high-performance jet aircraft. As the Mercury Project progressed, NASA's first seven astronauts became widely recognized by the American public as members of that elite cadre of steel-nerved jet pilots who had "the right stuff."

With the exception of Donald "Deke" Slayton, all the original Mercury Project astronauts piloted the very tiny project space capsule—the first man-rated spacecraft developed and successfully flown by the United States in the early 1960s. Slayton was originally scheduled to pilot the

The original seven Mercury Project astronauts are shown here inspecting a model of the Mercury–Atlas launch vehicle. Seated in the front row (from left to right) are Virgil I. "Gus" Grissom, M. Scott Carpenter, Donald "Deke" Slayton, and L. Gordon Cooper, Jr. Standing in the back row (from left to right) are Alan B. Shepard, Jr.; Walter M. Schirra, Jr.; and John H. Glenn, Jr. This April 1959 photograph was taken at NASA's Langley Research Center. *(NASA)*

Mercury–Atlas 7 mission (the second manned orbital mission) but was relieved of this assignment due to a minor heart condition (an irregular heartbeat) that physicians discovered in August 1959. Determined to fly in space, he stayed within NASA's astronaut corps and eventually traveled into orbit as a member of the American crew that participated in the 1975 Apollo-Soyuz Test Project—the world's first cooperative international rendezvous and docking space mission.

The Mercury Project began soon after the start of the Space Age. No human being had ever flown in outer space before, and some physicians and life scientists expressed serious doubts about whether the human body could survive in extended periods of microgravity (weightlessness). Consequently, NASA's aerospace engineers had the enormous challenge of designing a space vehicle that would protect a human being from many anticipated, but not yet sufficiently quantified, hazards—including temperature extremes, vacuum conditions, and the newly discovered trapped radiation environment that permeated near-Earth space. Added to these technical challenges was the need to keep an astronaut cool during the fiery, high-speed reentry of the spacecraft through the atmosphere. Factoring in the relatively limited thrust capabilities of the early man-rated American space launch vehicles (which were initially modified military ballistic missiles), the Mercury Project engineers came up with a wingless "capsule" design that proved acceptable for the task. Their engineering efforts produced the tiny one-person Mercury Project spacecraft (or space capsule), which had a maximum orbiting mass of about 3,200 pounds (1,454 kg). Beginning with Alan Shepard's *Freedom 7* suborbital flight (May 5, 1961), the Mercury Project astronauts were allowed to name their own spacecraft. To acknowledge the teamwork that characterized the original seven astronauts, each man added a "7" to his particular spacecraft's name.

Shaped somewhat like a bell or a gumdrop, this small spacecraft was just 74.5 inches (189 cm) wide across the bottom and about nine feet (2.7 m) tall. A solid-propellant rocket-powered astronaut escape tower added another 17 feet (5.2 m), creating for an overall spacecraft length of approximately 26 feet (8 m) at launch. This escape tower would yank the space capsule and its occupant away from a malfunctioning booster and deliver the capsule to a high-enough altitude so its parachutes could deploy and then return the astronaut safely to Earth. The blunt (bottom) end of the Mercury space capsule was covered with an ablative shield to protect the capsule and its passenger (an astronaut or astrochimp) against the searing heat of atmospheric reentry.

Two boosters were chosen: the U.S. Army's Redstone intermediate range ballistic missile with its 78,000 pounds-force (346,944 N) thrust for the suborbital flights and the U.S. Air Force's Atlas intercontinental ballistic missile with its 360,000 pounds-force (1,601,280 N) thrust for the orbital missions. Prior to the manned flights, NASA conducted several unmanned tests of the booster and the space capsule, including the astrochimp flights of the chimpanzees Ham and Enos (see chapter 3).

✧ Mercury–Redstone 3

On May 5, 1961, NASA launched astronaut Alan B. Shepard, Jr., from Complex 5 at Cape Canaveral Air Force Station, using a Redstone booster. This flight was the first American manned space mission. Shepard named his space capsule *Freedom 7*.

Though only a brief suborbital mission that lasted just over 15 minutes, Shepard's flight proved that an astronaut could survive a flight in space. The historic flight also demonstrated to the estimated 45 million Americans who observed the launch as it was carried live on television from Cape Canaveral that the United States was definitely in the manned spaceflight business. The cold war–era space race was heating up because the former Soviet Union (Russia) had sent cosmonaut Yuri A. Gagarin into orbit around Earth a little less than a month earlier. (See chapter 1.)

The flight of Shepard inside his *Freedom 7* space capsule was essentially a "cannon shot." After launch, Shepard followed a ballistic trajectory from Cape Canaveral and splashed down in the Atlantic Ocean approximately 302 miles (486 km) downrange from the launch site. During the brief suborbital flight, he reached a maximum altitude of 116 miles (187 km), a maximum velocity of 5,135 miles per hour (8,260 km/h), and experienced a maximum g-force of six during the booster acceleration phase of the mission. (A g-force of one corresponds to the normal acceleration due to gravity at the surface of Earth.)

The Redstone booster performed well during launch, although there were some vibrations. After the *Freedom 7* capsule separated from the rocket booster, Shepard maneuvered his spacecraft using hand controllers that pitched, yawed, and rolled the tiny space capsule with its small thrusters. He found the ride into space smoother than expected and reported no discomfort during his five minutes of weightlessness. Although this first Mercury Project space capsule lacked a window, Shepard was able to observe the Atlantic coastline through a periscope. The view was in black and white, because he had inadvertently left a gray filter in place while waiting on the launch pad for liftoff.

After splashdown in the Atlantic Ocean, the *Freedom 7* space capsule and its pilot were recovered and returned by helicopter to the aircraft carrier USS *Lake Champlain.* Just three weeks following Shepard's successful mission, President Kennedy made his historic address to the U.S. Congress in which he gave the United States a goal of sending American astronauts to the Moon and safely returning them to Earth before the end of the 1960s.

✧ Mercury–Redstone 4

On July 21, 1961, another Redstone booster hurled astronaut Virgil I. "Gus" Grissom through the second and last suborbital flight in the *Liberty Bell 7* Mercury space capsule. Grissom's suborbital mission was essentially a repeat of Shepard's flight, although the *Liberty Bell 7* spacecraft had a few minor improvements, including improved hand controllers, a window, and an explosively activated side hatch. The astronauts had requested the addition of an explosive side hatch for easier capsule escape in case of an emergency.

Learning from Shepard's overly busy suborbital flight, NASA mission managers intentionally reduced Grissom's duties and gave him more time to view Earth during the brief flight. Grissom's flight was 15 minutes and

37 seconds in duration. His spacecraft attained a maximum velocity of 5,140 miles per hour (8,270 km/h) and reached a maximum altitude of 117 miles (189 km). At the end of the suborbital mission, the *Liberty Bell 7* capsule splashed into the Atlantic Ocean about 300 miles (483 km) downrange from the launch site at Cape Canaveral.

From liftoff of the Redstone booster rocket to capsule reentry, the operational sequences during Grissom's Mercury–Redstone 4 mission were quite similar to those of Shepard's first suborbital flight. Like Shepard, Grissom reported no ill effects due to the five minutes of weightlessness he encountered. The major anomaly that happened during this mission took place after a successful splashdown in the Atlantic Ocean. While Grissom waited inside his floating space capsule to be picked up by the helicopter rescue and recovery team, the *Liberty Bell 7*'s side hatch somehow activated prematurely, and the tiny space capsule began to fill with seawater. Grissom safely exited the capsule and waited in the water for the arrival of the helicopter. As his space suit began to fill with water, he was plucked from the ocean and delivered unharmed to the recovery ship USS *Randolph*. Unfortunately, despite efforts by another helicopter crew to save the *Liberty Bell 7* space capsule, it filled with water and sank.

Discounting the postflight loss of the *Liberty Bell 7* space capsule, NASA officials regarded Grissom's flight as another successful mission and moved forward with plans to send the first American astronaut into orbit around Earth. However, the sinking-capsule incident resulted in a change in recovery procedures. Astronauts would now be required to keep the hatch's firing safety pin in place until after the recovery helicopter's hook was firmly attached and tension applied to the recovery cable.

Following a number of unsuccessful salvage attempts in 1992, 1993, and 1999, on July 20, 1999, a recovery expedition finally succeeded in raising the *Liberty Bell 7* from the bottom of the Atlantic Ocean. The successful deepsea salvage took place about 520 miles (830 km) northwest of Grand Turk Island. Still attached to the space capsule was the recovery line from the helicopter, whose crew had tried to prevent the *Liberty Bell 7* from sinking in 1961. Also among the artifacts discovered inside the space capsule were some of astronaut Grissom's flight gear and several Mercury head dimes, which were carried into space as souvenirs of this flight.

✧ Mercury–Atlas 6

Following two successful suborbital missions, NASA mission managers advanced the Mercury Project to the next important phase, the Mercury–Atlas series of orbital missions. A major milestone in the United States civilian space program was reached on February 20, 1962, when astronaut

The Mercury Atlas 6 rocket lifts off from Cape Canaveral Air Force Station in Florida on February 20, 1962. Inside the *Friendship 7* Mercury Project space capsule was Astronaut John H. Glenn, Jr.—the first American to travel around Earth in a spacecraft. *(NASA)*

John H. Glenn, Jr., became the first American in orbit and circled the Earth three times in the *Friendship 7* spacecraft.

The Mercury–Atlas 6 mission was the first orbital flight of an American spacecraft with a human passenger. The pilot was John Glenn, a U.S. Marine Corps officer and aviator, who named his capsule *Friendship 7*. NASA planners had set several objectives for this historic mission. The first objective was to evaluate the performance of this new man-rated spacecraft in a three-orbit duration mission. The second objective was to evaluate the effects of several hours of spaceflight on a human being. The third objective was to obtain the astronaut's personal evaluation of the suitability of the Mercury space capsule and its supporting systems for other, more extensive manned missions.

Originally scheduled for launch in late January 1962, the Mercury–Atlas 6 mission was postponed twice—once (on January 27) because of adverse weather and once (on January 30) because of a fuel leak in the Atlas rocket. Finally, on February 20, as an estimated 60 million people viewed the launch via live television coverage, Atlas rocket successfully lifted-off from Cape Canaveral and placed the *Friendship 7* spacecraft into orbit around Earth. During the flight, the spacecraft attained a maximum velocity in excess of 17,400 miles per hour (28,000 km/h) and an orbital altitude of approximately 160 miles (260 km).

Glenn's orbital flight lasted almost five hours, during which time he circled Earth three times and observed everything from city lights in Australia to a large dust storm in Africa. He was the first American to see a sunrise and sunset from the vantage point of space. On his own initiative, Glenn also became the first space traveler to photograph Earth from space. He accomplished this by purchasing a 35-millimeter camera in a local pharmacy in Cocoa Beach, Florida, and then taking this camera along on his mission.

Prior to Glenn's orbital flight, aerospace medicine experts were concerned about the possible physiological effects on an astronaut's body due to prolonged conditions of weightlessness and exposure to space radiation. Glenn reported that the microgravity conditions he experienced were actually "very handy" in performing his tasks. He also described his almost five hours of weightlessness as "exhilarating"—except perhaps for a moment when he sampled less than tasty space food. His postflight medical evaluation revealed that he had received less than half the expected amount of ionizing radiation dosage. This important fact indicated that the walls of the Mercury space capsule could provide a reasonable amount of radiation shielding—at least during short-duration flights in low Earth orbit.

There was a very curious observation that happened during Glenn's flight. When he entered the sunrise portion of an orbit, he reported seeing what he described as "fire flies" outside his space capsule. This unusual phenomenon remained a mystery until the Mercury–Atlas 7 mission. During that mission, astronaut M. Scott Carpenter accidentally tapped his hand against the interior wall of the *Aurora 7* space capsule. Carpenter's action released a swarm of so-called fire flies. NASA engineers quickly recognized that the source of the interesting phenomenon was just frost from the space capsule's reaction control jets.

Several major problems also occurred during Glenn's flight. First, a yaw attitude control jet became clogged, forcing the astronaut to abandon use of the automatic control system in favor of the manual-electrical fly-by-wire system and manual-mechanical system. Then, there was a signal in the heat shield circuit indicating that the clamp, which held the heat shield in place, had prematurely released. This caused a good deal of concern, suggesting that the capsule's heat shield was loose while the astronaut was still traveling in orbit. A loosened heat shield would not withstand the rigors of reentry heating, and the mission could end in disaster. As a safety measure, NASA mission controllers instructed Glenn not to jettison the retrorocket pack hardware prior to his reentry maneuver. This action would hold the heat shield in place in the event that the heat shield had somehow actually loosened during flight. As the capsule plunged into the atmosphere, Glenn saw some spectacular pyrotechnic displays as chunks of the retrorocket pack burned up and flew past the window. At one point he even thought his heat shield was burning up and breaking away. But, fortunately, that was not the case. Postflight investigations of the space capsule showed that a faulty switch had erroneously generated this alarming heat shield signal.

Finally, Glenn ran out of fuel for the capsule's attitude control system as he tried to stop the spacecraft's bucking motion as it descended through the atmosphere on reentry. Despite these difficulties, he splashed down safely in the Atlantic Ocean some 810 miles (1,300 km) southeast of Bermuda and about 40 miles (64 km) short of his intended target area. After bobbing about 21 minutes in the water, the *Friendship 7* space capsule with its pilot still inside, was plucked from the ocean and returned by helicopter to the destroyer USS *Noa*—the nearest surface ship in the recovery task force. Having achieved the Mercury Project's primary goal with his orbital mission, Glenn became a national hero.

Glenn retired from the U.S. Marine Corps on January 1, 1965, and won election to the U.S. Senate from Ohio in November 1974. He served in the U.S. Senate until January 1999. Decades after his historic Mercury–Atlas 6 flight, Glenn returned to space as a payload specialist on board NASA's space shuttle *Discovery* during the STS-95 mission, which took place between October 29 and November 7, 1998. With this space mission, John Glenn became the oldest human being to travel in space in the 20th century.

✧ Mercury–Atlas 7

On May 24, 1962, Astronaut M. Scott Carpenter completed a three-orbit flight in the *Aurora 7* spacecraft. The Mercury–Atlas 7 was the second manned orbital flight of the Mercury Project. NASA had originally selected Donald K. "Deke" Slayton to be the pilot for this mission. However, Carpenter received the assignment after a medical examination of Slayton revealed an irregularity of his heartbeat.

The objectives of Carpenter's mission were similar to those of John Glenn's mission (Mercury–Atlas 6). Since Glenn's mission had already demonstrated the efficacy of human spaceflight in the Mercury space capsule, NASA managers decided to include some science experiments as part of Carpenter's five-hour-long Aurora 7 mission. Consequently, Carpenter's flight plan contained the first study of the behavior of liquids under the microgravity conditions of an orbiting spacecraft, astronaut-conducted photography of Earth, and the inflation and deployment of a balloon to measure the drag of Earth's residual atmosphere and the influence of solar pressure on objects in low Earth orbit. The balloon experiment failed when the device did not properly inflate on deployment, but the liquid experiment behaved as generally anticipated. With this flight, NASA made astronaut-conducted photography of Earth an integral part of the American human spaceflight program.

During the flight, the *Aurora 7* spacecraft attained a maximum velocity in excess of 17,400 miles per hour (28,000 km/h) and a maximum altitude of about 165 miles (267 km). Carpenter encountered only one critical component malfunction. A random failure of the circuitry associated with the pitch horizon scanner, which provided a reference point to the space capsule's attitude control gyros, occurred. NASA mission controllers also had concern about excessive attitude control propellant consumption, primarily caused by the astronaut's extensive use of the high-thrust control rockets and his inadvertent use of two control systems simultaneously. To compensate for this circumstance, NASA flight controllers allowed the *Aurora 7* capsule to drift in attitude control for an additional 77 minutes beyond the time already built into the flight plan.

At sunrise on the third and final orbit, Carpenter inadvertently bumped his hand against the inside wall of the *Aurora 7*'s cabin and solved the mystery of the so-called fire flies that were seen by Glenn on the previous orbital flight. The resulting bright shower of particles outside the space capsule—what Glenn had called "fire flies"—turned out to be nothing more than ice particles shaken loose from the space capsule's exterior.

Partly because he had been distracted by watching the so-called fire flies and partly because of his very busy flight schedule, Carpenter overshot his planned reentry mark and splashed down in the Atlantic Ocean about 250 miles (400 km) beyond the planned impact point. After the space capsule's retrorockets fired, computers at NASA's Goddard Space Flight Center successfully predicted the area of splashdown, and naval ships and aircraft rapidly deployed to the new location, which was about 125 miles (200 km) northeast of Puerto Rico.

Some 39 minutes after the *Aurora 7* capsule splashed down in the ocean, a U.S. Navy amphibian aircraft was the first search-and-rescue craft to establish visual contact with the spacecraft. The USS *Farragut* was the first surface ship to reach the impact area. After almost three hours in the water, Carpenter was picked up from the floating *Aurora 7* capsule and returned by helicopter to the aircraft carrier USS *Intrepid*. The astronaut experienced no adverse physical or biomedical effects due to this flight. The *Aurora 7* space capsule was not retrieved until about six hours later, when the USS *John R. Pierce* arrived and used its special onboard equipment to retrieve the spacecraft from the surface of the ocean.

✧ Mercury–Atlas 8

By orbiting Earth six times, Astronaut Walter M. Schirra, Jr., doubled the American flight time in space. He concluded the Mercury–Atlas 7 mission by landing his *Sigma 7* space capsule in the planned Pacific Ocean recovery area. All previous Mercury Project water landings had been in the Atlantic Ocean.

Originally scheduled for launch in early September 1962, the Mercury–Atlas 8 mission was postponed twice to provide NASA engineers additional time for flight preparation. After Carpenter's flawed reentry, the flight emphasis of the Mercury Project returned to aerospace engineering tasks rather than science experiments. To highlight this renewed emphasis on engineering, Schirra even named his spacecraft *Sigma 7*—with sigma (Σ) representing the engineering and mathematical symbol for summation. NASA successfully launched Schirra's mission on October 3 from Cape Canaveral. The launch was the first to be relayed live (via the *Telstar* communications satellite) to television audiences in western Europe.

NASA engineers made two significant modifications to Schirra's *Sigma 7* spacecraft to eliminate the difficulties encountered during the two previous orbital flights. The first involved an alteration of the Mercury space capsule's reaction control system to disarm the high-thrust jets and to permit use of low-thrust jets only in manual operation. This change helped conserve attitude control system propellant during orbital flight. The second modification involved the addition of two high-frequency antennae (mounted on the space capsule's retrorocket package) to assist and maintain communications between the spacecraft and ground throughout the flight.

The six-orbit mission lasted nine hours and 13 minutes, during which Schirra mostly spent in what he termed *chimp configuration.* He was referring to a free drift flight mode that tested the Mercury space capsule's autopilot system—as done during the November 1961 orbital flight test with astrochimp Enos as the passenger. Schirra tried "steering" the *Sigma 7* spacecraft by the stars but found that task quite difficult. He also used a 70-millimeter Hasselblad camera with various filters to collect imagery of Earth from space. After his flight, NASA scientists began assembling these images into a catalog of astronaut-collected Earth photography for comparison with similar images obtained by (uncrewed) Earth-orbiting spacecraft.

During Schirra's flight, the *Sigma 7* spacecraft attained a maximum velocity of 17,448 miles per hour (28,092 km/h) and a maximum altitude of approximately 176 miles (283 km)—the highest altitude achieved in the Mercury Project. After completing six orbits, the Mercury space capsule reentered Earth's atmosphere and splashed down in the Pacific Ocean, about 273 miles (440 km) northeast of Midway Island. To complete his essentially textbook mission, Schirra landed the *Sigma 7* spacecraft in the water just five miles (8 km) away from the prime recovery ship, the USS *Kearsarge.*

✦ Mercury–Atlas 9

Between May 15 and 16, 1963, Astronaut L. Gordon Cooper, Jr., performed a 22-orbit mission around Earth in the *Faith 7* spacecraft. His flight triumphantly concluded the first American human spaceflight program. Because Cooper's flight went so well, Mercury Project officials decided to cancel a planned seventh human-crewed flight and proceed with the two-person spacecraft Gemini Project instead.

To support Cooper's long-duration mission, NASA engineers made a number of alterations in his *Faith 7* space capsule. These modifications included an increase in the capacity of several life support system components, such as additional oxygen and water, increased urine collection capacity, and increased condensate removal capacity. The space capsule was also given a larger supply of attitude control propellant and two larger capacity electric batteries. To accommodate the mass increase due to these additions while keeping the spacecraft's overall mass within tightly controlled limitations, NASA engineers removed several backup components that they deemed unnecessary for the mission.

During the Mercury–Atlas 9 mission, Cooper became the first astronaut to sleep in space. As he circled Earth, he also released a tiny minisatellite—a flashing beacon used to test the astronaut's ability to visually track objects in space. Cooper participated in an additional visual acquisition and perception study, during which he was able to spot a powerful 44,000-watt xenon lamp shining up at him from the ground.

The first significant malfunction of concern on the Mercury–Atlas 9 mission occurred during the 19th orbit, when the indicator light for an instrument sensitive to subtle changes in the spacecraft's microgravity level (on the order of 0.05-g to be precise) erroneously came on. This light normally appeared only during the reentry phase of an orbital flight. Cooper immediately checked the spacecraft's other instruments. Everything else appeared normal for an orbital flight. In addition, all telemetry being sent by *Faith 7* to mission support personnel on the ground indicated that Cooper's spacecraft was in the correct orbit. So NASA mission managers concluded that the indicator light was an erroneous signal. However, because of this instrument malfunction, NASA engineers also determined that the potential existed for a partial or total failure in the spacecraft's automatic system for reentry. They therefore advised Cooper to use the manual mode for reentry. He became the first astronaut to use this method exclusively.

During the flight, the spacecraft attained a maximum velocity of 17,438 miles per hour (28,075 km/h) and a maximum altitude of 165 miles (265 km). After completing 22 orbits, the *Faith 7* space capsule reentered Earth's atmosphere under the manual control of its human pilot. Cooper landed the spacecraft about 81 miles (130 km) southeast of Midway Island in the Pacific Ocean. In a testament to his superb skill as a spacecraft pilot, *Faith 7* splashed down in the ocean about four miles (6 km) from the prime recovery ship, USS *Kearsarge*.

Space Walks and the Gemini Project

This chapter summarizes the major activities and accomplishments of NASA's Gemini Project (1964–66). The Gemini Project was the second American crewed space program and the beginning of sophisticated human spaceflight. The project was announced to the public on January 3, 1962, after the Apollo Project was already well underway. The Gemini Project expanded and refined the scientific and technological endeavors of the Mercury Project and prepared the way for the technically more sophisticated Apollo Project, which carried American astronauts to the lunar surface.

Gemini means "twins" in Latin. Searching for a spacecraft (and project) name late in 1961, NASA officials selected the name Gemini in lieu of "Mercury Mark II"—the working designation for the follow-on Mercury space capsule model which was capable of carrying two astronauts. The choice proved quite popular. In astronomy, Gemini is the third constellation of the zodiac—a constellation characterized by the twin stars Castor and Pollux. In Greek mythology, Castor and Pollux are inseparable brothers who agree to share immortality by spending half their time on Earth and half in Olympus with the immortal gods.

The one-person Mercury space capsules that were launched between 1961 and 1963 did not provide sufficient spaceflight experience for the great endeavor of landing human beings on the Moon under the Apollo Project. The Gemini Project added a second crewmember and a maneuverable spacecraft. The project's objectives included demonstrations of the following: rendezvous and docking techniques with orbiting spacecraft, extravehicular activity (EVA) or a "walk in space," long-duration flight, and guided spacecraft reentry. The Gemini Project served as the essential technical bridge between the initial demonstration that human beings could travel in space and the ability of American astronauts to walk on the

Moon. It was in the Gemini Project that NASA gathered the major portion of the spaceflight experience necessary to accomplish President Kennedy's vision of the Apollo Project's Moon landings.

The first two Gemini missions were unmanned tests of spacecraft systems. NASA launched the first crewed Gemini orbital flight, called the Gemini 3 mission, from Cape Canaveral Air Force Station, Florida, on March 23, 1965. During a 20-month period between 1965 and 1966, 10 two-person launches occurred, successfully placing 20 astronauts in orbit and returning them safely to Earth. As part of the last five missions, manned Gemini spacecraft rendezvoused and docked with unmanned Agena vehicles that had been previously placed in orbit as docking targets.

The requirement to rendezvous and dock with an orbiting spacecraft often resulted in very short launch windows for the Gemini missions. Constrained by such windows, the efficiency of NASA's ground support operations at the Kennedy Space Center improved dramatically during the Gemini Project. During this project, NASA officials also made a decision to move the mission control activities for human spaceflight from Cape Canaveral, Florida, to Houston, Texas. Starting in the Gemini Project, once a manned rocket lifted off from its launch pad at Cape Canaveral, NASA immediately transferred control of the flight from launch personnel at Kennedy Space Center in Florida to human spaceflight personnel assigned to the Mission Control Center at the Johnson Space Center in Texas.

✧ Gemini Spacecraft

Shaped like a truncated cone, the Gemini spacecraft was a significant improvement over the Mercury Project space capsule in both size and capability. With an overall length of 18.8 feet (5.7 m), a maximum diameter (at the base) of 10 feet (3 m), and a maximum mass of approximately 8,400 pounds (3,820 kg), the Gemini spacecraft was a physical enlargement of the Mercury space capsule. Yet, despite having more than twice the mass of the Mercury capsule, the Gemini spacecraft remained quite cramped, since it had only 50 percent more passenger cabin space and had to carry twice as many people.

The Gemini spacecraft consisted of two components: a reentry module and an adapter module. The reentry module was mainly the pressurized cabin that held the two astronauts. It was a truncated cone that decreased in diameter from 7.5 feet (2.3 m) at the base to about 3.2 feet (0.9 m) at the upper end. The reentry module was topped by short cylinder (also 3.2 feet in diameter) and then another truncated cone, which decreased to a diameter of 2.4 feet (0.74 m) at the flat top. The reentry module had a total height of 11.3 feet (3.45 m).

At its base, a curved, ablative heat shield separated the reentry module from the retrorocket section of the adapter. The narrow top of the reentry module contained the cylindrical reentry control system section. Engineers placed the spacecraft's rendezvous and recovery section (including reentry parachutes) above the reentry control system section. The pressurized cabin had about 79 cubic feet (2.3 m³) of volume and contained two seats (each equipped with an emergency ejection device), instrument panels, life support equipment, and equipment stowage compartments. Finally, the pressurized cabin had two large hatches with small windows. There was one hatch over each seat. Both of these hatches opened outward to accommodate astronaut ingress and egress on Earth and (on certain missions) the performance of an extravehicular activity by one of the Gemini astronauts while the spacecraft was in orbit.

The adapter module made up the base of the Gemini spacecraft. It was a truncated cone 7.5 feet (2.3 m) high, 10 feet (3 m) in diameter at the base, and 7.3 feet (2.3 m) in diameter at the upper end, where the adapter module attached to the base of the reentry module. The adapter module consisted of an equipment section at the base and a retrorocket section at the top. The equipment section held fuel and propulsion systems. The retrorocket section contained four solid-propellant rocket motors that were used to bring the astronauts in the reentry module back to Earth from orbit.

While the Gemini spacecraft resembled an enlarged version of the Mercury space capsule, there were some significant differences between the two spacecraft. The Mercury capsule's rocket-propelled escape tower was replaced by ejection seats. Aerospace engineers also gave the Gemini spacecraft more storage space to accommodate the crew needs during the longer-duration flights that took place as part of this project. To support onboard electric power needs during the long-duration missions, NASA engineers installed fuel cells in place of batteries.

Engineers made the Gemini spacecraft more serviceable. This design approach greatly accelerated the performance of ground support activities prior to each launch. For example, onboard oxygen, fuel, and other consumable supplies were carried in an adapter module that fitted to the rear of the spacecraft and was jettisoned before reentry. This design approach not only improved spacecraft checkout and servicing but also enhanced end-of-mission safety, since the compartments containing potentially hazardous-on-impact consumables were separated from the main crew capsule prior to reentry and splashdown (landing).

The Gemini spacecraft was far more maneuverable than its predecessor, the Mercury space capsule. Unlike the Mercury capsule, which only needed to change its orientation in space, the Gemini spacecraft needed a more robust maneuvering capability if the two-man crew was to perform precise rendezvous and docking operations with another orbiting spacecraft. Gemini astronauts had to move their spacecraft forward, backward, and sideways with respect to its orbital path, as well change the spacecraft's orbit. The demands and complexities of an orbital rendezvous and dock-

A two-stage Titan II rocket lifts off from Complex 19 at Cape Canaveral, Florida, on July 18, 1966, at the start of the Gemini 10 mission. On board the *Gemini 10* spacecraft are astronauts John W. Young and Michael Collins, who will successfully rendezvous and dock with the unmanned Gemini Agena target vehicle that was launched two days earlier by an Atlas rocket from Complex 14 at Cape Canaveral. *(NASA)*

ing operation required the presence of two astronauts on board the space-craft. Gemini astronauts had to perform more piloting than required or even possible with the Mercury space capsule. The Gemini spacecraft was the first to carry onboard computers. These early computer systems helped the astronauts by calculating some of the data they needed, while perform-ing complicated rendezvous and docking maneuvers.

The Gemini spacecraft ascended into orbit on top of a powerful Titan II rocket. The target for the orbital rendezvous operations was an unmanned Agena upper-stage rocket, usually lifted into space a few days

AGENA

The Agena was a versatile, upper-stage rocket that supported numerous American military and civilian space missions in the 1960s and 1970s. One special feature of this liquid-propellant system was its in-space engine restart capability. The U.S. Air Force originally developed the Agena for use in combina-tion with Thor or Atlas rocket first stages. Agena A, the first version of this upper stage, was followed by Agena B, which had a larger fuel capacity and engines that could restart in space. The later Agena D was standard-ized to provide a launch vehicle for a variety of military and NASA payloads. For example, NASA used the Atlas-Agena vehicles to launch large Earth-orbiting satellites as well as lunar and interplanetary space probes; Thor-Agena vehicles launched scientific satellites, such as the *Orbiting Geophysical Observatory (OGO)*, and applications satellites, such as *Nimbus* meteorological satellites. In the Gemini Project, the Agena D vehicle, modified to suit special-ized requirements of space rendezvous and docking maneuvers, became the Gemini Agena Target Vehicle (GATV).

Aerospace engineers configured the GATV to be launched into Earth orbit as the upper stage of an Atlas launch vehicle configuration—

prior to a Gemini Project mission. Once it reached the planned orbit, Gemini astronauts used the GATV for rendezvous and docking practice. The GATV had a docking cone at the forward end into which the nose of the Gemini spacecraft could be inserted and held with docking latches. The GATV was a 19.7-foot- (6-m-) long cylinder with a diameter of 16.4 feet (4.9 m). The primary and secondary propulsion systems were located at the back end of the target vehicle along with the attitude control gas tanks and the main propellant tanks. The docking cone was connected to the front end of the GATV by shock-absorbing dampers. NASA engineers installed acquisition running lights and target vehicle status-display indica-tors on the front end of the vehicle to assist the Gemini astronauts as they performed orbital docking maneuvers. With respect to commu-nications, the GATV had a 6.9-foot- (2.1-m-) long retractable L-band boom antenna, which extended from the side of the target docking adapter cylinder near the front. Tracking and command of the GATV were also assisted by a rendezvous beacon, two spiral L-band antennae, two tracking antennae (C-band and S-band), two VHF telemetry antennae, and a UHF command antenna.

earlier by an Atlas rocket. After rendezvousing with the Agena target vehicle, the astronauts would precisely maneuver the Gemini spacecraft so as to fit its nose into a special docking collar on the Agena.

As previously mentioned, the Gemini spacecraft had two hatches, one for each astronaut. This design feature allowed American astronauts to make their first forays outside the spacecraft in an activity officially called an extravehicular activity (EVA) but more popularly referred to as space walking. In the Gemini Project, such space walks proved more challenging and difficult than previously anticipated. Following astronaut Edward H. White II's successful solo EVA outside the *Gemini 4* spacecraft, it was not until the last Gemini mission (namely, Gemini 12) that things would again go as smoothly as planned. As discussed later in this chapter, other Gemini astronauts who made solo ventures outside their orbiting spacecraft encountered a variety of problems. On the long-duration Gemini Project missions, the astronauts had to learn to sleep and perform housekeeping tasks in crowded quarters. Both of these new human spaceflight experiences also proved quite difficult.

✧ Gemini 3 Mission

The first manned mission of the Gemini Project was the Gemini 3 mission. It was launched on March 23, 1965, at 9:24 A.M. from Complex 19 at Cape Canaveral Air Force Station by a Titan II rocket. The Titan II rocket placed the 7,121-pound (3,237-kg) *Gemini 3* spacecraft into an initial 100-mile (161-km) perigee by 139-mile (224-km) apogee orbit with a period of 88.3 minutes and at an inclination of 32.6 degrees. Astronaut Virgil I. Grissom flew the mission as the spacecraft commander, and astronaut John W. Young served as pilot.

With reference to the hit Broadway show *The Unsinkable Molly Brown*, Grissom gave the *Gemini 3* spacecraft its nickname, "Molly Brown." By choosing this nickname, he was humorously alluding to what happened at the end of the second Mercury Project suborbital flight, when his *Liberty Bell 7* space capsule blew a hatch and sank to the ocean bottom after splashdown. Grissom was also implying that there would not be a repeat performance at the end of this Gemini mission. (As a historic note, NASA designated all subsequent flights in the Gemini Project with Roman numerals and then used these numerical designations for spacecraft names, as for example the *Gemini VI* spacecraft. For the sake of clarity and editorial continuity, however, Arabic numerals are used in this book to describe the Gemini spacecraft—a practice consistent with the style found in several contemporary NASA databases and reports.)

This mission had several major objectives, the most important of which was the evaluation and qualification of NASA's new two-man spacecraft. Other goals included testing of the worldwide tracking network, demonstrating the capability of the spacecraft's orbit attitude and maneuver system, evaluating the recovery procedures, and demonstrating controlled reentry and (ocean) landing. Grissom and Young completed three orbits of Earth in the *Gemini 3* spacecraft and then began a manually controlled reentry at the end of the third orbit.

Splashdown in the Atlantic Ocean occurred some 19 minutes later in the vicinity of Grand Turk Island. Because the spacecraft experienced a less than expected lift force during reentry, *Gemini 3* landed some 69 miles (111 km) short of the target point. The ocean landing took place on March 23, at 2:16 P.M. (EST). While awaiting retrieval from the ocean, both astronauts became seasick. At approximately 3:00 P.M. (EST), they decided to remove their space suits and climb out of the spacecraft, which was bobbing vigorously in the waves. Some 30 minutes later, a helicopter picked the astronauts up and delivered them to the recovery ship, USS *Intrepid*. A preliminary postflight medical examination indicated both Grissom and Young were in good condition. Later that afternoon, the *Gemini 3* spacecraft was also recovered from the surface of the ocean. NASA records the orbital duration of the Gemini 3 mission as approximately four hours and 53 minutes. With all major objectives met (save for the demonstration of a precision reentry and landing), NASA considered the first crewed flight of the Gemini spacecraft a success. Although the spacecraft was supposed to have sufficient lift during reentry to support a precision landing, preflight wind tunnel predictions simply did not match physical reality.

✧ Gemini 4 Mission

The Gemini 4 mission (also designated by NASA as Gemini IV) was the second crewed flight of the Gemini Project. Astronaut James A. McDivitt served as spacecraft commander, and astronaut Edward H. White II served as pilot. The 7,863-pound (3,574-kg) *Gemini 4* spacecraft was launched on June 3, 1966, at 10:16 A.M. from Cape Canaveral AFS's Complex 19 by a Titan II rocket. The Titan II placed the *Gemini 4* spacecraft into an initial 101-mile (162-km) perigee by 175-mile (282-km) apogee orbit with a period of 88.9 minutes and at an inclination of 32.5 degrees.

The Gemini 4 mission had several major objectives: to evaluate the consequence of prolonged spaceflight on human beings; to evaluate the performance of the spacecraft and its subsystems on an extended (four-day duration) flight; and to evaluate procedures for crew rest and work cycles, eating schedules, and real-time flight planning. The mission had two secondary goals: to perform the first extravehicular activity (EVA) by an American astronaut and to make the first attempt at an orbital rendezvous,

On June 3, 1965, astronaut Edward H. White II became the first American to step outside of an orbiting spacecraft to perform an extravehicular activity (EVA), or space walk. For 22 minutes, White, while attached to a 23-foot- (7-m-) long tether, floated outside of the *Gemini 4* spacecraft. To maneuver while floating in space, he used a handheld "zip gun"—a small propulsion device officially called the handheld self-maneuvering unit. White's traveling companion, astronaut James A. McDivitt, photographed the historic EVA from inside the *Gemini 4* spacecraft. *(NASA)*

in this case using the second stage of the Titan II launch vehicle as a target, and then to practice stationkeeping maneuvers.

Following orbital insertion, the astronauts raised their spacecraft's orbit in an attempt to rendezvous with the Titan II rocket's second stage. But as McDivitt and White thrust their spacecraft toward the orbital target, the *Gemini 4* spacecraft only moved farther away. This exercise in rendezvous and stationkeeping was canceled early in the second orbit, after the astronauts had depleted about 42 percent of the spacecraft's propellant supply. Despite the failure, NASA engineers and mission planners learned an important lesson about the complications of orbital mechanics. On subsequent rendezvous missions, astronauts on board the chaser spacecraft would first drop to a lower, faster orbit and then, at the right moment, rise the chaser spacecraft to the higher orbital altitude, where the target was.

The next activity during the Gemini 4 mission was far more successful and spectacular. At approximately 2:33 P.M. astronaut White donned special gear and pressurized his space suit to 3.7 psi (25.5 kilopascals). McDivitt (also wearing a space suit) then depressurized the crew cabin. Within a minute, White opened his hatch. Then, two minutes later, he stood up and exited the *Gemini 4* spacecraft, becoming the first American astronaut to perform an extravehicular activity. White used a handheld gas gun to help him walk in space. He remained attached to the spacecraft by means of a 26-foot- (8-m-) long tether. After the gas gun's propellant supply became exhausted (about three minutes into the EVA), White pulled on the tether and twisted his body to maneuver in space around the *Gemini 4* craft. The historic EVA lasted 23 minutes, after which White pulled himself back into the spacecraft. He had difficulty sealing the hatch but, working together with McDivitt, managed to close it properly. Following the EVA, the astronauts repressurized the cabin. They then let the spacecraft fly in a drifting mode for the next 30 hours to conserve propellant.

A malfunction occurred in the spacecraft's computer during the 48th orbit, making it impossible to conduct the planned computer-controlled reentry. Instead, at the start of the 62nd orbit, the astronauts initiated a ballistic reentry—similar to the reentry procedures used during the Mercury Project. The *Gemini 4* spacecraft splashed down 16 minutes later (at 12:12 P.M.) in the western portion of the Atlantic Ocean. When the space capsule hit the water, it was only 50 miles (81 km) from the planned impact point. A helicopter quickly recovered McDivitt and White, who were then flown to the aircraft carrier USS *Wasp*. About an hour later, the *Gemini 4* spacecraft was successfully recovered. With the exception of a rendezvous with the Titan II rocket's second stage and a computer-controlled reentry, the Gemini 4 mission achieved all its major objectives. The astronauts also performed several experiments as the spacecraft traveled in orbit for 97 hours and 56 minutes. This mission provided aerospace medicine experts an increased level of confidence that human beings could survive and function for extended periods of time in space—bringing the Apollo Project a step closer to reality.

✧ Gemini 5 Mission

The Gemini 5 mission (also designated by NASA as Gemini V) was the third crewed flight of the Gemini Project. Astronaut L. Gordon Cooper served as spacecraft commander, and astronaut Charles (Pete) Conrad, Jr., served as pilot. The *Gemini 5* spacecraft was launched on August 21, 1965, at 8:59 A.M. from Cape Canaveral AFS's Complex 19 by a Titan II rocket. The Titan II placed the 7,931-pound (3,605-kg) *Gemini 5* spacecraft into an initial 101-mile (162-km) perigee by 218-mile (350-km) apogee orbit with a period of 89.6 minutes and at an inclination of 32.6 degrees.

The Gemini 5 mission had several major objectives: to evaluate the spacecraft's guidance and navigation system to support rendezvous and controlled reentry guidance, to demonstrate a long-duration (eight-day) crewed flight, to evaluate the consequences of long-term exposure to microgravity, to evaluate the spacecraft's new fuel-cell power system, and to test rendezvous capabilities and maneuvers using the radar evaluation pod (REP). The REP was a 76-pound (34.5-kg) optical and electronic duplicate of the Gemini Agena Target Vehicle planned for use in later missions.

During the second orbit (about two hours after liftoff), the astronaut crew deployed the REP. Some 36 minutes into the evaluation of the spacecraft's rendezvous system, Cooper and Conrad noticed the pressure in the fuel cell's oxygen supply tank was dropping. Although the tank's pressure was still above the allowed minimum, NASA flight controllers instructed the astronauts to cancel the REP exercise and to power down the spacecraft. After engineers at the Johnson Space Center performed an analysis of the fuel-cell problem, NASA mission control instructed the astronauts to begin a powering-up procedure. This procedure took place during the seventh orbit around Earth. For the rest of the mission, the pressure slowly rose in the fuel cells, and sufficient power was available for the remainder of the mission. As would occur on many future crewed flights that encountered a hardware problem, NASA personnel on the ground would perform analyses and tests on identical mission hardware, isolate the problem, and then mission control would send repair or workaround procedures to the astronauts traveling in space. A workaround procedure is one that isolates or bypasses the hardware or software problem, allowing the mission to continue—sometimes at a reduced level of activity or performance.

On the second day in space (during orbit number 14), the crew conducted rendezvous radar tests, followed on the third day by a simulated rendezvous with a phantom Agena. During orbit 120 (on August 29), the crew fired the spacecraft's retrorockets. (The retrofire procedure was performed one orbit early due to the threat of a tropical storm near the planned landing area.) About 28 minutes later, the *Gemini 5* spacecraft splashed down in the western Atlantic Ocean at a point about 105 miles (169 km) short of the intended target area. The "short" landing was due to a ground-based computer program error. Within 90 minutes, astronauts Cooper and Conrad were safely on board the recovery ship, the aircraft carrier USS *Lake Champlain*. The *Gemini 5* spacecraft was recovered about two hours later.

During the Gemini 5 mission, Cooper and Conrad had traveled in space for 190 hours and 55 minutes. One significant result of this mission was the demonstration of the human body's ability to adapt to microgravity conditions over an extended period and then to readapt to normal (Earth) gravity. All other major objectives of the mission were also achieved, except rendezvous with the REP.

✧ Gemini 6A Mission

The Gemini 6A mission (also called Gemini VI-A by NASA) was the fifth crewed Earth-orbiting spacecraft of the Gemini Project. NASA launched the *Gemini 6A* spacecraft on December 15, 1965, after the *Gemini 7* spacecraft was already in orbit, so that both spacecraft could rendezvous in space and practice close proximity orbital flight.

The original Gemini 6 mission had been scheduled for launch on October 25, 1965, but was canceled 101 minutes before launch when the planned Gemini 6 Agena Target Vehicle (vehicle GATV 5002) blew up in space and disappeared. NASA managers decided to reschedule the Gemini 6 mission (renaming it the Gemini 6A mission) and to use the *Gemini 7* spacecraft as a cooperative orbital target in order to demonstrate orbital rendezvous and stationkeeping.

NASA launched the Gemini 6 Agena Target Vehicle from Cape Canaveral on October 25, 1965, using an Atlas-Agena D launch vehicle configuration. After liftoff and ascent through the atmosphere, the Agena successfully separated from the Atlas rocket, and all telemetry signals were normal. Then, 376 seconds into the flight, as the GATV tried to fire its primary propulsion system to achieve orbital insertion, the vehicle's telemetry and radar beacon signals ceased. All attempts to communicate with or track the vehicle using ground-based radar systems failed. Just before the loss of contact with the Agena, telemetry indicated that the

On December 15, 1965, American astronauts successfully completed the first rendezvous mission of two crewed spacecraft. This photograph was taken from the *Gemini 7* spacecraft and shows the *Gemini 6* spacecraft in orbit 160 miles (257 km) above Earth. Once in formation, the two spacecraft flew around each other, coming within one foot (0.31 m) of each other but not making physical contact. The two spacecraft stayed in close proximity for five hours, clearly demonstrating one of the Gemini Project's primary goals—orbital rendezvous. *(NASA)*

vehicle was experiencing a marked rise in pressure in its liquid propellant tanks. Shortly afterward, U.S. Air Force radar systems reported detecting at least five pieces of debris at the point in space where the Agena should have been. Since the Gemini 6 Agena Target Vehicle apparently exploded in space, NASA officials immediately canceled the *Gemini 6* spacecraft launch, minutes before the astronaut crew was to be sent into space.

But the saga of the Gemini 6A mission was just beginning. NASA had rescheduled the Gemini 6A launch to December 12, 1965. Astronaut Walter M. Schirra, Jr., served as spacecraft commander, and astronaut Thomas P. Stafford served as pilot. But their launch was aborted one second after engine ignition because an electrical umbilical separated prematurely.

Remaining calm, Schirra and Stafford did not eject from the fully fueled Titan II rocket. Afterward, Schirra reported that he did not order an emergency ejection, even though the countdown clock was ticking, because he felt no motion and trusted his own senses, training, and judgment. He was right, and his steadfast course of action prevented significant delays in the Gemini Project. Three days later, these two brave men rode flawlessly into orbit on the very same Titan II rocket vehicle.

The *Gemini 6A* spacecraft was successfully launched on December 15, 1965, at 8:37 A.M. from Cape Canaveral AFS's Complex 19 by a Titan II rocket. The Titan II placed the 7,800-pound (3,545-kg) *Gemini 6A* spacecraft into an initial 100-mile (161-km) perigee by 161-mile (259-km) apogee orbit with a period of 89.6 minutes and at an inclination of 28.9 degrees. When the *Gemini 6A* spacecraft carrying Schirra and Stafford reached orbital altitude, it trailed the *Gemini 7* spacecraft, carrying astronauts Frank Borman and James A. Lovell, Jr., by about 1,180 miles (1,900 km). Schirra and Stafford then performed four major thruster burns, and the *Gemini 6A* spacecraft began to catch up with the *Gemini 7* spacecraft. After a final braking maneuver, the two spacecraft achieved rendezvous and remained in zero relative motion at a distance of 361 feet (110 m).

The stationkeeping maneuvers continued for five hours and 19 minutes. At one point the two spacecraft came within one foot (0.3 m) of each other but did not make physical contact. During these close proximity maneuvers, the two Gemini spacecraft circled each other, and all four astronauts (that is, the crews of *Gemini 6A* and *Gemini 7*) participated in precision-formation flying activities. At the end of these stationkeeping operations, the astronauts on board the *Gemini 6A* spacecraft fired their thrusters and moved to a position roughly 31 miles (50 km) away from the *Gemini 7* spacecraft.

On December 16, near the end of the 15th orbit, Schirra and Stafford fired the *Gemini 6A* spacecraft's retrorockets, and approximately 35 minutes later, the two astronauts successfully splashed down in the Atlantic Ocean only eight miles (13 km) from the planned landing location. This was the first successful controlled reentry to a predetermined point in the U.S. manned spaceflight program. The *Gemini 6A* spacecraft and its crew were plucked from the ocean and safely delivered to the aircraft carrier the USS *Wasp*. The two astronauts had logged 25 hours and 51 minutes in orbit and participated in an important rendezvous and stationkeeping demonstration.

✧ Gemini 7 Mission

The Gemini 7 mission (also designated by NASA as Gemini VII) was the fourth crewed flight of the Gemini Project. Astronaut Frank Borman served as spacecraft commander, and astronaut James A. Lovell, Jr., served as pilot. The *Gemini 7* spacecraft was launched on December 4, 1965, at 2:30 P.M. from Cape Canaveral AFS's Complex 19 by a Titan II rocket. As explained in the previous section, the *Gemini 7* spacecraft was actually launched before the *Gemini 6* spacecraft and then served as a convenient substitute target spacecraft during rendezvous operations. The Titan II placed the 8,059-pound (3,663-kg) *Gemini 7* spacecraft into an initial 100-mile (161.6-km) perigee by 204-mile (328-km) apogee orbit with a period of 89.6 minutes and at an inclination of 28.9 degrees.

Immediately after separation from the second stage of the launch vehicle, the crew of the *Gemini 7* spacecraft began stationkeeping operations with the expended Titan II stage. The stationkeeping exercise lasted for about 17 minutes, during which Borman and Lovell kept their spacecraft at distances ranging from 19.7 feet (6 m) to 50 miles (80 km) of the Titan II second stage. On the third orbit, the astronauts fired the spacecraft's thrusters to raise the perigee from 100 miles to 143 miles (230 km). This maneuver ensured that their spacecraft had an orbital lifetime of at least 15 days.

The Gemini 7 mission had several major objectives. The first goal was to perform a 14-day orbital flight and to evaluate the consequences of such an extended mission on the crew. The second goal was to serve as a target for the *Gemini 6A* spacecraft during rendezvous operations. The third goal was to evaluate a new lightweight pressure suit on extended flight conditions.

On December 6 at about 45 hours into the mission, astronaut Lovell removed his space suit to experience and evaluate the shirtsleeve environment of the Gemini spacecraft's pressurized crew cabin. Lovell put his space suit back on some 140 hours into the mission (on December 9), and astronaut Borman subsequently removed his space suit in order to evaluate the cabin's shirtsleeve environment. Twenty hours later, Lovell again removed his space suit, and both astronauts performed the remainder of the mission, except during rendezvous operations with the *Gemini 6A* spacecraft and reentry, without space suits.

The *Gemini 6A* spacecraft was launched on December 15, soon caught up with the *Gemini 7* spacecraft, achieved rendezvous, and began stationkeeping operations. At 2:33 P.M. (on December 15), the two Gemini spacecraft were in a state of zero relative motion with respect to each other at a distance of 361 feet (110 m). For the next five hours and 19 minutes, the two astronauts on board each spacecraft took turns piloting, as the vehicles flew in tight orbital formation. After three and a half orbits of Earth, the *Gemini 6A* spacecraft fired its thrusters and moved to an orbital position approximately 31 miles (50 km) away from the *Gemini 7* spacecraft, which was now placed in a drifting flight to allow Borman and Lovell to have a sleep period. The *Gemini 6A* crew returned to Earth on December 16, while the *Gemini 7* crew continued their long-duration mission.

On December 18, Borman and Lovell fired their spacecraft's retrorockets at the end of the 208th orbit and initiated the reentry sequence. Approximately 37 minutes later, they successfully splashed down in the Atlantic Ocean southwest of Bermuda. The impact point of the *Gemini 7* spacecraft was only 7.6 miles (12.2 km) from the target point. Astronauts Borman and Lovell were quickly retrieved from the ocean by helicopter and taken to the aircraft carrier USS *Wasp*. The *Gemini 7* spacecraft was also recovered about an hour later. The astronauts had accumulated 330 hours and 35 minutes in orbit and were pronounced to be in "better than expected" physical condition after their two-week flight—a flight that represents the longest-duration crewed mission of either the Gemini Project or Apollo Project. (As described in the next chapter, the longest-duration mission during the Apollo Project, namely *Apollo 17,* involved a total flight time of slightly less than 302 hours, including lunar landing activities.)

✧ Gemini 8 Mission

The Gemini 8 mission (also designated by NASA as Gemini VIII) was the sixth crewed flight of the Gemini Project. Astronaut Neil A. Armstrong served as spacecraft commander, and astronaut David R. Scott served as pilot. The *Gemini 8* spacecraft was launched on March 16, 1966, at 10:41 A.M. from Cape Canaveral AFS's Complex 19 by a Titan II rocket. The Titan II placed the 8,336-pound (3,789-kg) *Gemini 8* spacecraft into an initial 99.4-mile (160-km) perigee by 169-mile (272-km) apogee orbit with a period of 88.8 minutes and at an inclination of 28.9 degrees. The primary objective of the Gemini 8 mission was to perform rendezvous and docking tests with an Agena target vehicle. Astronaut Scott was to perform an extravehicular activity.

On March 16, 1966, about 100 minutes before the *Gemini 8* spacecraft was scheduled to liftoff, NASA launched the Gemini 8 Agena Target Vehicle (GATV 8) from Cape Canaveral, using an Atlas rocket. As planned, the uncrewed GATV entered a nearly circular 186-mile- (300-km-) altitude orbit around Earth and awaited the arrival of the crewed *Gemini 8* spacecraft.

Following launch from Cape Canaveral, the *Gemini 8* spacecraft performed maneuvers over the next six hours to rendezvous with GATV 8. The rendezvous phase ended with the two spacecraft approximately 148 feet (45 m) apart, with zero relative motion. Armstrong and Scott then performed stationkeeping and other maneuvers for about 30 minutes. Then, during orbit five, the astronauts carefully guided the *Gemini 8* spacecraft toward the Agena target vehicle and accomplished the first orbital docking.

But this moment of accomplishment would soon turn perilous, almost ending in tragedy save for the quick action of the crew.

About 27 minutes after docking, the combined (cojoined) spacecraft began rolling continuously. This unusual problem was not one for which the astronauts had trained in simulators back at the Johnson Space Center. Relying on overall pilot experience, the astronauts immediately undocked from the Agena vehicle. However, the problem did not go away. Instead, it became more pronounced and more violent. Soon the *Gemini 8* spacecraft rolled and tumbled even faster, at a rate of about one revolution per second, causing Armstrong and Scott to become dizzy. With only seconds to respond before they would black out, Armstrong and Scott realized that one of the attitude control thrusters was firing continuously (later determined to be roll thruster number 8), but they could not figure out which thruster was the problem. In a last-ditch effort to save their lives, the astronauts deactivated the entire orbit attitude and maneuver system and then fired all 16 thrusters that made up the spacecraft's reentry control system (RCS) to dampen the spacecraft's violent tumbling. Their quick thinking worked, and the RCS stabilized the previously out of control spacecraft.

But there was a price to pay for this lifesaving decision. The RCS thrusters had consumed 75 percent of the propellant supply to stabilize the spacecraft's errant motion. Due to the premature use of the reentry control system, NASA flight-safety rules required that the astronauts land immediately. Any further docking operations with GATV 8 and Scott's planned space walk were canceled. On the seventh orbit, Armstrong and Scott fired the RCS thrusters again, this time to initiate the mandatory emergency reentry procedures. Some 37 minutes later, the *Gemini 8* spacecraft safely splashed down in the Pacific Ocean about 500 miles (800 km) west of Okinawa. Despite the emergency nature of the reentry procedure, the astronauts landed in the water about one mile (2 km) from the target point. Within minutes, U.S. Air Force personnel parachuted from a C-54 rescue plane, entered the water, and placed a flotation collar around the spacecraft. Three hours later, the recover ship USS *Mason* picked up the two astronauts. Armstrong and Scott were nauseated and disappointed, but they were also alive and would continue to participate actively in the space program. While Armstrong and Scott traveled in space for only 10 hours and 41 minutes on this perilous mission, within a few years they both became Moon walkers during different Apollo Project missions.

The *Gemini 8* spacecraft was also recovered from the ocean. NASA engineers performed a postflight inspection of the spacecraft as part of NASA's overall investigation of the near-fatal mission. The violent tumbling of the *Gemini 8* spacecraft was due to the continuous firing of roll thruster number 8. Why the thruster misbehaved remains a mystery. Apparently, the thruster experienced a short circuit while being used to maneuver the docked and cojoined *Gemini 8*–GATV spacecraft and stuck open. By continuously firing, this thruster caused the combined vehicle to rotate and then, once the *Gemini 8* spacecraft undocked, caused that spacecraft to rotate and tumble in a more rapid and violent fashion. After the astronauts undocked the *Gemini 8* spacecraft from the Agena target vehicle, ground control personnel successfully performed further tests with GATV 8 and then parked the target vehicle in a near-circular 236-mile- (380-km-) altitude orbit. The crew of the *Gemini 10* spacecraft would use GATV 8 as a passive target for rendezvous in July 1966.

✧ Gemini 9A Mission

The Gemini 9A mission (also designated by NASA as Gemini IX-A) was the seventh crewed flight of the Gemini Project. Astronaut Thomas P. Stafford served as spacecraft commander, and astronaut Eugene A. Cernan served as pilot. NASA had originally scheduled this mission for launch on May 17, 1966, calling it the Gemini 9 mission at the time. But NASA managers had to postpone the launch because the Agena target vehicle (or GATV) failed to achieve orbit earlier in the day due to a booster failure. On June 1, a replacement docking vehicle, called the Augmented Target Docking Adapter (ATDA), was launched successfully from Cape Canaveral. However, after the ATDA vehicle reached orbit, telemetry indicated that the protective shroud had failed to jettison properly, making any attempt to dock with this defective target vehicle highly unlikely. So NASA mission planners modified the goals of the already once postponed Gemini 9A mission, removing docking as a primary objective. Although the *Gemini 9A* spacecraft was ready to launch on June 1, the flight was postponed until June 3 due to problems with launch-critical ground-support equipment.

The *Gemini 9A* spacecraft was finally launched on June 3, 1966, at 8:39 A.M. from Cape Canaveral AFS's Complex 19 by a Titan II rocket. The Titan II placed the 8,250-pound (3,750-kg) spacecraft into an initial 99-mile (159-km) perigee by 166-mile (267-km) apogee orbit with a period of 88.8 minutes and at an inclination of 28.9 degrees. The primary objective of the Gemini 9A mission was now to rendezvous with the ATDA vehicle and to demonstrate maneuvers that simulated those planned for use during the Apollo Project. Astronaut Cernan was to perform an extravehicular activity, demonstrating the use of the astronaut maneuvering unit (AMU).

On the third orbit of Earth following launch, the *Gemini 9A* spacecraft successfully rendezvoused with the ATDA vehicle, coming within 26 feet (8 m) of it. The astronauts confirmed that the ATDA's protective shroud had failed to jettison, which made it look, in astronaut Stafford's words, like an "angry alligator." Since docking with the defective ATDA vehicle was not possible, the crew performed several passive rendezvous maneuvers, including a rendezvous-from-above maneuver that simulated the rendezvous of an Apollo command module with a lunar excursion module after abort from the Moon.

On June 5, the crew depressurized the spacecraft's cabin, and Cernan opened the hatch to begin an extravehicular activity, which included testing the astronaut maneuvering unit. The 151-pound (69-kg) rocket-propelled backpack was mounted on the rear of the *Gemini 9A* spacecraft's adapter section. This unit had a form-fitting seat, a 148-foot (45-m) nylon tether, manual and automatic stabilization systems, a self-contained life support system, and communications and telemetry systems. The AMU's propulsion system consisted of 12 small thrusters that were mounted on the corners of the pack and used hydrogen peroxide fuel. As somewhat unrealistically envisioned by NASA mission planners, a space-walking astronaut would travel outside the crew cabin and reach the pack located at the back end of the spacecraft. The astronaut would then put on this backpack, disconnect the oxygen supply line and its companion 26-foot (8-m) tether (which came from within the cabin) before using the AMU. That was, in theory, how things should happen smoothly.

At 10:02 A.M., Cernan opened the hatch and left the *Gemini 9A* spacecraft. For safety during this space walk, he was connected to a 26-foot tether, as well as a companion line attached to the spacecraft's oxygen supply. Outside of the spacecraft, he first retrieved a micrometeorite impact detector. But he soon experienced great difficulty maneuvering and maintaining his orientation as the safety tether extended to its full length. After taking some photographs of the extended tether, Cernan moved to where the astronaut maneuvering unit was mounted on the back of the spacecraft. The task of donning the AMU required about five times more work than anticipated, and the physical exertion overwhelmed the astronaut's environmental control system—so much so that the faceplate of his space suit fogged up, seriously restricting his visibility. In addition, the AMU's communications system was misbehaving and sending garbled transmissions. Stafford, the spacecraft commander, assessed these problems, made an immediate decision to cancel the EVA, and recalled Cernan back to the spacecraft's cabin. Cernan reentered the *Gemini 9A* spacecraft at 12:05 P.M., and five minutes later, the astronauts sealed the hatch and repressurized the cabin. Despite all the difficulties encountered, Cernan had performed a 128-minute-duration space walk—a significant feat for the early days of human spaceflight.

At the end of the 45th orbit (on June 6), the astronauts began their reentry procedure by firing the spacecraft's RCS thrusters. About 34 minutes later, they safely splashed down in the Atlantic Ocean 342 miles (550 km) east of Cape Canaveral. In a superb example of precision reentry, the

Gemini 9A spacecraft landed in the ocean just 0.4 mile (0.7 km) away from the target point. The astronauts were so close to the recovery ship that they stayed inside their spacecraft while it was brought aboard the aircraft carrier USS *Wasp*. Stafford and Cernan traveled in space a total of 72 hours and 21 minutes during this Gemini Project mission. Because of Cernan's difficulties with the AMU, NASA decided not to test the EVA mobility device again in space until *Skylab* in the 1970s.

✧ Gemini 10 Mission

The Gemini 10 mission (also designated by NASA as Gemini X) was the eighth crewed flight of the Gemini Project. Astronaut John W. Young served as spacecraft commander, and astronaut Michael Collins served as pilot.

The *Gemini 10* spacecraft was launched on July 18, 1966, at 5:20 P.M. from Cape Canaveral AFS's Complex 19 by a Titan II rocket. The Titan II placed the 8,278-pound (3,763-kg) spacecraft into an initial 99-mile (160-km) perigee by 167-mile (269-km) apogee orbit with a period of 88.8 minutes and at an inclination of 28.9 degrees. The primary objective of the Gemini 10 mission was to conduct rendezvous and docking tests with the Gemini 10 Agena Target Vehicle (GATV 10). Other important objectives of this mission were to conduct two extravehicular activities (EVAs) and to rendezvous with dormant GATV 8, now in a parking orbit following the hastily terminated Gemini 8 mission.

NASA successfully launched the GATV 10 from Complex 14 at Cape Canaveral AFS on July 18, 1966, using an Atlas rocket. Following launch, GATV 10 went into a nearly circular orbit around Earth at an altitude of 186 miles (300 km). The *Gemini 10* spacecraft lifted off about 100 minutes later and, during its fourth orbit of Earth, rendezvoused with GATV 10. Successful docking took place some 20 minutes later. Due to an out-of-plane error in the *Gemini 10* spacecraft's initial orbit, the astronauts had to use about 60 percent of the spacecraft's fuel to conduct the rendezvous operation. The propellant consumption was about twice the planned amount. As a result, NASA managers had to revise the mission plan and canceled several practice docking operations.

To conserve propellant, the astronauts kept the *Gemini 10* spacecraft docked to GATV 10 for the next 39 hours and used the Agena's propulsion system to perform any necessary orbital maneuvers. This was the first successful demonstration of using the thrust of a fueled spacecraft to move a combined (docked) space-vehicle configuration. The astronauts used a 14-second firing of the GATV 10's primary propulsion system on July 18 to raise the apogee of the docked space-vehicle configuration to 475 miles (764 km). Another firing of the GATV 10's propulsion system (on 19 July) brought the combined *Gemini 10*/GATV

10 space vehicle into the same orbit as the expended GATV 8 from the Gemini 8 mission.

On July 19, astronaut Collins started his first EVA. He opened the hatch and stood up for about three minutes to perform some ultraviolet astrophotography. While this EVA was underway, both Collins and Young began experiencing extreme eye irritation. Young ordered cancellation of the EVA. Collins sat back down in the *Gemini 10* spacecraft and closed the hatch. The astronauts then used a high flow rate of oxygen to purge the environmental control system of the unknown eye irritant.

The *Gemini 10* spacecraft separated from GATV 10 and initiated a series of thrusting maneuvers on July 20, which brought the crewed spacecraft within 48 feet (15 m) of the drifting, dormant GATV 8. Later that day, Collins began his second EVA, an ambitious space walk in which the tethered astronaut left the *Gemini 10* spacecraft and floated over to the derelict GATV 8. Despite some difficulties due to a lack of handholds on this Agena target vehicle, Collins managed to retrieve a micrometeorite detection package that was mounted on the outside of the vehicle. But, while he was floating back to the *Gemini 10* spacecraft, he somehow lost his camera, which apparently had worked itself free and drifted away. Collins also retrieved a micrometeorite experiment mounted on the outside of the *Gemini 10* spacecraft. This experiment package also drifted away from the astronaut as he was attempting to get back within the crew cabin. The astronauts moved their *Gemini 10* spacecraft away from GATV 8 and jettisoned several more items into space in preparation for reentry.

On July 21, during their 43rd orbit of Earth, Young and Collins fired the *Gemini 10* spacecraft's retrorockets and initiated reentry procedures. About 37 minutes later, the astronauts safely splashed down in the Atlantic Ocean, 544 miles (875 km) east of Cape Canaveral. They landed in the ocean just 3.9 miles (6.3 km) from the planned target point. The astronauts were then taken by helicopter to the recovery ship USS *Guadalcanal*. During the Gemini 10 mission, Young and Collins flew in space for a total of 70 hours and 47 minutes.

✧ Gemini 11 Mission

The Gemini 11 mission (also designated by NASA as Gemini XI) was the ninth crewed flight of the Gemini Project. Astronaut Charles (Pete) Conrad, Jr., served as spacecraft commander, and astronaut Richard F. Gordon, Jr., served as pilot. With the Apollo Project close on NASA's technical horizon, the primary objective of *Gemini 11* was to demonstrate an ability to rendezvous with an orbiting spacecraft immediately after launch. This was a critical demonstration because the procedure was precisely what would have to be done during the Apollo Project, when the Apollo command module orbited around the Moon and the upper portion of the lunar excursion module blasted off from the Moon's surface. There were no time-outs, reruns, or backup plays. Rendezvous quickly in lunar orbit or get stranded in space. It was as simple as that. The *Gemini 11* astronauts would lead the way by demonstrating that this type of time-constrained orbital linkup was possible.

The *Gemini 11* spacecraft was launched on September 12, 1966, at 9:42 A.M. from Cape Canaveral AFS's Complex 19 by a Titan II rocket. The Titan II placed the 8,357-pound (3,798-kg) spacecraft into an initial 100-mile (161-km) perigee by 173-mile (279-km) apogee orbit with a period of 88.9 minutes and at an inclination of 28.8 degrees. The primary objective of the Gemini 11 mission was to achieve a first orbit rendezvous and docking with the Gemini 11 Agena Target Vehicle (GATV 11). Other important objectives of this mission included the performance of two extravehicular activities (EVAs) and a completely automated (computer-controlled) reentry.

NASA successfully launched the GATV 11 from Complex 14 at Cape Canaveral AFS on September 12, 1966, using an Atlas rocket. Following launch, GATV 11 (also identified in NASA's literature as GATV 5006 and Agena Target Vehicle 11) went into a nearly circular orbit around Earth at an altitude of 186 miles (300 km). The *Gemini 11* spacecraft lifted off about 100 minutes later and, during its first orbit of Earth, successfully rendezvoused and docked with the Agena target vehicle. Following this successful demonstration, each astronaut conducted two docking exercises with GATV 11. A subsequent set of orbital maneuvers brought the combined (docked) spacecraft configuration into a 178-mile- (287-km-) by 189-mile- (304-km-) altitude orbit. Following this very busy day in space, Conrad and Gordon enjoyed a well-deserved sleep period while their spacecraft orbited Earth in a docked configuration with GATV 11.

On September 13, astronaut Gordon started his planned 107-minute EVA. One task involved detaching one end of an approximately 100-foot- (30-m-) long tether from the Agena target vehicle and attaching it to the docking bar on the *Gemini 11* spacecraft. This EVA task proved far more strenuous and demanding than ground simulations predicted. After attaching the tether, Gordon stopped to rest astride the GATV. As heavy perspiration seriously obscured his vision, Conrad ordered him to cancel a planned power tool evaluation and return immediately to the cabin.

Following another sleep period, the astronauts used a 25-second firing of the GATV 11's primary propulsion system on September 14 to raise the docked spacecraft configuration to an apogee of 854 miles (1,374 km). Until American astronauts circumnavigated the Moon (during the Apollo

8 mission in December 1968), this apogee represented the highest distance human beings had traveled above Earth's surface. As of January 2007, the *Gemini 11* altitude record still stands for Earth-orbiting, human-crewed spacecraft. (As a technical note, astronauts and cosmonauts are not usually placed in orbits around Earth with altitudes much above 250 miles [400 km] to minimize their cumulative exposure to ionizing radiation from the inner portions of Earth's trapped radiation belts.)

After two orbits of Earth, the astronauts again fired the Agena target vehicle's primary propulsion system—this time for 22.5 seconds to lower the combined space vehicle to a 178-mile (287-km) by 189-mile (304-km) orbit. Once at this lower altitude, the crew (wearing space suits) depressurized the *Gemini 11* spacecraft's cabin, and Gordon opened his hatch to begin a two-hour-long standup EVA. During this procedure, while tethered, Gordon stood up in his seat, looked out beyond the spacecraft's cabin, and then performed a variety of photographic experiments. At the conclusion of this standup EVA, Gordon secured the hatch, after which Conrad repressurized the cabin.

They then slowly undocked the *Gemini 11* spacecraft from the GATV 11 target vehicle. Conrad then carefully piloted the spacecraft to a distance that precisely corresponded to the length of the 100-foot- (30-m-) long tether, which connected the two orbiting vehicles. Although he had a little difficulty in keeping the tether taut, Conrad was able to then start a slow rotation of the *Gemini 11* spacecraft about GATV 11. This maneuver helped keep the tether between the two vehicles taut and allowed the two vehicles to remain a constant distance apart. Conrad then increased the rotation rate. This action caused some temporary oscillations that eventually damped out. The astronauts observed that (as predicted by physics) the circular motion of the tethered spacecraft combination was producing a slight amount of artificial gravity within their spacecraft. After three hours, the astronauts released the end of tether attached to the *Gemini 11* spacecraft. This test sequence represents the first time the rotation of two tethered spacecraft was used to produce artificial gravity—a technique now viewed with considerable interest for possible use on long-duration human flights to Mars and beyond.

At the end of the 44th orbit (on September 15), Conrad and Gordon prepared the *Gemini 11* spacecraft for an automatic reentry. In an American space program technology first, the astronauts allowed onboard computers to command the firings of the thrusters of the reentry control system. All went well, and about 35 minutes later, Conrad and Gordon splashed down in the western Atlantic Ocean just three-miles (4.9 km) from the target point. Within 30 minutes of splashdown, a helicopter picked up the astronauts and delivered them to the recovery ship USS *Guam*. The *Gemini 11* spacecraft was recovered within an hour after landing in the ocean. Conrad and Gordon had spent a total of 71 hours and 17 minutes in space.

Gemini 12 Mission

The Gemini 12 mission (also designated by NASA as Gemini XII) was the 10th and final crewed flight of the Gemini Project. Astronaut James A. Lovell, Jr., served as spacecraft commander, and astronaut Edwin E. (Buzz) Aldrin, Jr., served as pilot. Up until this point, Gemini Project missions had demonstrated all the technical steps needed for the Apollo Project, save for one—the ability of an astronaut to efficiently perform useful work during EVA. NASA engineers added new hand and foot restraints to the outside of the *Gemini 12* spacecraft to support the EVA work demonstration that became a major goal for this final mission. Other goals included continued demonstrations of rendezvous and docking operations and automatic reentry. NASA managers included orbital maneuvering experiments, involving the behavior of tethered vehicles, along with a number of photographic activities and science experiments.

The *Gemini 12* spacecraft was launched on November 11, 1966, at 3:46 P.M. from Cape Canaveral AFS's Complex 19 by a Titan II rocket. The Titan II placed the 8,277-pound (3,762-kg) spacecraft into an initial 100-mile (161-km) perigee by 168-mile (271-km) apogee orbit with a period of 88.9 minutes and at an inclination of 28.9 degrees. A major objective of the Gemini 12 mission was to achieve quick orbit rendezvous and docking with the Gemini 12 Agena Target Vehicle (GATV 12). Other important objectives of this mission included the performance of three extravehicular activities and a completely automated (computer-controlled) reentry.

NASA successfully launched the GATV 12 from Complex 14 at Cape Canaveral AFS on November 11, 1966, using an Atlas rocket. Following launch, GATV-12 (also identified in NASA's literature as GATV-5001 and Agena Target Vehicle 12) went into a nearly circular orbit around Earth at an altitude of 186 miles (300 km). The *Gemini 12* spacecraft lifted off 100 minutes later and successfully rendezvoused and docked with the Agena target vehicle about four hours after the Titan II inserted the crewed spacecraft into orbit.

Two anomalies impacted the conduct of this mission. Due to problems with the spacecraft's onboard radar, astronauts Lovell and Aldrin relied heavily upon visual sightings to accomplish docking. During the insertion of the Agena target vehicle into orbit, ground controllers noticed an anomaly in the primary propulsion system. So NASA mission planners canceled the plan to use GATV 12's primary propulsion system to send a docked vehicle combination to a higher-altitude orbit, as was done during the Gemini 11 mission. Instead, the Agena vehicle's secondary propulsion system was used to support additional rendezvous operations.

During the Gemini 12 mission, Aldrin successfully performed three EVAs—the first and the third being stand-up EVAs, in which Aldrin stood on his seat with the upper portion of his (space suited) body extending out the hatch and took photographs and collected a micrometeorite experiment. During his second EVA (on November 13), Aldrin left the depressurized crew cabin, moved along the spacecraft's handrails, and then used foot restraints and tethers to position himself in front of a work panel that engineers had mounted on the rear of the spacecraft's adapter. At this work panel, he successfully performed a variety of simple manual tasks. While tethered to the *Gemini 12* spacecraft, Aldrin then moved over to the adapter portion of the GATV 12 vehicle and conducted another series of work-related tasks, including using a torque wrench. Based on the difficulties encountered during previous Gemini Project EVAs, NASA mission planners provided Aldrin a dozen or so two-minute rest periods. These mandatory breaks prevented the astronaut from becoming exhausted or from overtaxing his space suit's life support system. Finally, Aldrin attached one end of the 100-foot- (30-m-) long tether (which was stowed in the GATV 12's adapter) to the *Gemini 12* spacecraft's adapter bar. After spending a little over two hours spacewalking, Aldrin climbed back inside the *Gemini 12* spacecraft and secured the hatch. His second EVA accomplished the important mission goal of demonstrating the efficient performance of work during extravehicular activity.

About two hours after Aldrin's second EVA, the astronauts undocked the *Gemini 12* spacecraft from GATV 12 and slowly backed away until the 100-foot tether became taut. They then performed several orbital experiments with the tethered spacecraft configuration, reporting the tendency of the tether to become slack during circular rotational activities. After four hours of these tethered vehicle experiments, the crew released the tether from the *Gemini 12* spacecraft. The next day (on November 14), Aldrin performed his third EVA of the mission—the second stand-up EVA. The third EVA lasted 55 minutes and included photography and the jettisoning of unused equipment.

On November 15 (at the end of the 59th orbit), Lovell and Aldrin prepared the spacecraft for its automatically controlled reentry sequence. Splashdown took place about 35 minutes later in the western Atlantic Ocean just three miles (4.8 km) away from the target point. A helicopter picked up the astronauts and brought them aboard the recovery ship the USS *Wasp*. During the Gemini 12 mission, Lovell and Aldrin had flown in space a total of 94 hours and 35 minutes. This mission brought to a successful conclusion NASA's Gemini Project. The space technology pathway was now established for the Moon-landing missions of the Apollo Project.

✦ Blue Gemini

The Dyna-Soar (Dynamic Soaring), or X-20, Project was an early U.S. Air Force space plane development effort that occurred from 1958 to 1963. The central concept for this military man-in-space project was a crewed, boost–glide orbital vehicle that could be sent into orbit by an expendable launch vehicle (like the Titan rocket), perform its military mission, and return to Earth. When the orbital mission was completed, the military pilot would control the sleek glider through atmospheric reentry and then land it on a runway, much like a conventional jet fighter.

The term *blue Gemini* is the informal, working name given to an unofficial (that is, not funded) concept, which emerged within the U.S. Air Force about February 1962. The blue Gemini concept paralleled two hardware development projects: NASA's Gemini Project and the U.S. Air Force's Dyna-Soar Project, which was an attempt to develop a piloted military vehicle that explored human flight in the hypersonic and orbital regimes. The central idea of blue Gemini was to develop rendezvous, docking, and orbital transfer capabilities to support military activities in space, using a Gemini-type spacecraft. The concept wandered through the Department of Defense and eventually caught the attention of Secretary of Defense Robert S. McNamara.

McNamara not only welcomed the idea of possible cooperation between NASA's emerging Gemini Project and the U.S. Air Force blue Gemini concept, he even suggested combining the civilian and military Gemini spacecraft efforts within the Department of Defense. His suggestion proved too much for NASA officials, who mounted a strong political counteroffensive within all accessible power circles of the federal government.

For their part, senior U.S. Air Force officials were just as surprised at McNamara's suggestion. According to published historic reports, the senior leadership with the U.S. Air Force shared NASA's distaste for a "military takeover" of the civilian Gemini Project. But, the military leaders had quite different reasons. The primary concern within the U.S. Air Force was that responsibility for a combined civilian and blue (military) Gemini Project could jeopardize ongoing activities within the Dyna-Soar Project. The military leaders reasoned that large quantities of defense money would be drained from the military space plane (X-20) project in return for only a relatively few blue Gemini flights.

By January 1963, U.S. Air Force and NASA leadership had provided convincing arguments to McNamara and other senior government officials, and transfer of NASA's Gemini Project to the Department of Defense was deemed inappropriate. NASA proceeded with its civilian Gemini Project, and the Department of Defense, while still pursuing the Dyna-Soar Project, also began examining other military man–in-space options.

The blue Gemini idea that started circulating within the U.S. Air Force in early 1962 was actually part of a much more ambitious, long-term (10-year) visionary plan for the development of military space technology. The concept matured somewhat in June 1962, when personnel at the U.S. Air Force's Space Systems Division in Los Angeles started a study on how to use Gemini hardware as the first step in a proposed new U.S. Air Force man-in-space program, called the Manned Orbital Development System, or MODS. As first conceived, MODS would be a type of military space station (or orbiting facility) with blue Gemini spacecraft serving as ferry vehicles for the military astronauts.

Much of the conceptual work on MODS and the so-called blue Gemini came to a temporary halt when the U.S. Air Force canceled the Dyna-Soar Project in December 1963. Even before a prototype vehicle could be constructed and flown, the project had proven extremely expensive. Despite termination of this project, the concept of a manned military space program was kept alive. On December 10, 1963, McNamara gave a speech in which he announced an intention to explore the requirements for military man in space. The U.S. Air Force was given the lead responsibility in this study effort. The original MODS concept was reborn and soon emerged with the name Manned Orbiting Laboratory (MOL).

In August 1965, in the midst of NASA's Gemini Project activities, President Lyndon Johnson publicly announced that the U.S. Air Force Manned Orbiting Laboratory program would proceed. Johnson's decision essentially fulfilled the promise that McNamara made to the U.S. Air Force in December 1963, following cancellation of the Dyna-Soar (X-20) space plane project. Most of the technical details concerning the MOL program were shrouded in secrecy. However, some facts about this program have now been publicly revealed.

MOL was to incorporate a modified Gemini spacecraft that rode into space attached to a cylindrical laboratory. Both pieces of military space hardware would ride into space as a single unit on top of a powerful Titan IIIC booster vehicle. Several highly qualified armed services officers were selected and began training for duty as MOL military astronauts. The publicly announced purpose of the MOL program was to learn about space, to test equipment, and to conduct experiments. Behind the veil of secrecy, the U.S. Air Force regarded MOL as an orbiting platform from which to conduct strategic reconnaissance, to gather all types of intelligence (using telescopes, radar systems, radio-frequency (ELINT) receivers, and the like), and even to covertly examine foreign satellites.

As part of the overall MOL program effort, the McDonnell Aircraft Company modified its NASA Gemini spacecraft design to produce a military spacecraft design, called the Gemini B. The name of this modified Gemini spacecraft (Gemini B) is often confused with the term *blue Gemini,* but the two terms are not synonymous. The Douglas Aircraft Company was assigned the task of constructing a 42-foot- (12.8-m-) long and 10 foot (3 m) in diameter cylindrical laboratory module that would be attached to the manned Gemini B spacecraft. As originally envisioned, MOL would be capable of supporting a minimum of four military astronauts, their reconnaissance equipment, and a variety of laboratory experiments.

As the United States became more deeply mired in the war in Southeast Asia, funding support for MOL steadily eroded, primarily because within the overall defense program this very costly effort did not offer immediate war-fighting benefits. Soon the MOL program could not sustain itself. The budget became inadequate to meet the program's major objectives. Funding stretch-outs caused serious schedule slips. In addition to severe fiscal pressures, the U.S. Air Force could never really formulate a convincing military man–in-space mission—that is, one of high national defense importance—that readily justified such enormous expenditures. Fiscal planners in the Department of Defense needed only to look at the huge expense NASA was incurring as a result of the Apollo Project and draw a simple comparison to MOL as a similar resource-consuming military astronaut program. To make matters worse, unmanned reconnaissance satellites, such as the *Corona* spacecraft, were collecting superb images from space, so a manned military platform doing the same thing would be quite redundant and fiscally irresponsible.

Consequently, like its predecessor the Dyna-Soar (X-20) Project, the Manned Orbiting Laboratory program never reached fruition. Despite a great deal of development work, President Richard M. Nixon decided to abruptly cancel the MOL program on June 10, 1969. His decision was made as part of a sweeping government effort to reprogram defense funds to pay for the war in Vietnam. As will become apparent in the next chapter, the manned military space program was not the only space program casualty of this war-related, budget-cutting frenzy. The global celebrations over the triumphant Apollo Project lunar landings had hardly subsided when Nixon's administration slashed the budget for the last three planned, but never accomplished, lunar-landing missions: *Apollo 18, 19,* and *20.*

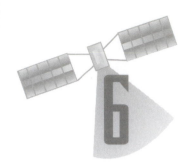

Moonwalks and the Apollo Project

On July 20, 1969, more than 500 million people around the world heard: "Houston, Tranquility Base here—the *Eagle* has landed." Whether listening on radios or watching on television sets, a major portion of the world's population witnessed one the greatest triumphs of modern technology—the first manned landing on the Moon in the Apollo 11 mission. This incredible feat was accomplished in less than a decade as part of NASA's cold war–era human spaceflight program.

One of the unexpected impacts of the Apollo Project was the surge of environmental awareness stimulated by the many inspiring, long-distance images of Earth, taken by the astronauts during their translunar flights. By showing the human's home planet as a uniquely beautiful "blue marble" that traveled around the Sun through the vast emptiness of space, these pictures dramatically reinforced the Copernican hypothesis and heightened environmental consciousness.

The Apollo 8 mission (December 1968) was the first time in history that human beings could look back across the interplanetary void and personally view the entire Earth as a majestic, complex system bursting with life. As they observed Earth from their spacecraft, the three *Apollo 8* astronauts could not help but compare the planet's dynamic and bountiful biosphere with the barren and lifeless lunar landscape below them. Mission commander Frank Borman recalls the powerful, almost spiritual, impact of glimpsing Earth above the lunar landscape: "We were the first humans to see the Earth in its majestic totality, an intensely emotional experience for each of us. And it was the most beautiful, heart-catching sight of my life. And I thought: 'This must be what God sees.'" Similarly, *Apollo 11* astronaut Michael Collins said, "As viewed from the Moon, the Earth is the most beautiful object I have ever seen."

These Earth images caused millions of people to recognize and appreciate the fragile, interconnected nature of their home planet. A person did not have to be an astronaut or a rocket scientist to grasp the fact that the Earth's biosphere—its oceans, clouds, atmosphere, snow-covered polar regions, and great variety of landmasses—some lush with green vegetation and others quite barren—was a delicately interwoven system, capable of supporting life in an incredible number of forms.

For many scientists and historians, this is the principal legacy of the Apollo Project. The human-crewed missions to the Moon provided humans with a dramatic new cosmic perspective of Earth, its unique life-sustaining characteristics, and—perhaps most shocking of all—the planet's relatively insignificant physical size and location in an immense universe. The cold-war geopolitical advantage earned by the American Moon-landing missions is now fading into history. But once the Apollo astronauts walked on another world, the entire human race "came of age" in the universe. Future generations of space-faring human beings will celebrate the triumph of the Apollo Project as the most definitive technical milestone in human history. It is the singularly special event when intelligent life finally emerged from the cradle of Earth and cautiously first ventured into the cosmos.

✧ Origins of the Apollo Project

In July 1960, NASA officials announced that they were preparing to implement a long-range human spaceflight program beyond the Mercury Project. This new effort, called "Apollo Project," was publicly introduced that month in Washington, D.C., during a NASA/Industry Program Plans Conference. As originally presented, the project would involve a manned circumlunar mission—that is, a human-crewed flight around the Moon. The proposed spaceflight scenario was even somewhat reminiscent of the fictional flight in Jules Verne's famous story *From the Earth to the Moon* (without the giant cannon, of course).

The precedent for naming human-crewed space projects after mythological gods and heroes had been set within NASA by the Mercury Project. In keeping within this tradition, headquarters officials selected the name "Apollo" for the civilian space agency's ambitious new project. In Greek mythology, Apollo was born on the tiny island of Delos—the son of Zeus and Leto (Latona). He was the Greek god of archery, prophecy, music, and poetry. His full name, Phoebus Apollo, meant the "shining" or "brilliant" one. Throughout history, Apollo has (somewhat inaccurately) also been referred to as the "Sun god." This minor misconception has occurred because within Greek myth, Apollo used his horse-drawn, golden chariot to pull the Sun in its course across the sky each day. However, according to classicists and mythology experts, the ancient Greek Sun god was actually

Helios, an offspring of the Titan Hyperion. Misconception aside, Apollo proved a proper and fitting name for this exciting new NASA project.

The fledgling Apollo Project took a dramatic new turn in 1961, thanks to the geopolitical pressures of the cold war and a beleaguered young president's need to demonstrate American technical superiority on the global stage. On May 25, 1961, President John F. Kennedy proposed before a joint session of Congress that the United States establish as a national goal landing astronauts on the Moon and then returning them safely back to Earth by the end of the decade. Responding to this incredible, technically demanding presidential initiative, NASA refocused the primary objective of the fledgling Apollo Project. The technical effort would be preceded by both the Mercury and Gemini Projects (discussed in chapters 4 and 5, respectively). It is important to remember that when Kennedy proposed this daring and bold initiative, no American astronaut had yet traveled around Earth in an orbiting spacecraft. To fulfill Kennedy's vision, the Apollo Project became a human spaceflight program consisting of a series of three-person flights, leading to the landing of men on the Moon. The rendezvous and docking of Apollo spacecraft components in lunar orbit became vital techniques for the intricate flight to and return from the Moon.

A giant new rocket, the Saturn V, would be needed to send astronauts and their equipment safely to the lunar surface. All three stages of the colossal Saturn V used liquid oxygen (LO_2) as the oxidizer. The fuel for the first stage was kerosene, while the fuel for the upper two stages was liquid hydrogen (LH_2). The Saturn V vehicle, with the Apollo spacecraft and its emergency escape rocket on top, stood 363 feet (111 meters) tall and developed 7.76 million pounds-force (34.5 million N) of thrust at liftoff. The Saturn V first stage used a cluster of five F-1 engines to generate this enormous liftoff thrust. The second stage used a cluster of five J-2 engines that developed a combined thrust of 1 million pounds-force (4.4 million N). The third stage of the gigantic "Moon rocket" used a single J-2 engine and had a 200,000 pound-force (889,600 N) thrust capability.

The Saturn V rocket was the brainchild of Wernher von Braun—the famous German-American rocket scientist who had two decades earlier developed the V-2 rocket during World War II. To provide some idea of the true enormity of the Moon rocket, imagine eight, individual, 46-foot- (14-m-) long V-2 rockets, standing end on end. This postulated configuration would approximately equal the height of just one Saturn V launch vehicle with its Apollo spacecraft payload perched on top.

✧ Space Robots Scout the Moon

In the early 1960s, scientists did not know very much about conditions on the surface of the Moon. Some scientists even speculated that there could be primitive lunar life-forms (discussed in the next section) that might represent a biological danger to Earth when the astronauts returned soil examples. To resolve most of these uncertainties, NASA designed and flew three families of robot spacecraft. Their collective mission was to gather data in direct support of the planned human-landing missions. These trailblazing space robots were the Ranger, Surveyor, and Lunar Orbiter spacecraft.

The Ranger spacecraft were the first U.S. robot spacecraft sent toward the Moon in the early 1960s to pave the way for the anticipated Apollo Project's human-landing missions. The Ranger Project involved a series of fully attitude-controlled spacecraft designed to photograph the lunar surface at close range before impacting. *Ranger 1* was launched on August 23, 1961. Its mission was to set the stage for the other Ranger missions by testing the spacecraft's ability to navigate. The *Ranger 2* through *9* spacecraft were launched from November 1961 through March 1965. All of the early Ranger missions (*Ranger 1* through *6*) were problem-plagued and suffered setbacks of one type or another. Only the *Ranger 7, 8,* and *9* spacecraft succeeded with flights that returned many thousands of images (before impact) and greatly advanced scientific knowledge about the lunar surface.

NASA's highly successful Surveyor Project began in 1960. It consisted of seven unmanned lander spacecraft that were launched between May 1966 and January 1968, as an immediate precursor to the Apollo Project's human expeditions to the lunar surface. These robot lander craft were used to develop soft-landing techniques, to survey potential Apollo mission–landing sites, and to improve scientific understanding of the Moon.

The *Surveyor 1* spacecraft was launched on May 30, 1966, and soft-landed in the Ocean of Storms region of the Moon. It found the bearing strength of the lunar soil was more than adequate to support the Apollo Project lander spacecraft (called the lunar module, or LM). This contradicted the then-prevalent hypothesis that the LM might sink out of sight in the fine lunar dust. The *Surveyor 1* spacecraft also telecast many pictures from the lunar surface.

The *Surveyor 3* spacecraft was launched on April 17, 1967, and soft-landed on the side of a small crater in another region of the Ocean of Storms. This robot spacecraft used a shovel attached to a mechanical arm to dig a trench and discovered that the load-bearing strength of the lunar soil increased with depth. *Surveyor 3* also transmitted many pictures from the lunar surface.

The *Surveyor 5* spacecraft was launched on September 8, 1967, and soft-landed in the Sea of Tranquility. An alpha particle–scattering device on board this craft examined the chemical composition of the lunar soil and revealed a similarity to basalt on Earth.

The *Surveyor 6* was launched on November 7, 1967, and soft-landed in the Sinus Medii (Central Bay) region of the Moon. In addition to performing soil analysis experiments and taking many images of the lunar surface, this spacecraft also performed a critical "hop experiment." NASA engineers back on Earth remotely fired *Surveyor 6*'s vernier rockets to launch it briefly above the lunar surface. The spacecraft's launch did not create a dust cloud and resulted only in shallow cratering. This important demonstration indicated that the Apollo astronauts could safely lift off from the lunar surface with their rocket-propelled craft (upper portion of the LM) when their surface exploration mission was completed.

Finally, the *Surveyor 7* spacecraft was launched on January 7, 1968, and landed in a highland area of the Moon, near Crater Tycho. Its alpha particle–scattering device showed that the lunar highlands contained less iron than the soil found in the mare regions (lunar plains). Numerous images of the lunar surface also were returned.

Despite the fact that the *Surveyor 2* and *4* spacecraft crashed on the Moon, the overall Surveyor Project was extremely successful.

Finally, five Lunar Orbiter missions were launched by NASA in 1966 and 1967 to perform detailed mapping of the Moon's surface prior to the landings by the Apollo astronauts. All five missions were highly successful, photographing 99 percent of the lunar surface with an approximately 200-foot (61-m) spatial resolution or better. The first three Lunar Orbiter missions were dedicated to imaging 20 potential Apollo landing sites that had been preselected, based on telescopic observations of the Moon's nearside from Earth. The fourth and fifth missions were committed to broader scientific objectives and were flown in high-altitude polar orbits around the Moon. *Lunar Orbiter 4* photographed the entire nearside and 95 percent of the farside, and *Lunar Orbiter 5* completed the farside coverage and acquired medium- (66-foot [20-m]) and high- (6.6-foot [2-m]) resolution images of 36 preselected areas. These probes were sent into orbit around the Moon to gather information and then purposely crashed at the end of each mission to prevent possible interference with future projects.

✦ The Life-on-the-Moon Debate and the Issue of Extraterrestrial Contamination

Scientists define extraterrestrial contamination as the contamination of one world by life-forms, especially microorganisms, from another world. Using Earth and its biosphere as the reference, the planetary-contamination process is called *forward contamination* if the alien world (or material sample returned from that world) is contaminated by contact with terrestrial organisms, and it is called *back contamination* if alien organisms are released into the Earth's biosphere.

At the start of the Space Age, scientists were keenly aware of the potential problem of extraterrestrial contamination—in either direction. Based on international discussions and agreements, the scientific community drew up various quarantine protocols (procedures) to avoid the forward contamination of other worlds by outbound, unmanned spacecraft. These protocols were also intended to safeguard Earth against the possible problem of back contamination, primarily when lunar samples were returned as part of the Apollo Project.

Quarantine is basically a forced isolation to prevent the movement or spread of a contagious disease. Historically, quarantine was the period during which a ship suspected of carrying persons or cargo (such as produce or livestock) infected with contagious diseases was detained at its port of arrival. The length of the quarantine, generally 40 days, was considered sufficient to cover the incubation period of most highly infectious terrestrial diseases. If no symptoms appeared at the end of the quarantine, officials allowed the travelers to disembark or the cargo to be unloaded.

In modern times, the term *quarantine* has obtained a new meaning—namely, that of holding an infected person (or a suspicious organism) in strict isolation until the person or the organism is no longer capable of transmitting the disease. With the Apollo Project and the advent of the NASA's lunar quarantine program, the term acquired elements of both meanings. As responsible members of the scientific community, NASA scientists began a planetary quarantine program at the beginning of the American civilian space program in the late 1950s. This quarantine program was conducted in a spirit of openness and international cooperation and was intended to prevent, or at least minimize, the possibility that early space probes would contaminate other planetary bodies.

At that time, scientists were especially concerned with forward contamination—the potential process whereby terrestrial microorganisms, "hitchhiking" on early planetary probes and landers, would spread to another world, thrive, and possibly destroy any native life-forms, life-precursors, or perhaps even fossil remnants of past life-forms. If forward contamination occurred, it would compromise future scientific attempts to search for and identify extraterrestrial life-forms that had arisen independently of the Earth's biosphere.

The human-crewed Apollo Project missions to the Moon stimulated a great deal of debate about forward and back contamination. Early in the 1960s, scientists began asking in earnest: Is there life on the Moon? Some of the most heated, technical exchanges that took place during the Apollo Project concerned this particular question. If life existed on the Moon, no matter how primitive or microscopic, scientists would want to examine it carefully and compare it with life-forms here on Earth. But any scientific search for these postulated microscopic lunar life-forms would be made very difficult and expensive because of the forward-contamination problem. To avoid this problem, all equipment and materials landed on the Moon would need rigorous sterilization and decontamination procedures. But this cautious approach represented a costly and time-consuming process. In a politically charged era in which the unofficial "superpower race to the Moon" had gained enormous international notoriety, senior NASA officials generally viewed such extra steps as programmatically unnecessary and scientifically unwarranted.

However, there was also the nagging uncertainty about back contamination. If some form of microscopic life did exist on the Moon (however remote the possibility), did it represent a serious hazard to the terrestrial biosphere? Because of the prevalent uncertainty about the extraterrestrial contamination issue, some members of the scientific community still urged that NASA undertake time-consuming and expensive quarantine procedures for all robot spacecraft bound for the Moon.

On the other side of this contamination argument were those scientists who pointed out the anticipated extremely harsh lunar conditions—virtually no atmosphere, probably no water, temperature extremes ranging from 248°F (120°C) at lunar noon to −238°F (−150°C) during the frigid lunar night; and unrelenting exposure to lethal doses of ultraviolet, charged particle and X-ray radiations from the Sun. No life-form, these scientists argued, could possibly exist under such hostile conditions. But other scientists countered this line of reasoning by speculating that trapped water and moderate temperatures might be found below the lunar surface. If such subsurface environmental niches existed, perhaps they could sustain primitive life-forms.

And so the great extraterrestrial contamination debate of the Apollo Project raged back and forth, until finally the Apollo 11 expedition departed on the first lunar-landing mission. As a compromise, *Apollo 11* traveled to the Moon with careful precautions set up against back contamination but with only a limited effort to protect the Moon from forward contamination by terrestrial organisms.

The Lunar Receiving Laboratory (LRL) at NASA's Johnson Space Center in Houston, Texas, provided quarantine facilities for returned lunar soil and rock samples, as well as isolation facilities for the *Apollo 11, 12,* and *14* astronauts. What scientists learned from operating the LRL serves as a convenient starting point for planning any future sample quarantine activities. Quarantine facilities will be needed to accept, handle, and test extraterrestrial materials from Mars and other solar system bodies of interest in the scientific search for alien life-forms (present or past).

During the Apollo Project, no evidence was discovered that native alien life was then present or had ever existed on the Moon. NASA scientists at the LRL performed a careful search for carbon, since terrestrial life is carbon-based. One hundred to 200 parts per million of carbon were found in the lunar samples. Of this amount, only a few tens of parts per million are considered indigenous to the lunar material, while the bulk of carbon appears to have been deposited by the solar wind. Scientists therefore concluded that none of this carbon was derived from indigenous biological activity on the Moon. Since the Moon appeared completely devoid of all life, NASA officials abandoned the back contamination quarantine procedures after the first three Apollo expeditions to the lunar surface. These procedures had involved physically isolating the returning Apollo astronauts for approximately three weeks after their return to Earth.

Is the life-on-the-Moon debate over? Quite possibly it is not, because of discoveries made within the last decade or so. The suspected presence of lunar water ice in permanently shadowed craters found in the polar regions of the Moon (discussed in chapter 10) may revive some modest portion of the "microscopic lunar life" debate of the early 1960s. Furthermore, here on Earth, scientists have discovered various extremophiles—very hardy microorganisms capable of living in extremely harsh environmental conditions. If these hardy life-forms have been discovered in the strangest places on Earth, what about elsewhere in the solar system in places where there is water?

✦ The Apollo Spacecraft

The Apollo spacecraft consisted of a command module, service module, and lunar module. The command and service module (CSM) was collectively a single spacecraft but was separable into two components: the command module and the service module. The CSM environmental control system regulated cabin atmosphere, pressure, temperature, carbon dioxide levels, and ventilation, including the removal of odors and particles. The environmental system also controlled the temperature range of the spacecraft's electronic equipment. The command module served as the crew's quarters and flight control section. The service module contained propulsion and spacecraft support systems. The lunar module—sometimes referred to as the lunar excursion module (until about 1966)—carried two Apollo astronauts down to the lunar surface, supported them while they remained on the lunar surface, and then returned them to the command and service module, which was orbiting the Moon with the third astronaut on board.

COMMAND MODULE

The conical command module (CM) was compact, solid, sturdy, and crammed with some of the most complex equipment ever sent into space (up to that time). It was 12 feet (3.65 m) high, had a maximum diameter of 12.8 feet (3.9 m) at its base, and was protected by an ablative (charring) heat shield. Engineers had designed the CM with one overriding criterion: to survive the fiery heat of reentry as the command module jettisoned the service module and slammed back into Earth's atmosphere at the tremendous speed of about 25,000 miles per hour (40,225 km/h). From a physical perspective, the speed of reentry from a mission to the Moon is nearly one and one-half times as fast as a mission returning from orbit around Earth. So aerospace engineers had to design the CM to dissipate great amounts of energy as the capsule made its high-speed plunge into Earth's atmosphere. Their successful design approach included the use of an ablative heat shield that charred and slowly burned away while protecting all that the shield surrounded, including the three astronauts. The crew compartment comprised most of the volume of the CM, approximately 218 cubic feet (6.2 m^3) of space. The three Apollo astronauts had to live inside the cramped CM for most of the lunar journey, and one of them for all of it. Their couches were surrounded by instrument panels, radios, navigation equipment, life support subsystems, and small reaction engines to keep the capsule stable during reentry.

The three astronaut couches were lined up facing forward in the center of the crew compartment. There was a large access hatch located above the center couch. A short access tunnel led to the docking hatch in the CM nose. The CM had five windows: one in the access hatch, one next to each astronaut in the outer two seats, and two forward-facing rendezvous windows. Five silver/zinc oxide batteries provided electric power after the CM and service module detached. Three batteries supported reentry and post-landing activities, while two of these batteries were for vehicle separation and parachute deployment. The CM had 12 94.4-pound-force (420 N) nitrogen tetroxide/hydrazine reaction control thrusters. Finally, the forward compartment in the nose of the cone held three 83.3-foot- (25.4-m-) diameter main parachutes and two 16.4-foot- (5-m-) diameter drogue parachutes for a soft-landing on Earth at the end of the reentry operation.

SERVICE MODULE

Packed with tanks and plumbing, the service module (SM) served as the constant companion to the command module (CM) until just before reentry. NASA engineers reasoned that they would place all the components needed for the trip to the Moon—but not for reentry operations—in the SM. This avoided having to provide thermal protection against the searing conditions of reentry heating for all this equipment. So engineers designed the SM to carry the crew's oxygen for most of the journey, fuel cells to generate electricity (along with the supply of oxygen and hydrogen to run them), four identical banks of four 100 pound-force (450 N) reaction control thrusters for attitude control, and one large 20,460-pound-force (91,000 N) restartable hypergolic liquid-propellant rocket engine to propel the command and service module (CSM) spacecraft into and out of orbit around the Moon. The SM was a cylinder 12.8 feet (3.9 m) in diameter and 24.9 feet (7.6 m) in length. The front end of the SM was attached to the back of the CM. The two components remained joined throughout the mission until minutes before reentry—at which point the astronauts jettisoned the service module and returned to Earth in just the command module. The SM's large, gimballed restartable rocket motor was mounted at the back (or aft) end of the SM. When the CM and SM modules were attached, the CSM spacecraft had a combined length of 36.1 feet (11 m) and a maximum diameter of 12.8 feet (3.9 m).

The mass of the combined command service module varied somewhat from mission to mission. The *Apollo 11* CSM had a launch mass of 63,362 pounds (28,801 kg), including propellants and expendables. Of this total CSM launch mass, the *Apollo 11* CM had a mass of 12,225 pounds (5,557 kg), and the SM had a mass of 51,137 pounds (23,244 kg). By way of contrast, during the last lunar-landing mission, the *Apollo 17* CSM had a total launch mass of 66,704 pounds (30,320 kg), including propellants and expendables. Of this total CSM launch mass, the *Apollo 17* CM had a mass of 13,112 pounds (5,960 kg), and the SM had a mass of 53,592 pounds (24,360 kg).

Telecommunications included voice, television, data, and tracking and ranging subsystems for communications among astronauts, the command module, the lunar module, and the NASA mission control center on Earth. Voice contact was provided by an S-band uplink and downlink system. Tracking was accomplished by means of a unified S-band transponder.

LUNAR MODULE

The lunar module (LM) was a two-stage space vehicle specifically designed for operations near and on the Moon. NASA engineers designed the LM to operate only in the vacuum conditions of outer space and not in a planetary atmosphere, making the craft the first crewed spaceship. Resembling a giant spider with its spindly legs, the lunar module's mission was to carry two astronauts from lunar orbit to the surface of the Moon, provide them a temporary base while they were on the Moon, and then send its upper half back into lunar orbit to rendezvous with its mother ship. The mother ship, in this case, was the command and service module (CSM) spacecraft, which patiently orbited the Moon with the third Apollo astronaut on board. The ascent and descent stages of the lunar module operated as a single unit, until staging took place on the lunar surface. At that point, the ascent stage separated from the LM's descent (lower) stage by firing its rocket engine to lift off from the Moon's surface. After abandoning the

LM's lower stage at their particular lunar-landing site, the two Moonwalking astronauts carefully piloted and navigated the LM (ascent stage) so it could rendezvous and dock with the orbiting CSM. Once successfully docked, the two astronauts transferred equipment from the LM ascent stage into the CSM and then rejoined the third mission astronaut. To prepare for the journey back to Earth, the astronauts jettisoned the LM ascent stage from the CSM and then waited patiently in lunar orbit to perform the transearth injection maneuver by firing the CSM's rocket engine at the precise time.

On some missions (such as Apollo 17), the LM ascent stage had sufficient residual propellant for the spacecraft to fire its reaction-control system rockets and crash the vehicle into the Moon. Taking this action at the end of an Apollo lunar-landing mission avoided any possibility (however remote) of an on-orbit collision between the jettisoned LM ascent stage and the CSM—during the ongoing mission or during some future one. On other missions (such as Apollo 11), the LM ascent stage was simply abandoned in orbit. The derelict spacecraft then experienced orbital decay for one to several months before finally crashing into the Moon at some unknown site. On the aborted Apollo 13 mission, the entire LM (upper and descent stages combined) served as a miraculous lifeboat. While joined to the disabled CSM, the LM's environmental subsystems kept the three astronauts alive as they perilously swung around the Moon and headed back to Earth.

Built by Grumman Aerospace Corporation (now Northrop Grumman) in Bethpage, Long Island, New York, the two-part lunar module had a total overall height (including deployed landing legs) of approximately 23 feet (7 m) and an overall width of 31 feet (9.4 m) with the landing legs extended and deployed. The descent stage was 10.6 feet (3.23 m) tall and the ascent (upper) stage 12.3 feet (3.75 m) tall.

The total liftoff mass of the LM varied slightly from lunar-landing mission to lunar-landing mission. The *Apollo 11* LM had a total mass of 33,143 pounds (15,065 kg), including astronauts, propellants, and expendables. The dry mass of the ascent stage was 4,796 pounds (2,180 kg), and it held 5,806 pounds (2,639 kg) of propellant. The descent (lower) stage had a dry mass of 4,475 pounds (2,034 kg) and carried an initial supply of 18,066 pounds (8,212 kg) of propellant. The LM used in the Apollo 17 mission had a total mass of 36,186 pounds (16,448 kg), with approximately 26,400 pounds (12,000 kg) of this total liftoff mass being propellants. The fully fueled mass of the ascent stage was 10,967 pounds (4,985 kg) and the descent stage 25,219 pounds (11,463 kg). As previously mentioned, the lunar module's ascent and descent stages operated as a single unit, until the two Moonwalking astronauts departed from the lunar surface. At that point, the descent stage served as a platform for launching the ascent stage, which carried the astronauts into lunar orbit for rendezvous and docking with the CSM mother ship.

The descent stage comprised the lower part of the lunar module spacecraft. This part of the LM contained the landing rocket—a deep-throttling ablative rocket with a maximum thrust of approximately 10,120 pounds-force (45,000 N). Engineers mounted this rocket engine on a gimbal ring in the center of the descent stage. The LM's descent stage also had two tanks of aerozine 50-rocket fuel, two tanks of nitrogen tetroxide oxidizer, water, oxygen, helium tanks, and storage space for lunar equipment and experiments. For the Apollo 15, 16, and 17 missions, the LM's descent stage also carried the Apollo lunar rover. Since the LM's descent (or lower) stage served as a launch platform for the LM ascent stage, it was left behind at the lunar-landing site.

Each abandoned LM descent stage has a plaque attached to the ladder, commemorating the particular lunar-landing mission. The plaque on the *Apollo 11* lunar module (called the *Eagle*) is inscribed as follows: "Here men from the planet Earth first set foot upon the Moon July 1969 A.D. We came in peace for all mankind." This plaque also has the engraved names and signatures of the three *Apollo 11* astronauts (Neil A. Armstrong, Edwin E. "Buzz" Aldrin, and Michael Collins) as well as the president of the United States (Richard M. Nixon). The plaque attached to the ladder of the *Challenger* LM, which was left behind on the Moon at the conclusion of the Apollo 17 mission, is inscribed: "Here man completed his first exploration of the Moon, December 1972 A.D. May the spirit of peace in which we came be reflected in the lives of all mankind." The *Apollo 17* plaque has the engraved names and signatures of the three mission astronauts (Eugene A. Cernan, Ronald E. Evans, and Harrison H. "Jack" Schmitt) as well as the president of the United States (Richard M. Nixon). The plaques on the other lunar modules left behind on the Moon (during the Apollo 12, 14, 15, and 16 missions, respectively) are a bit more simplified and only contain the mission number, the name of the LM, the month and year, and the names and signatures of the three astronauts. All Apollo lunar module plaques also have engraved at the top a symbolic map of Earth, divided into the Western Hemisphere and Eastern Hemisphere.

What happened to the plaque on the *Aquarius* lunar module? The plaque on the *Apollo 13* lunar module (*Aquarius*) burned up in Earth's atmosphere along with the entire LM, after the spacecraft had done a magnificent job of serving as a lifeboat for three stranded astronauts. They jettisoned *Aquarius* moments before reentry, after they had boarded and reactivated the disabled *Apollo 13* command and service module.

The LM descent stage was an octagonal prism 13.8 feet (4.2 m) across and 5.6 feet (1.7 m) tall. It had four landing legs with round footpads mounted on the sides. The landing legs kept the bottom of the descent stage about 4.9 feet (1.5 m) above the surface. A 3.3-foot- (1-m-) long conical descent engine skirt protruded from the bottom of this stage. The distance between the ends of the footpads on opposite landing legs was 31 feet (9.4 m). One of the legs had a small egress platform and ladder for the astronauts to descend to the lunar surface.

The ascent stage of the lunar module mounted on top of the descent stage. It was an irregularly shaped unit approximately 9.2 feet (2.8 m) high and 13.1 feet (4.0 m) by 14.1 feet (4.3 m) in width. The ascent stage housed the astronauts in a pressurized crew compartment, which had a volume of approximately 235 cubic feet (6.65 m^3) and functioned as the base of operations for lunar surface activities. The LM's ascent stage had an ingress-egress hatch at one side and a docking hatch for connecting to the CSM on top. Engineers also mounted a parabolic rendezvous radar antenna; a steerable, parabolic, S-band antenna; and two in-flight VHF antennae on the top of the ascent stage. There were two triangular windows located above and to either side of the ingress-egress hatch—the hatch that provided each pair of Moonwalking astronauts with access to the lunar surface.

Engineers placed the ascent stage's fixed, constant-thrust 3,370-pounds-force (15,000-N) rocket engine at the base of the unit. The LM ascent stage also contained an aerozine 50-fuel tank, an oxidizer tank, and other storage tanks for helium, liquid oxygen, gaseous oxygen, and the reaction-control system fuel. Maneuvering was achieved by means of the reaction-control subsystem, which consisted of the four thrust module assemblies mounted around the sides of the LM ascent stage. Each thrust module assembly consisted of four 100-pound-force (450-N) thrust chambers and nozzles pointing in different directions.

The crew accommodations in the LM were quite spartan. For example, the LM contained no seats, and astronauts had to sleep on the floor (as best they could) during any rest periods that occurred during lunar surface operations. While they occupied the LM's crew compartment, instruments and control panels surrounded the astronauts. In function, if not in looks, conditions inside the LM was pretty much like those found in the cramped CSM. There was a control console mounted in the front of the crew compartment above the ingress-egress hatch and between the two triangular windows; there were two more control panels mounted on the sidewalls. The LM instruments and control panels allowed the astronauts to communicate, navigate, and rendezvous.

Telemetry, television, voice, and range communications with Earth from the LM were all accomplished by means of the LM's S-band antenna. VHF was used for communications between the astronauts and the LM, and between the LM and the orbiting CSM mother ship. There were also redundant transceivers and equipment for both S-band and VHF communications. The LM had an environmental control system that recycled oxygen and maintained the temperature in the crew cabin and in the electronics subsystems. The LM obtained its electric power from six silver-zinc batteries. Guidance and navigation were accomplished by a radar ranging system, an inertial measurement unit (consisting of gyroscopes and accelerometers), and by the Apollo guidance computer.

APOLLO SPACECRAFT NAMES

Beginning with the flight of *Apollo 9,* NASA allowed the astronauts who were to fly on each mission to select code names for both the command and service module (CSM) and the lunar module (LM). The following combinations of code names were chosen: *Apollo 9—Gumdrop* for the CSM and *Spider* for the LM; *Apollo 10–Charlie Brown* (CSM) and *Snoopy* (LM); *Apollo 11–Columbia* (CSM) and *Eagle* (LM); *Apollo 12–Yankee Clipper* (CSM) and *Intrepid* (LM); *Apollo 13–Odyssey* (CSM) and *Aquarius* (LM); *Apollo 14–Kitty Hawk* and *Antares* (LM); *Apollo 15–Endeavor* (CSM) and *Falcon* (LM); *Apollo 16–Casper* (CSM) and *Orion* (LM); and *Apollo 17–America* (CSM) and *Challenger* (LM).

✧ A Summary of the Apollo Missions

NASA's Apollo Project included a large number of unmanned test missions and 11 manned missions: two Earth-orbiting missions (with *Apollo 7* and *9*), two lunar-orbiting missions (with *Apollo 8* and *10*), a lunar swingby (with the in-flight aborted *Apollo 13*), and six lunar-landing missions (with *Apollo 11, 12, 14, 15, 16,* and *17*). The Apollo-Soyuz Test Project (ASTP), which took place in July 1975, is often referred to as the 12th manned Apollo Project mission, especially since the American spacecraft in this international rendezvous and docking mission was called *Apollo 18.* (The ASTP mission is discussed in the next chapter, as is the *Skylab* Project, which also used spacecraft and launch vehicles leftover when NASA terminated the Apollo Project after the Apollo 17 mission.)

Two astronauts from each of the six lunar-landing missions walked on the Moon. This group of astronauts became popularly known as the "Moonwalkers." Members included Neil A. Armstrong and Edwin E. "Buzz" Aldrin, Jr. (*Apollo 11*); Charles "Pete" Conrad, Jr., and Alan L. Bean (*Apollo 12*); Alan B. Shepard, Jr., and Edgar D. Mitchell (*Apollo 14*); David R. Scott and James B. Irwin (*Apollo 15*); John W. Young and Charles M. Duke, Jr. (*Apollo 16*); and Eugene A. Cernan and Harrison H. "Jack" Schmitt (*Apollo 17*). So far, these are the only human beings who have set foot on another body in the solar system. As of July 20, 2006, nine Moonwalkers still serve as a living testimonial to the great technical achievements of the Apollo Project. The other three—Charles Conrad, Jr.; James B. Irwin; and Alan B. Shepard, Jr.—are now deceased. To help preserve the important legacy of Apollo, this section provides a concise summary of each of the project's manned missions. This section also includes the tragic *Apollo 1* launch pad accident, which claimed the lives of astronauts Virgil (Gus) Grissom, Edward White II, and Roger B. Chaffee, and the so-called missing Apollo 18, 19, and 20 missions, which were planned for, but then canceled, in 1970 due to budget constraints.

From October 1961 through April 1968, NASA also conducted a variety of unmanned flight tests involving the Saturn IB and V launch vehicles and the Apollo spacecraft/lunar module. These unmanned tests included the Apollo 4, 5, and 6 missions (November 1967 to April 1968)—missions that involved Earth orbital trajectories.

THE *APOLLO 1* TRAGEDY

On January 27, 1967, tragedy struck the American space program when a flash fire erupted inside the command module spacecraft for the Apollo 204 (AS-204) mission during ground testing at Complex 34, Cape Canaveral Air Force Station in Florida. The fire resulted in the deaths of astronauts Virgil (Gus) Grissom, Edward White II, and Roger B. Chaffee, who were to be the crew of the first manned mission of the Apollo Project. At the time, NASA was planning to place the spacecraft that suffered the fire into orbit around Earth on February 21, 1967, by the Saturn IB rocket from Complex 34.

NASA officials had originally designated this first manned Apollo mission as the AS-204 mission, meaning the fourth launch in the Apollo Saturn IB series. The earlier, unmanned Apollo Saturn IB AS-201, AS-202, and AS-203 missions had not been assigned official "Apollo" flight numbers. The unmanned AS-201 mission and the AS-202 missions involved the Saturn IB rocket with the Apollo spacecraft aboard. In the AS-203 mission, the Saturn IB carried only the aerodynamic nose cone into orbit for testing. Even though this tragic accident took place on the launch pad during a preflight test, NASA historians reported the fatal event as follows: "First manned Apollo Saturn flight—failed on ground test."

On that fateful day, astronauts Grissom, White, and Chaffee were seated in the command module, moving through the countdown of a simulated launch of the Saturn IB rocket. At T minus 10 minutes, tragedy struck without warning. In a public statement, Major General Samuel C. Phillips (USAF) described how ground crew members saw a flash fire break through the spacecraft shell and envelop the spacecraft in smoke. Rescue attempts failed. It took a tortuous five minutes to get the spacecraft's hatch open from the outside. Long before that the three astronauts were dead

The *Apollo 1* astronauts (left to right) Virgil I. (Gus) Grissom, Edward H. White II, and Roger B. Chaffee pose in front of Launch Complex 34 at Cape Canaveral Air Force Station, Florida, on January 17, 1967. The launch complex (in the background) housed their Saturn 1 launch vehicle. Ten days later (on January 27), the three astronauts died in a tragic fire on the launch pad during a preflight test. The first manned Apollo mission was to have been launched on February 21, 1967. *(NASA)*

from asphyxiation. This was the first fatal accident in the American human spaceflight program.

Shock and disbelief swept across the United States and the world. As befitting fallen heroes, the deceased astronauts were laid to rest with full national honors. As a permanent tribute to their sacrifice, NASA officially proclaimed in spring of 1967 that the mission originally scheduled for Grissom, White, and Chaffee would be known as Apollo 1.

By April 1967, a review board reached the conclusion that the fire on *Apollo 1* had apparently been started by an electrical short circuit that ignited the crew cabin's oxygen-rich atmosphere and fed on combustible materials in the spacecraft. Although the precise wire that caused the tragedy was never identified, the accident forced NASA to make major modifications in command module spacecraft prior to its first crewed mission in space. NASA officials postponed any manned launches until safety experts and aerospace engineers reviewed the changes in the interior of the command module and deemed the improved, extensively reworked spacecraft ready for flight. For example, aerospace engineer redesigned the command module hatch to open out instead of in, because the old hatch had been a significant factor in trapping Grissom, White, and Chaffee inside their burning craft that dreadful day in January. The spacecraft engineers also rewired the spacecraft, rerouted wire bundles, and used better, fire-resistant insulation for the wires. NASA safety experts also examined all potentially combustible materials being used inside the spacecraft and replaced many of these nonmetallic materials with less combustible substitutes. Overall, thousands of engineers and technicians across the United States helped make the redesigned Apollo command module a much better and safer spacecraft. Changes were also made to the lunar module to avoid the occurrence of any similar accident in the crew cabin area.

NASA's administrative formula for numbering Apollo missions was significantly altered as a result of the accident on *Apollo 1*. No missions or flights were ever designated as Apollo 2 or Apollo 3. The Apollo 4 mission was the name given to the first (unmanned) flight of the Saturn V rocket on November 9, 1967. This mission is also referred to as the AS-501 mission. In the Apollo 5 mission, the original AS-204 Saturn IB rocket vehicle (the one atop which the *Apollo 1* fire occurred) placed an unmanned lunar module (LM) spacecraft into Earth orbit on January 22, 1968. The LM was enclosed in a spacecraft–lunar module adapter and topped by an aerodynamic nose cone in place of the Apollo command and service modules. Finally, the Apollo 6 mission involved another unmanned test of the Saturn V launch vehicle. The Apollo 6 (or AS-502) mission took place on April 4, 1968. This unmanned mission was plagued with problems. Two of the J-2 rocket engines that made up the five-engine second-stage configuration of the Saturn V launch vehicle shut down prematurely during the ascent to orbit. In addition, the J-2 engine in the giant rocket's third stage would not restart after achieving orbit around Earth. By carefully examining large quantities of telemetry data, engineers were eventually able to trace the main problem to failed igniter lines on the faulty J-2 engines. Under intense schedule pressure, the rocket engineers improvised fixes, and NASA managers declared the giant Saturn V launch vehicle ready to carry human beings into space.

APOLLO 7

Apollo 7 served as a confidence-building mission, allowing NASA personnel involved in the Apollo Project to recover from the *Apollo 1* tragedy. Liftoff of the first crewed Apollo launch took place on October 11, 1968. A Saturn IB vehicle (designated AS-205) left Complex 34 at Cape Canaveral Air Force Station (AFS) and carried astronauts Walter Schirra (commander), Donn Eisele (command module pilot), and R. Walter

Cunningham (lunar module [LM] pilot) into orbit around Earth. As an historic note, even though this particular mission did not involve the flight of a lunar module spacecraft, Cunningham was nevertheless assigned the title of LM pilot to maintain crew-position continuity within the overall project. All subsequent Apollo missions used the Saturn V launch vehicle and were flown from Complex 39 at the NASA Kennedy Space Center, which is adjacent to Cape Canaveral AFS.

The principal objectives of this 11-day Earth-orbiting mission were to demonstrate the suitability of the redesigned and remodeled command and service module (CSM) spacecraft, the use of the Saturn IB launch vehicle with a human crew, and the rendezvous capability of the CSM spacecraft. Following a successful launch, the *Apollo 7* spacecraft traveled in an orbit around Earth characterized by a perigee of 144 miles (232 km), an apogee of 185 miles (297 km), a period of 89.8 minutes, and an inclination of 31.6 degrees.

After launch ascent, the combined Saturn IB upper stage/Apollo spacecraft payload configuration, designated as the S-IVB/CSM, was placed in a 142-mile (228-km) by 175-mile (282-km) orbit around Earth. NASA mission managers then vented the residual propellants from the S-IVB upper stage—an action which gradually raised the orbital altitude of the S-IVB/CSM configuration to 144 miles by 192 miles (309 km) over the next three hours. The astronauts then separated their spacecraft (the CSM) from the Saturn IB upper stage and used the S-IVB as a target vehicle with which to practice orbital rendezvous maneuvers for two days.

THE SATURN IB VEHICLE

The Saturn IB rocket served as the junior partner to the gigantic Saturn V rocket used by NASA in the Apollo Project. The Saturn IB launch vehicle embodied significant advances in the propulsion hardware and operational techniques—advances that would be needed to send astronauts to the Moon. The Saturn IB rocket stages contained a number of modifications that increased overall launch vehicle performance, as compared to the performance of the earlier series of Saturn I vehicles. For example, NASA propulsion engineers improved the thrust of the vehicle's eight H–1 rocket engines from 188,000 pounds-force (836,000 N) to 200,000 pounds-force (890,000 N) each. Of perhaps greater overall significance to the Apollo Project, the Saturn IB missions provided NASA engineers an early opportunity to flight-test some of the Saturn V hardware. Specifically, the Saturn IB's upper stage, called the S-IVB stage, had a single J-2 rocket engine that was nearly identical to the upper (third) stage carried on the Saturn V launch vehicle. The instrumentation unit enjoyed a similar design commonality.

During the 11-day orbital mission of *Apollo 7*, the CSM's rocket, called the service propulsion system (SPS), made eight nearly perfect firings. The successful demonstrations were of critical importance to the entire Apollo Project, because the SPS was the rocket engine that would place the Apollo astronauts in and out of orbit around the Moon. Other Apollo spacecraft hardware and operational procedures worked equally well, and there were no significant problems. This extended mission clearly demonstrated the space-worthiness of the redesigned CSM spacecraft and served as a major technical milestone on the pathway to the Moon.

However, one unintentional biological glitch did occur on the extended flight. Although the Apollo spacecraft provided a somewhat larger crew cabin for the three astronauts, living under cramped conditions in microgravity for 11 days eventually took its toll. Basic human-factor issues were compounded when, shortly after liftoff, Schirra developed a really bad head cold. The next day, the other two astronauts (Eisele and Cunningham) also developed colds. The microgravity environment of their orbiting spacecraft exacerbated the cold conditions, because normal drainage of fluids from the head did not occur. Despite taking medication, the colds caused all three astronauts extreme discomfort throughout the lengthy mission. Their personal discomfort hampered the performance of some scheduled duties. In addition, during reentry operations the astronauts did not wear their space suit helmets so they could clear their throats and unblock their ears more effectively.

On October 22, the *Apollo 7* crew jettisoned the service module and prepared the command module for reentry operations, which started about 10 minutes later. The spacecraft splashed down in the Atlantic Ocean near Bermuda, about eight miles (13 km) north of the recovery ship USS *Essex*. Schirra, Eisele, and Cunningham had logged 260 hours and nine minutes of flight time during this extended orbital mission, in which they made 163 revolutions of Earth. The *Apollo 7* command module is now on display at the Canada Science and Technology Museum in Ottawa.

APOLLO 8

Apollo 8 astronauts were the first human beings to venture beyond low Earth orbit, escape from Earth's gravity, and visit another world. Astronaut Frank Borman served as commander; James A. Lovell, Jr., as command module pilot; and William A. Anders as lunar module pilot. Although a functional lunar module was not used on this mission, a lunar module test article (LMTA) was included. NASA engineers mounted the LMTA in the spacecraft/launch vehicle adapter as ballast for mass balance purposes.

The mission began on December 21, 1968, when a powerful Saturn V rocket (AS-503), successfully lifted off from Complex 39-A at the Kennedy

Space Center in Florida. This mission was the first human-crewed mission to employ the Saturn V rocket and the first human-crewed mission of any nation to circumnavigate the Moon. Over a total period of 147 hours, the *Apollo 8* spacecraft took its crew on a faultless, 497,200-mile (800,000-km) journey to and around the Moon—an historic flight that included 10 revolutions of the Moon.

The mission achieved important operational experience and tested the Apollo command module systems, including communications, tracking, and life support, in both cislunar space as well as in lunar orbit. NASA mission planners and human-factor specialists were also provided their first opportunity to evaluate the performance of an astronaut crew in orbit around the Moon. The *Apollo 8* crew photographed both the nearside and farside of the lunar surface and collected additional information about nearside topography and surface features in anticipation of future landing missions.

The crew also performed six live television transmission sessions, including an inspirational Christmas Eve broadcast in which each astronaut read from the Book of Genesis as they observed the stark lunar landscape below their spacecraft and saw Earth in all its colorful majesty rise above the Moon's horizon.

After completing a total of 10 orbits around the Moon, the astronauts successfully performed the all-important translunar injection burn of the command and service module's rocket on December 25. The *Apollo 8* spacecraft then left lunar orbit and headed back to Earth. When the astronauts splashed down in the Pacific Ocean on December 27, they were about 1,000 miles (1,600 km) south-southwest of Hawaii and only three miles (5 km) from the recovery ship, the aircraft carrier USS *Yorktown*. The *Apollo 8* command module is now on display at the Chicago Museum of Science and Industry in Illinois.

APOLLO 9

The Apollo 9 mission took place in March 1969 and was the first *all-up* flight of the Apollo/Saturn V space vehicle configuration—that is, all lunar mission equipment was flown together for the first time, including a flight article lunar module instead of the lunar module test article (LMTA) used in the Apollo 8 mission. This mission was the third crewed flight in the Apollo Project. Astronaut James A. McDivitt served as commander, David R. Scott as command module pilot, and Russell Schweickart as lunar module (LM) pilot. Starting with this mission, the astronauts received permission from NASA headquarters to give names to both the command and service module (CSM) and the lunar module spacecraft. For this mission, the CSM was called *Gumdrop*, and the LM was called *Spider*. After the service module was jettisoned from the combined CSM spacecraft at the end of the mission (just prior to reentry), the command module retained the name *Gumdrop.*

The primary objective of the Apollo 9 mission was to test all aspects of the lunar module in Earth orbit. The astronauts demonstrated that the LM could function properly as an independent, self-sufficient spacecraft. They also performed a series of rendezvous and docking maneuvers with *Gumdrop* and *Spider.* These demonstration tests provided all the evidence NASA managers needed to remain confident that both Apollo Project spacecraft, namely the CSM and LM, were up to the rigors of a lunar mission—a mission that included orbital rendezvous and docking while the spacecraft traveled around the Moon.

A giant Saturn V rocket (AS-504) lifted the astronauts, the CSM, and the LM into orbit around Earth from Complex 39-A at the Kennedy Space Center on March 3, 1969. After launch, the combined S-IVB upper stage and the adapter-LM-CSM-payload were inserted into an almost circular 119-mile- (192-km-) altitude orbit around Earth. NASA mission control personnel then vented the propellant tanks on the S-IVB upper stage rocket, changing the orbit to approximately 123 miles (198 km) by 127 miles (204 km). Next, almost three hours after launch, the astronauts separated the CSM from the S-IVB upper stage rocket. They also jettisoned the adapter panels, exposing the LM, which was mounted on the S-IVB. Several minutes later, the astronauts turned the CSM around and docked with the LM. About one hour later (or some four hours after launch), the astronauts separated the S-IVB upper stage rocket from the docked CSM-LM spacecraft combination. NASA mission controllers then supervised a 62-second firing of the S-IVB's J-2 rocket engine—a propulsive action that raised the expended rocket vehicle's apogee to 1,896 miles (3,050 km).

Over the next few days, the *Apollo 9* astronauts fired the CSM's service propulsion system (SPS) five times to change the orbit of the combined CSM-LM configuration, to prepare for rendezvous maneuvers, and to test the dynamics of the cojoined CSM and LM under thrust. On March 5, the LM descent engine was also fired for 367 seconds. On March 6, two of the *Apollo 9* astronauts performed a simultaneous extravehicular activity (EVA). Schweickart performed a 37.5-minute EVA on the LM porch to test the astronaut's portable life support system and the extravehicular mobility unit. Schweickart's use of the new Apollo space suit was the first time an American astronaut's space suit had its own (independent) life support system rather than being dependent on an umbilical connection to the spacecraft. While Schweickart was on an EVA on the LM "front porch," Scott (with a life support umbilical connection) performed an EVA from the CSM side hatch.

Test operations in Earth orbit got even more interesting on March 7, when McDivitt and Schweickart climbed into the *Spider* LM and separated from the *Gumdrop* CSM. As part of the test maneuvers, they jettisoned the LM descent stage and fired the LM ascent stage's rocket for the first time. These orbital activities culminated with the simulation of the LM returning from a lunar (landing) mission, rendezvousing with the orbiting CSM, and then docking. After *Spider* successfully docked with *Gumdrop,* McDivitt and Schweickart transferred back to the CSM. The astronauts then jettisoned the LM ascent stage (for continuity, this spacecraft was still called *Spider* even though its descent stage was previously discarded). *Spider's* ascent engine was then commanded to fire until propellant depletion. This placed the LM ascent stage into a 146-mile (235-km) by 4,332-mile (6,970-km) orbit around Earth. *Spider* eventually decayed from its orbit and burned up in Earth's atmosphere on October 23, 1981. The jettisoned LM descent stage experienced orbital decay much more rapidly and reentered Earth's atmosphere on March 22, 1969.

While in Earth orbit, the *Apollo 9* astronauts had successfully demonstrated key docking and rendezvous maneuvers between the CSM spacecraft and the lunar module. A landmark tracking exercise was also accomplished during this mission. After logging 241 hours in space, the three astronauts returned to Earth on March 13. The *Gumdrop* CM splashed down in the Atlantic Ocean about 180 miles (290 km) east of the Bahamas and within sight of the recovery ship USS *Guadalcanal.* The *Apollo 9* command module *Gumdrop* is now on display at the Michigan Space and Science Center in Jackson.

APOLLO 10

The Apollo 10 mission in May 1969 successfully accomplished the second human flight that orbited the Moon. In a dress rehearsal for the actual lunar-landing mission (which occurred in Apollo 11), the *Apollo 10* astronauts came within 8.9 miles (14.3 km) of the lunar surface and spent nearly 62 hours (31 revolutions) in lunar orbit.

Astronaut Thomas P. Stafford served as commander of the mission, John W. Young as command module (CM) pilot, and Eugene A. Cernan as lunar module (LM) pilot. The crew named the command and service module (CSM) *Charlie Brown* and the LM *Snoopy,* after the popular *Peanuts* cartoon characters created by Charles M. Schulz.

NASA launched the Apollo 10 mission with another powerful Saturn V rocket vehicle (AS-505), which lifted off from Complex 39-A at the Kennedy Space Center on May 18, 1969. The primary objectives of the mission were to demonstrate the interactions of crew, space vehicle, and mission support facilities throughout a complete manned mission to the Moon. The astronauts also performed a variety of operations that evaluated the performance of the *Snoopy* lunar module in both cislunar and lunar orbit environments. Apollo 10 was a "dry run" for the planned Apollo 11 lunar-landing mission.

During Apollo 10, the Apollo Project hardware was thoroughly tested, and all operations, except the actual lunar landing, were performed.

One occasionally raised speculation is whether or not the *Apollo 10* crew, in coming so close (within 8.9 miles [14.3 km]) to the Moon's surface, might have decided to go ahead and land. Perhaps the lunar module crew might have had the mental desire to be the first to touch down on the Moon's surface, which lay tantalizingly just beneath their spacecraft. But the LM crew consisted of astronauts Stafford and Cernan, who were well-disciplined, highly trained military officers on loan to NASA and specially selected for this important mission precisely because they knew how to follow orders dependably and how to respond decisively to any unplanned events or equipment malfunctions. Furthermore, their lunar module *Snoopy* was actually an early design, test article that was simply too massive for any successful lunar landing and subsequent ascent flight back up to the CSM. While *Snoopy* was suitable for this dress rehearsal mission, it was not capable of returning any unofficially "landed astronauts" to the orbiting CSM for the journey back to Earth.

After launch, the *Apollo 10* spacecraft was inserted into a 118-mile (190-km) by 115-mile (185-km) nearly circular parking orbit around Earth. After completing one and one-half orbits of their home planet, the astronauts accomplished the translunar injection maneuver. The *Charlie Brown* (CSM) separated from the Saturn V rocket's third stage (known in aerospace shorthand as the S-IVB stage), transposed, and docked with the *Snoopy* (LM). After a three-day cruise through cislunar space, the astronauts used a 356-second firing of the CSM's service propulsion system (SPS) on May 21 to enter an initial 196-mile (316-km) by 68.6-mile (110-km) orbit around the Moon. The crew then fired the SPS for 19.3 seconds. This second SPS firing placed the *Apollo 10* spacecraft into a nearly circular 70-mile (112-km) altitude lunar orbit.

On May 22, astronauts Stafford and Cernan entered the *Snoopy* LM, and Young fired the service module's reaction control thrusters to separate the LM and the CSM spacecraft. Stafford and Cernan then placed the lunar module into an orbit that allowed low-altitude passes over the lunar surface. Their closest approach brought them within 8.9 miles of the Moon's surface. The astronauts tested all systems on the lunar module during this period of separation from the CSM. The tests included communications, propulsion, attitude control, and radar. They also snapped numerous high-resolution photographs of the Moon's surface, especially pictures of candidate landing sites for future Apollo missions. At the end of these tests, timed to simulate the upcoming Apollo 11 landing mission, the astronauts jettisoned the LM descent stage into lunar orbit and returned to the CSM using *Snoopy*'s ascent stage. Early in the morning on May 23, Stafford and Cernan rendezvoused and docked with the *Charlie Brown* (CSM). They had spent about eight hours testing the LM. All systems worked well, and the tests went smoothly, except for a momentary gyration in the lunar module's motion, caused by a faulty switch setting.

Later in the day, the *Apollo 10* astronauts jettisoned the LM ascent stage (that is, what remained of *Snoopy*) into a heliocentric orbit. On May 24, after completing 31 revolutions of the Moon, the astronauts fired the CSM's SPS to achieve transearth injection. Late in the afternoon on May 26, the crew jettisoned the service module and prepared the command module (*Charlie Brown*) for reentry. About 30 minutes later, the *Apollo 10* CM splashed down in the Pacific Ocean at a point some 400 miles (644 km) west of American Samoa and just 3.4 miles (5.5 km) away from the recovery ship USS *Princeton*.

As an historic note, the Apollo astronauts took great pride in the accuracy of their landings during reentry; they even had an informal pool that went to the crew whose capsule landed closest to the planned target point. Early in the project, the Apollo spacecraft's computer could pinpoint the target location far better than the recovery ships. Eventually, the recovery ships began using navigation equipment comparable in accuracy to that provided by the Apollo spacecraft's digital computer. Nevertheless, the captains of the recovery ships still cautiously kept them about a mile (2 km) or so away from the targeted splashdown point to avoid any possibility of a collision.

During this highly successful rehearsal mission, the *Apollo 10* astronauts clearly demonstrated that all was ready for the long-anticipated lunar landing. Stafford, Young, and Cernan had logged a little more than 192 hours in space and made 31 revolutions of the Moon. Another feature of this mission was the first live color television broadcast from a spacecraft. The *Apollo 10* command module (*Charlie Brown*) is currently on display at the Science Museum in London.

APOLLO 11

The Apollo 11 mission achieved the national goal set by President Kennedy in 1961—namely, landing human beings on the surface of the Moon and returning them safely to Earth within the decade of the 1960s. On July 20, 1969, astronauts Neil Armstrong (commander) and Edwin (Buzz) Aldrin (lunar module [LM] pilot) flew the lunar module (called *Eagle*) to the surface of the Moon, touching down safely in the Sea of Tranquility. While Armstrong and then Aldrin became the first two persons to walk on another world, their fellow astronaut, Michael Collins (the command module [CM] pilot), orbited above in the *Columbia* command and service module (CSM).

Neil Armstrong's own words capture best the feeling of excitement on July 16, 1969, when a gigantic Saturn V rocket (AS-506) flawlessly lifted off its pad at Complex 39-A of the Kennedy Space Center and began the most profound journey in human history. In a NASA Apollo Project summary report (published in 1975), Armstrong noted that: "As we ascended in the elevator to the top of the Saturn on the morning of July 16, 1969, we knew that hundreds of thousands of Americans had given their best effort to give us this chance. Now it was time for us to give our best."

The *Apollo 11* spacecraft (CSM-LM combined) followed a similar mission profile to the Moon, as had its immediate predecessor, the *Apollo 10* spacecraft. Following launch, *Apollo 11* entered orbit around Earth. After completing one and one-half orbits of Earth, the third stage of the Saturn V (the S-IVB upper stage) reignited for an approximately six-minute, translunar injection burn that placed the spacecraft on a course for the Moon. Thirty-three minutes later, the *Apollo 11* astronauts separated the *Columbia* CSM from the S-IVB stage, turned the spacecraft around, and docked with the *Eagle* LM. About 75 minutes later, they released the S-IVB and injected this spent rocket stage into heliocentric orbit. While the docked CSM-LM spacecraft configuration coasted through cislunar space for its rendezvous with history, the *Apollo 11* astronauts made a live color television broadcast back to Earth.

On July 19, the astronauts performed a 358-second, retrograde firing of the CSM's service propulsion system (SPS) in order to achieve insertion into lunar orbit. This orbit insertion burn was accomplished, while the spacecraft was behind the Moon and out of contact with Earth. A second, much shorter duration (17-second) SPS burn circularized the spacecraft's orbit around the Moon.

The next day (July 20), Armstrong and Aldrin entered the *Eagle* to perform a final checkout before traveling in the LM down to the lunar surface. As the LM and CSM separated, Collins made a visual inspection of the *Eagle* from the *Columbia* (CSM). Armstrong and Aldrin then fired the LM's descent engine for 30 seconds. Their actions put the *Eagle* in a descent orbit, which had a closest approach to the Moon's surface of nine miles (14.5 km). The two astronauts fired the LM descent engine once again, this time for 756 seconds, and they began the final phase of their historic descent to the Moon's surface. Although Armstrong piloted the LM to a safe touchdown on the lunar surface, when he finished he had less than 30 seconds of propellant supply remaining. The problem the astronauts encountered was finding a suitable landing site. Despite all the previous photographic reconnaissance that was performed during the site selection process, the original landing site chosen in Mare Tranquilitatis (the Sea of Tranquility) was actually populated with a large number of small craters and rocks—landing wrong near any one of which could have spelled disaster for the mission. Searching for a suitable landing spot as the fuel supply for the *Eagle*'s descent engine approached exhaustion,

On July 16, 1969, American astronauts Neil Armstrong, Edwin (Buzz) Aldrin, and Michael Collins lifted off from Launch Pad 39A at the Kennedy Space Center in Florida, perched atop a mammoth–size Saturn V rocket. As the giant rocket slowly rumbled off its pad, the three *Apollo 11* astronauts began their historic journey to the Moon, climaxed by the first lunar–landing mission. *(NASA)*

Armstrong finally spotted a relatively smooth place and quickly set the spidery-looking spacecraft down at 4:17 P.M. (EDT) on July 20, 1969.

Throughout this harrowing search for a safe lunar-landing spot, personnel at NASA's mission controller center in Houston, Texas, were anxiously monitoring the depletion of the *Eagle*'s propellant supply. As soon as signals from the LM indicated some type of contact had been made with the surface, mission control sent the following short message: "We copy you down, *Eagle*." The modest time delay (a little more than two seconds) for radio signals to go back and forth between Earth and the Moon seemed like an eternity to everyone in the room that day. Then came Armstrong's famous reply: "Houston, Tranquility Base here. The *Eagle* has landed!" The response from Houston at this historic moment proved equally memorable: "Roger, Tranquility. We copy you on the ground. You' ve got a bunch of guys here about to turn blue. We' re breathing again. Thanks a lot." This simple dialogue marked the start of one of the greatest moments in exploration.

Six hours later, Armstrong opened the ingress-egress hatch on the LM and cautiously descended down the ladder. As his left foot made contact with the lunar soil, he reported back to Houston: "That's one small step for (a) man . . . one giant leap for mankind." About 19 minutes later, Aldrin followed and became the second human being to walk on the Moon. As he looked out at the lunar landscape and noticed the starkness of the shadows and the barren, almost desertlike, characteristics of the Moon's surface, Aldrin remarked: "Beautiful, beautiful. Magnificent desolation."

Like tourists everywhere, Armstrong and Aldrin began their visit to the Moon by snapping lots of pictures. They also collected souvenirs, some 47.7 pounds (21.7 kg) of soil and rock samples for the planetary scientists back home. Once their initial euphoria subsided, they began deploying instruments, such as the Early Apollo Surface Experiments Package (EASEP), near the lunar module. In a few hours, they traversed a total of about 820 feet (250 m) across the Moon's surface, gathered rock specimens, inspected the LM, positioned science instruments, and planted an American flag—not as a symbol of territorial claim but rather as a permanent symbol of the nation that accomplished the first human landing. They also removed the protective, thin metal plate that was covering the *Apollo 11* plaque mounted on the LM's ladder.

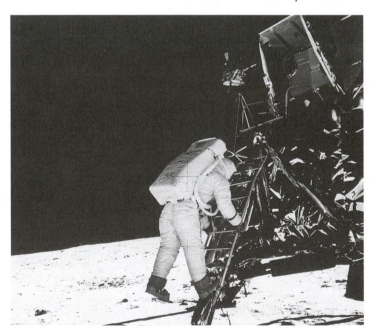

Apollo 11 astronaut Edwin (Buzz) Aldrin, Jr., descends the ladder of the lunar module (LM) *Eagle* and becomes the second human to walk on the Moon (July 20, 1969). *(NASA)*

At the conclusion of their EVA, the astronauts returned to the LM and closed the hatch. They were supposed to sleep for a few hours before attempting to blast off from the lunar surface and rejoining Michael Collins, who was orbiting above in the *Columbia* CSM. But they were unable to sleep. Clearly, there was far too much to do and see and too little time. The best "rest" Aldrin managed to accomplish on the floor of the cramped lunar module was (in his own words) a "couple of hours of mentally fitful drowsing." Armstrong simply stayed awake inside the tiny crew cabin filled with noisy pumps and bright warning lights.

On July 21, after spending 21 hours and 36 minutes on the lunar surface, the astronauts fired the LM's ascent engine and lifted off from the Moon's surface. As the upper half of the *Eagle* arose into lunar orbit, the lower half remained behind on the surface in the Sea of Tranquility at 00.6 degree N latitude, 23.5 degrees E longitude (lunar coordinates). The *Eagle* (as well as the other abandoned lunar module descent stages from *Apollo 12, 14, 15, 16,* and *17*) now serves as a permanent memorial to human's conquest of space. Armstrong and Aldrin then docked with *Columbia* and transferred the collection of lunar rocks and some equipment into the command module.

On July 22, in preparation for the journey back to Earth, the astronauts jettisoned the ascent stage of the lunar module into orbit around the Moon. The precise fate of the upper half of the *Eagle* LM is not known, but NASA mission managers assumed that it crashed into the Moon's surface within a month to four months after being abandoned in orbit. After completing 31 revolutions of the Moon, the *Columbia* prepared to return home to Earth. A two and one-half minute firing of the CSM's main rocket engine began the all-important transearth injection process.

On the morning of July 24, the command module made its programmed separation from the service module (SM), and its three occupants prepared for reentry. After a mission elapsed time of 195 hours, 18 minutes, and 35 seconds, Armstrong, Aldrin, and Collins splashed down in the middle of the Pacific Ocean about 15 miles (24 km) away from the recovery ship USS *Hornet.* U.S. Navy recovery crews arrived quickly by helicopter and tossed biological isolation garments into the spacecraft. After the suitably "cocooned" astronauts emerged from the *Columbia* command module, the team of recovery swimmers swabbed the spacecraft's hatch down with an organic iodine solution. Then the astronauts and recovery team personnel decontaminated each other's protective garments with a solution of sodium hypochlorite. The three astronauts were then plucked from the ocean's surface, transported by helicopter to the USS *Hornet,* and placed immediately in a special lunar quarantine trailer facility on the deck of the aircraft carrier. After a quick change of clothing inside the quarantine facility, the astronauts appeared at the window and received personal congratulations from President Richard M. Nixon (who had flown to the *Hornet*).

While confined in their quarantine trailer, Armstrong, Aldrin, and Collins—along with their now biologically isolated command module and its precious cargo of lunar rocks—traveled to the Lunar Receiving Laboratory in Houston, Texas. There they remained in quarantine until late in the evening on August 10. From a medical perspective, this period of quarantine proved totally uneventful for the astronauts. Showing no signs of any ill effects from exposure to lunar dust or to any "postulated" extraterrestrial microorganism that might have hitchhiked back to Earth on their space suits or equipment, NASA's biomedical experts decided to release the three astronauts from quarantine. They went home to their families and, after a much deserved period of privacy, embarked on a triumphant tour around the world. A publicly accessible lunar rock sample and the *Apollo 11* command module (*Columbia*) are on exhibit at the National Air and Space Museum in Washington, D.C. The descent stage of the *Eagle* awaits the next generation of lunar explorers at its original landing site in the Sea of Tranquility.

APOLLO 12

The Apollo 12 mission was the first of the H-series missions, during which scientific exploration of the Moon began in earnest. The primary mission goals for Apollo 12 involved completion of an extensive list of lunar exploration tasks, the deployment of the Apollo lunar surface experiments package (ALSEP), and the demonstration of the ability to remain and work on the surface of the Moon for an extended period. The astronaut crew accomplished all these goals in November 1969.

Astronaut Charles P. "Pete" Conrad, Jr., served as the mission commander; Richard F. Gordon, Jr., as the command module pilot; and Alan L. Bean as the lunar module (LM) pilot. The astronauts named the *Apollo 12* command and service module (CSM) *Yankee Clipper* and the LM *Intrepid*. Apollo 12 was the second mission in which human beings walked on the lunar surface and safely returned to Earth.

The successful mission got off to a somewhat dubious start on November 14, 1969, when NASA launched the Saturn V rocket (AS-507) from Complex 39-A at the Kennedy Space Center during a driving rainstorm. As the rocket ascended to orbit, its spacecraft payload was hit twice by lightning—at 36 seconds and 52 seconds after launch. The lightning strikes momentarily caused an electrical power failure in the spacecraft, and there was also a brief cessation of telemetry from the spacecraft back to the ground. However, power was automatically switched to battery backup, and this gave the crew time to examine the problem and to restore the primary (fuel cell provided) spacecraft power.

As the huge rocket continued to climb toward outer space, NASA engineers and technicians on the ground worked feverishly to help the three astronauts resolve any problems. In particular, while the engineering team on the ground coached and advised, the astronauts kept busy resetting circuits and checking all the spacecraft's operating systems. The crew wanted to make absolutely sure the surprise lightning flashes had not harmed the spacecraft or threatened their ability to safely accomplish the upcoming Moon landing.

After the mission was completed, NASA scientists performed a more thorough investigation of the cause of the sudden electric power outage. They concluded that the *Apollo 12* vehicle had created its own lightning. The scientists postulated that as the rocket passed up through the rain clouds, there was a buildup of static electricity on the ascending vehicle. Apparently, both flashes involved an accumulation of static charge, which then suddenly discharged—disrupting the spacecraft's electrical systems.

Fortunately, the *Apollo 12* spacecraft and its crew survived this electrifying experience unharmed. As the spacecraft entered orbit around Earth (some 11 minutes and 44 seconds after launch), the astronauts had restored primary power, and all spacecraft systems appeared to be functional. After one and one-half orbits around Earth, the S-IVB upper stage engine came to life again, flawlessly performed a five-minute and 45-second burn, and placed the *Apollo 12* spacecraft on its intended translunar trajectory. Following an uneventful cruise through cislunar space, the *Apollo 12* spacecraft arrived at the Moon and achieved orbit on November 18. The next day (November 19), astronauts Conrad and Bean boarded the LM *Intrepid;* bid a brief farewell to Richard Gordon, who remained on board the CSM *Yankee Clipper;* and safely landed on the lunar surface in Oceanus Procellarum (Ocean of Storms).

Their pinpoint landing was so precise that Conrad and Bean had set the *Intrepid* down within 600 feet (183 m) of the *Surveyor 3*—a robot spacecraft sent by NASA to the Moon in April 1967. Conrad became the third person to walk on the Moon. Bean followed and became the fourth member of the exclusive Moonwalker club.

During the Apollo 12 mission, Conrad and Bean performed two lunar surface extravehicular activities (EVAs)—one on November 19 and the other on November 20. During both these walks on the Moon, the two astronauts worked in approximately four-hour-long shifts, exploring each site of interest and carrying all their equipment, tools, and experiments by hand. They set up the ALSEP unit. They also walked over to the *Surveyor 3* spacecraft and retrieved the robot's TV camera and mechanical scoop, so scientists back on Earth could study the impact on spacecraft components that some 30 months of exposure on the lunar surface had caused. They also took a variety of photographs and gathered up about 75.7 pounds

(34.4 kg) of lunar rock and soil samples from the surrounding Oceans of Storms region.

After spending a total of 31 hours and 31 minutes on the lunar surface, Conrad and Bean prepared the ascent stage of the *Intrepid* for liftoff on November 20 and then fired the LM's ascent engine. The descent stage of the *Intrepid* remained on the Moon where the astronauts landed in the Sea of Storms at approximately 3.0 degrees S latitude, 23.4 degrees W longitude (lunar coordinates). Once in lunar orbit, the astronauts docked the *Intrepid* (upper stage) with the *Yankee Clipper* (CSM) and transferred themselves and the collection of lunar soil and rock samples to the command module. About two hours later, they jettisoned the upper stage of *Intrepid* and used the onboard computer to fire its rocket engine once again, deliberately crashing the spacecraft into the Moon at 3.94 S latitude, 21.20 W longitude. The impact of the LM's upper stage caused the first artificial Moonquake—an event recorded and reported by the seismic instruments in the ALSEP. The crew remained an extra day in lunar orbit to continue taking photographs. On November 21, after completing 45 revolutions of the Moon, the astronauts fired the CSM's main rocket engine and departed from lunar orbit on a transearth trajectory.

Following a midcourse correction on November 22, the astronauts arrived at Earth on November 24. The command module properly separated from the service module and reentered the atmosphere several minutes later. After a mission elapsed time of 244 hours and 36 minutes, Conrad, Bean, and Gordon splashed down in the Pacific Ocean near American Samoa. The astronauts impacted the water just 4.3 miles (6.9 km) from the recovery ship USS *Hornet*. They then underwent the same three-week quarantine procedures that had been followed by the returning *Apollo 11* astronauts. The *Apollo 12* command module *Yankee Clipper* is now on display at the Virginia Air and Space Center in Hampton. The *Surveyor 3* camera recovered from the Moon and returned to Earth by the *Apollo 12* astronauts is on exhibit at the National Air and Space Museum in Washington, D.C.

APOLLO 13

Rocket scientists, aerospace engineers, and astronauts are, by nature, persons who make decisions based on facts and mathematical logic. They understand that superstitions have no place in human spaceflight. However, even the most hard-nosed, number-crunching scientist or engineer cringes a bit when he or she reviews the cumulative misfortunes that struck NASA's third manned lunar-landing mission, Apollo 13. The good news, of course, is that the three affected astronauts refused to succumb to the life-threatening misfortunes that reached a climax some 200,000 miles (322,000 km) away from Earth. Their struggle to survive is an inspiring account of personal courage and professional behavior in the face of grave danger.

The *Apollo 13* spacecraft was the second of the Apollo H series. As originally planned, the primary purpose of this third lunar-landing mission was to explore the hilly upland Fra Mauro region of the Moon. The astronauts were also scheduled to deploy an ALSEP unit and to obtain photographs from orbit as well as at the landing site. Although the main objectives of this mission were not realized, the astronauts—despite the perilous nature of their journey around the Moon in a badly disabled spacecraft—did collect a limited amount of photographic data.

Astronaut James A. Lovell, Jr., served as the mission commander; John L. Swigert, Jr., as the command module (CM) pilot; and Fred W. Haise, Jr. as the lunar module (LM) pilot. The crew named the *Apollo 13* command and service module *Odyssey* and the lunar module *Aquarius*. The first omen of pending misfortune took place several days before the launch. The originally scheduled CM pilot was astronaut Thomas K. Mattingly II. But Mattingly's participation as a member of the prime crew had to be scrubbed at the last minute because of his inadvertent exposure to German measles—a relatively minor childhood disease against which he had no immunity. Astronaut John L. Swigert, Jr., was his substitute on this mission. He proved to be an excellent addition to the team even though he had only two days of training with the other two members of the prime crew.

On April 11, 1970, the mission's Saturn V rocket (AS-508) lifted off from Complex 39-A at the Kennedy Space Center at 13:13 (Houston time), carrying the *Apollo 13* spacecraft into orbit around Earth. Things appeared normal during the operation of the Saturn V's powerful first stage. But during the second-stage firing, the center engine of the S-II stage cut off 132 seconds early, causing the remaining four engines in that stage to burn 34 seconds longer than planned. Since the spacecraft's velocity after the S-II stage burn was lower than planned by 223 feet per second (68 m/s), the Earth orbital insertion burn of the Saturn rocket's third stage (S-IVB) had to take place for nine seconds longer than originally planned.

About an hour and one-half-later, when the S-IVB fired again, the *Apollo 13* spacecraft experienced translunar injection and left Earth orbit. Although the second-stage booster anomaly caused some concern among the NASA officials monitoring the flight in Houston, their concerns temporarily dissipated after the successful translunar injection burn. In fact, as the spacecraft traveled through cislunar space, activities on board the *Apollo 13* spacecraft appeared to be going well. The crew separated the command and service module (CSM) from the S-IVB, docked with the lunar module, and then sent the expended S-IVB on an impact trajectory into the Moon. (The S-IVB upper stage impacted into the lunar surface in the early morning of

April 14 [universal time] at 2.75 degrees S latitude, 27.9 degrees W longitude, with a velocity of 1.6 miles per second [2.58 km/s].)

At 55 hours and 46 minutes into the flight, Lovell, Swigert, and Haise completed a 49-minute television broadcast back to Earth extolling how they were living and working comfortably in microgravity conditions inside the command module. All that changed a few minutes later when Jack Swigert turned the fans on to stir the oxygen tanks in the service module portion of the *Odyssey* spacecraft. As determined by NASA's postflight accident review board, wires—damaged during preflight testing in oxygen tank number 2—had shorted, and their insulation had caught fire. The fire continued to spread within the tank, until—at 55 hours, 54 minutes, and 53 seconds mission elapsed time on April 13 (EST)—oxygen tank number 2 exploded, damaging oxygen tank number 1, and much of the interior of the service module. Swigert sent what is perhaps the most understated distress message in aerospace history when he calmly reported: "Houston. We've had a problem."

Problem indeed! With its oxygen supplies rapidly venting into space, the command module became unusable. The lunar-landing mission had to be aborted. In order to save their lives, the three astronauts had to quickly power down the *Odyssey* and then seek refuge in the lunar module *Aquarius*—the spiderlike spacecraft designed to take two astronauts to the surface of the Moon. As they abandoned the command module, each astronaut silently hoped that, when and if the time ever came, the *Odyssey* could be reactivated for reentry. But that moment was a long way off, since Earth was then about 200,000 miles away.

The crew cleverly used the lunar module's descent engine to make a critical midcourse correction that placed the disabled spacecraft (CSM-LM joined together) on a free-return trajectory. After rounding the Moon on April 15, the crew performed another LM descent engine burn for 263 seconds. This successful second firing gave the *Apollo 13* spacecraft a differential velocity of 859 feet per second (262 m/s) and shortened the return to Earth by about nine precious hours. Water and power supplies were getting dangerously low on board the LM, and there was a dangerous buildup of carbon dioxide (CO_2) in the atmosphere of the crew cabin. Water rationing, minimal use of onboard electric power, and the application of a clever, hastily improvised technique to use the "square" lithium canisters from the CM as substitutes for the expended "round" lithium hydroxide canisters that were removing CO_2 on the LM alleviated some of the most threatening problems and kept the cold, dehydrated, and uncomfortable astronauts alive.

On April 17 at 13:15 (universal time), the astronauts jettisoned the badly damaged service module. They had hauled the disabled spacecraft all the way to the Moon and back in order to protect the command module's heat shield from exposure to the frigid temperatures of outer space. They would need this heat shield in good working condition if they were to survive the fiery conditions of reentry. As the disabled service module separated from the CM-LM spacecraft configuration, the astronauts dutifully took pictures of it. These photographs later provided helpful information in determining the cause of the disastrous explosion. At about 16:43 (universal time) on April 17, the three beleaguered space travelers climbed into the cold, damp command module. They followed the carefully worked-out instructions provided to them by the team engineers at NASA's mission control center. Lovell, Swigert, and Haise slowly applied electric power to all the *Odyssey*'s systems that were absolutely necessary for reentry. No one knew how long the CM's batteries would last or if the accumulated moisture behind the instrument panels would short out a critical electrical circuit. As the *Odyssey* came back to life, the astronauts

The *Apollo 13* astronauts (left to right) Fred Haise, John Swigert, and James Lovell smile during the press conference held weeks after they recovered from their aborted lunar landing mission. They were about 200,000 miles (322,000 km) from Earth and traveling to the Moon on April 13, 1970, when an oxygen tank in the service module portion of their Apollo spacecraft *Odyssey* exploded. A model of the lunar module *Aquarius*—the spacecraft that served as a lifeboat and saved their lives—is on the table in front of Fred Haise. *(NASA)*

jettisoned their lifeboat, the *Aquarius,* and prepared for the last remaining hurdle in their race for survival—reentry.

As people around the world held a collective breath, the distressed *Apollo 13* spacecraft made its fiery plunge into Earth's atmosphere, and minutes later it splashed safely down in the Pacific Ocean southeast of American Samoa. Remarkably, the three astronauts—though exhausted, sick, and dehydrated—managed to land their wounded spacecraft in the ocean just four miles (6.5 km) from the recovery ship USS *Iwo Jima.* Its duty done, the lifeboat *Aquarius* burned up in Earth's upper atmosphere. Any surviving parts of the lunar module (including possibly the SNAP-27 radioisotope power supply for the ALSEP) fell harmlessly into a remote area of the Pacific Ocean, somewhere northeast of New Zealand. The *Apollo 13* astronauts had logged a total of 142 hours, 54 minutes, and 41 seconds in space during this drama-filled, aborted lunar-landing mission. In reporting the safe return of the *Apollo 13* astronauts, the *Christian Science Monitor* said: "Never in recorded history has a journey of such peril been watched and waited-out by almost the entire human race."

The quick decision to use their lunar module (*Aquarius*) as a lifeboat, and the skillful maneuvering of their disabled spacecraft around the Moon, brought astronauts Lovell, Swigert, and Haise safely back to Earth. Rescue credit is also due to the thousands of NASA engineers and managers who worked continuously throughout the in-space emergency to provide the stranded astronauts with the best technical advice on how to stretch dwindling resources and return their disabled spacecraft home. Today the *Apollo 13* command module *Odyssey* is on display at the Kansas Cosmosphere and Space Center in Hutchinson.

APOLLO 14

The Apollo 14 mission was retargeted to accomplish the lunar-landing mission planned for *Apollo 13.* This mission was the third of the Apollo H series. The primary mission goals included collecting lunar samples in a different (highland) region of the Moon, deploying an ALSEP unit and other scientific experiments, and photographing from the lunar surface and from lunar orbit.

Astronaut Alan B. Shepard, Jr., the first American to fly a Mercury Project capsule into outer space, served as the mission commander. Stuart A. Roosa served as the command module (CM) pilot and Edgar D. Mitchell as the lunar module (LM) pilot. For all practical purposes, NASA had chosen a rookie spaceflight team for this particular lunar-landing mission. Up until Apollo 14, Shepard's spaceflight experience consisted of one cannonball-like, suborbital flight on May 5, 1961, that lasted about 15 minutes. For both Roosa and Mitchell, the Apollo 14 mission was their first flight in space. Regardless of the lack of previously accumulated spaceflight experience, the *Apollo 14* astronauts were well trained and were able to accomplish all the primary goals of the mission with little or no difficulty. They selected the name *Kitty Hawk* for the command and service module (CSM) and *Antares* for the lunar module.

The Apollo 14 mission was launched on January 31, 1971, by a Saturn V rocket (AS-509) from Complex 39-A at the Kennedy Space Center. The ascent into parking orbit around Earth was flawless. Following translunar injection, the astronauts separated the CSM from the S-IVB upper stage rocket, which at the time still contained the lunar module. The astronauts then made five frustrating attempts to dock the CSM and the LM. However, all five attempts failed because the catches on the docking ring would not release. Fortunately, the astronauts persisted and made a sixth attempt at approximately 02:00 (universal time) on February 1. This docking attempt proved successful. No further problems with the CSM-LM docking mechanism occurred. The astronauts then jettisoned the S-IVB stage and sent it on a lunar impact trajectory. The expended S-IVB upper stage rocket crashed into the lunar surface on February 4 with a velocity of 1.58 miles per second (2.54 km/s). The lunar surface impact took place at 8.1 degrees S latitude and 6.0 degrees W longitude (lunar coordinates).

On February 4, the *Apollo 14* achieved insertion into lunar orbit. The next day (February 5), Shepard and Mitchell transferred from the CSM into the LM, separated from the CSM piloted by Roosa, and landed on the lunar surface in the hilly upland region about 14.9 miles (24 km) north of the rim of Fra Mauro crater. Specifically, the *Antares* lunar module had successfully touched down at 09:18 (universal time) in the Fra Mauro region at 3.6 degrees S latitude and 17.5 degrees W longitude—just 60 feet (18.3 m) from the targeted point. Because of the hilly nature of the highland site, the *Antares* came to rest on the slope of a small depression and tilted about eight degrees.

Shepard and Mitchell performed two lunar surface extravehicular activities (EVAs), totaling nine hours and 23 minutes. Their first walk on the Moon began on February 5 at 14:42 and ended at 19:30 (universal time). During the first EVA, the astronauts deployed the *Apollo 14* ALSEP unit and other surface experiments. They then returned to the *Antares* for a long rest period. During the second EVA on February 6, Shepard and Mitchell used a wagon-like contraption to help them move equipment and rock specimens across the lunar surface. NASA engineers had given this device a formal aerospace title, the modularized equipment transporter (MET), while the astronauts preferred to call it the "lunar rickshaw."

As part of their second EVA, Shepard and Mitchell attempted to walk all the way to the rim of nearby Cone crater. They collected samples along the traverse. However, the ruggedness and unevenness of the terrain made it difficult for them to navigate using distinctive surface features as landmarks. When they became somewhat disoriented, Shepard decided to abandon the quest, and the two Moonwalkers headed back to the *Antares*. NASA's postflight review of the mission (assisted by high-resolution orbital photography of the area) suggested that Shepard and Mitchell were actually only about 30 feet (10 m) below the rim when they turned back. However, because of the rough terrain, this was not readily apparent to the astronauts as they walked on the surface. On the way back to *Antares*, they continued to gather rock samples and to take numerous surface photographs. At the end of this long and arduous EVA, the space-suited Shepard paused briefly to make sports history. He connected a six-iron head to a contingency sampler tool and then used this makeshift club to hit two golf balls out of sight. Shepard and Mitchell (the sixth human to walk on the Moon) had traversed 2.1 miles (3.4 km) and collected 94.4 pounds (42.9 kg) of lunar samples.

After spending a total of 33 hours and 31 minutes on the lunar surface, the two astronauts fired the ascent engine of the *Antares* at 18:48 (universal time) on February 6 and climbed into orbit around the Moon. After docking with the CSM being piloted by Roosa, they transferred the lunar samples and other equipment to the *Kitty Hawk* and prepared the spacecraft for the return journey to Earth. At 22:48 (universal time) on February 6, the astronauts jettisoned the ascent stage of the *Antares* lunar module. The discarded spacecraft impacted the Moon's surface early in the morning of February 8 (universal time) at 3.42 degrees S latitude, 19.67 degrees W longitude (lunar coordinates), a location lying between the seismic stations in the *Apollo 12* and *Apollo 14* ALSEP units. The impact of a jettisoned spacecraft caused an artificial Moonquake that the sensitive seismometers in the deployed ALSEP units could record. Scientists used seismic data from these artificial Moonquakes and the well-known location of the ALSEP units to estimate some of the geophysical properties of the Moon's core and crust.

Astronauts Shepard, Roosa, and Mitchell fired the *Kitty Hawk*'s main rocket engine at 01:39 (universal time) on February 7, initiating the transearth injection maneuver. As they rapidly approached Earth on February 9, the *Apollo 14* command module automatically separated from the service module (SM). Reentry operations began a few minutes later, and the three astronauts splashed down safely at 21:05 (universal time) in the Pacific Ocean about 880 miles (1,417 km) south of American Samoa. The astronauts and the command module were plucked from the ocean by the recovery ship USS *New Orleans*. Upon recovery, Shepard, Roosa, and Mitchell entered quarantine.

Although neither the *Apollo 11* nor *Apollo 12* astronauts showed any signs of bringing back an exotic disease from the Moon, NASA officials cautiously decided to continue the quarantine protocol with the returning *Apollo 14* astronauts. The main reason for this decision was because the astronauts had collected deep core samples from another (highland) region of the Moon. When no evidence of diseases or native organic lunar material showed up, NASA discontinued the use of quarantine procedures for the returning *Apollo 15, 16,* and *17* astronauts.

The *Apollo 14* astronauts carried out numerous experiments and conducted successful exploration and specimen collections in the lunar highlands. With a total mission elapsed time of 216 hours, one minute, and 58 seconds, the crew became seasoned space-travel veterans. The *Kitty Hawk* command module is currently on exhibit at the Astronauts Hall of Fame in Titusville, Florida.

APOLLO 15

The Apollo 15 mission was the fourth successful lunar-landing mission and the first of the Apollo J-series missions, which involved longer expedition-style operations on the lunar surface. The J-series Apollo missions can be distinguished from the previous G (for example, *Apollo 11*) and H series missions (for example, *Apollo 12*) by the extended hardware capability, larger scientific payload capacity, and by the use of the lunar roving vehicle (LRV). Starting with the Apollo 15 mission, each lunar module carried a stowed electric vehicle, called the Apollo lunar roving vehicle, to the Moon's surface. Once deployed on the surface, this electric vehicle allowed the astronauts to explore larger areas of the Moon during each surface extravehicular activity (EVA).

Astronaut David R. Scott was the commander of the Apollo 15 mission, Alfred M. Worden served as command module pilot, and James B. Irwin was the lunar module pilot. The crew selected the name *Endeavor* for the command and service module (CSM) and *Falcon* for the lunar module.

NASA used the powerful Saturn V rocket (AS-510) to launch *Apollo 15* on July 26, 1976, from Complex 39-A at the Kennedy Space Center. The mission followed the profile of the previous successful lunar-landing missions—namely, insertion into Earth orbit, travel along a translunar trajectory, lunar orbit injection, and the use of the lunar module to reach and return from the Moon's surface. There was only one hardware anomaly encountered on the way to the Moon that had any mission-threatening significance. The astronauts discovered a short in the *Endeavor*'s service propulsion system, the rocket engine needed to achieve orbit around the Moon and then to leave lunar orbit and return safely to Earth. Working with NASA engineers and mission controllers at the Johnson Space Center, the *Apollo 15* astronauts developed contingency procedures for using this important rocket engine, and the lunar-landing mission proceeded.

The *Falcon* lunar module carrying Scott and Irwin landed at 26.1 degrees N latitude and 3.6 degrees E longitude in the Mare Imbrium (Sea of Rains) region of the Moon at the foot of the Apennine mountain range on July 30, 1971. The astronauts became the seventh and eighth persons to walk on the Moon. They spent a total of 66 hours and 55 minutes on the lunar surface, and their exploration activity consisted of three surface extravehicular activities totaling 18 hours and 35 minutes. During this time, they used the LRV to traverse a total of 17.3 miles (27.9 km), collected approximately 169 pounds (76.8 kg) of rock and soil samples, took an extensive number of photographs, set up the *Apollo 15* ALSEP, and performed other scientific experiments.

Scott and Irwin took their first Moonwalk on July 31, during which they unloaded, deployed, and drove the Apollo LRV for the first time. At the end of this traverse, they deployed the *Apollo 15* ALSEP. Their second EVA on the lunar surface occurred on August 1. They drove the LRV to a number of interesting sites and took a core sample from about 10 feet (3 m) below the surface. On August 2, they performed their third EVA, during which they drove the LRV to Hadley Rille and other craters in the area.

After they had completed their third and final EVA on the Moon's surface, Scott performed a televised physics demonstration, in which he imitated the famous cannonball drop experiment from the tower of Pisa—historically attributed to the great Italian scientist and astronomer Galileo Galilei (1564–1642). In the demonstration, astronaut Scott simultaneously released a feather and a hammer, which in the lunar vacuum then fell to the Moon's surface at the same rate. Scott's simple demonstration took place on the Moon a little more than 300 years after the death of Galileo and the birth of Sir Isaac Newton (1642–1727). It illustrated the universality of the physical laws that emerged from the great intellectual accomplishments of Galileo, Newton, and other scientists who looked up at the Moon and stars and wondered. One of most important contributions of Western civilization to the human race is the development of modern science and the emergence of the scientific method, which started in the 17th century. During this intellectually turbulent period, men of great genius and personal courage, like Galileo, developed those important physical laws and experimental techniques that helped human beings explain the operation and behavior of the physical universe and ultimately allowed human beings to walk on the Moon.

The use of the LRV by Scott and Irwin made the Apollo 15 mission the first of three great voyages of lunar surface exploration. On previous landing missions, such as Apollo 12 and Apollo 14, the astronauts, encumbered by the space suits, could not venture far from the lunar module. In contrast, on the last three lunar-landing missions (namely, Apollo 15, 16,

Apollo 15 astronaut James B. Irwin salutes the American flag while on the lunar surface in the Hadley Rille/Apennines region (July 31, 1971). The lunar module *Falcon* appears in the center (background) of this picture, and the Apollo lunar roving vehicle (first driven on the Moon by astronauts during this mission) appears on the right. *(NASA)*

and 17), the electric car allowed the space-suited explorers to travel miles from the lunar module.

Driving the LRV on all three of their EVAs, Scott and Irwin collected numerous rock samples from the vicinity of Hadley Rille and the Apennines mountain region of Mare Imbrium. The so-called space buggy, or Moon car, performed well, and the only restriction on its range was a safety requirement not to drive it more than about six miles (10 km) away from the lunar module on any given scientific traverse. NASA mission managers imposed this limit because they regarded that distance as the maximum EVA astronauts could travel by foot back to the lunar module should the electric car encounter a serious problem and break down.

After spending a total of almost 67 hours on the Moon's surface, Scott and Irwin fired the *Falcon*'s ascent stage engine on August 2 and returned to the orbiting CSM, which was being piloted by Worden. After docking with the *Endeavor,* the two astronauts transferred the lunar samples, some equipment, and themselves into the command module. Then all three astronauts prepared *Endeavor* for the long journey back to Earth. On August 3, they jettisoned the *Falcon*'s ascent stage, which then impacted on the Moon two hours later at 26.36 degrees N latitude, 0.25 degree E longitude (lunar coordinates). The *Falcon*'s ascent stage crashed into the Moon about 58 miles (93 km) west of the *Apollo 15* ALSEP site with an estimated impact velocity of 1.1 miles per second (1.7 km/s). The lower portion of the *Falcon* lunar module remains at the Mare Imbrium landing site along with the *Apollo 15*'s lunar roving vehicle.

After the *Apollo 15* CSM performed an orbit-shaping maneuver on August 4, the astronauts spring-launched a small scientific satellite from the scientific instrumentation module (SIM) at the rear of the service module. (A *spring-launch* is a mechanical technique in which the potential energy of a coiled spring is released and transforms into the kinetic energy of the object being ejected.) The tiny scientific subsatellite went into a 63.4-mile- (102-km-) by 87.8-mile- (141-km-) altitude orbit around the Moon. On the *Endeavor*'s next lunar orbit, the astronauts began the transearth injection maneuver by firing the CSM's main engine for 141 seconds. While the *Apollo 15* spacecraft was cruising back to Earth, Worden performed the first deep-space extravehicular activity when he exited the CM on August 5 and made three trips to the service module's SIM bay to retrieve film canisters and to check some equipment. His walk in deep space lasted a little over 38 minutes.

As the *Apollo 15* spacecraft approached Earth on August 7, the crew-carrying command module jettisoned the service module, and the astronauts prepared their spacecraft for reentry. During the last stages of atmospheric descent, one of the three main parachutes failed to open fully, so the command module experienced a terminal descent velocity of 21.8 miles per hour (35 km/h)—a velocity that was about 2.8 miles per hour (4.5 km/h) faster than planned. However, the astronauts survived their somewhat bumpy splashdown in the Pacific Ocean about 330 miles (531 km) north of Honolulu, Hawaii. The command module had landed in the water about six miles (10 km) away from the recovery ship USS *Okinawa*. During this mission, the *Apollo 15* astronauts accumulated a total of 295 hours, 11 minutes, and 53 seconds in space, including (for Scott and Irwin) about 67 hours on the Moon. The CSM (piloted by Worden) made a total of 74 revolutions of the Moon. The CM *Endeavor* is now on exhibit at the National Museum of the United States Air Force at Wright-Patterson Air Force Base, near Dayton, Ohio.

APOLLO 16

Apollo 16 was the fifth mission in which American astronauts walked on the Moon and the second of the Apollo J-series missions, involving extended exploration activities assisted by the lunar roving vehicle (LRV). The primary mission goals included exploring, surveying, and gathering sample materials in the Moon's Descartes highland region; placing and activating science experiments on the surface; and collecting of photographic images both on the surface and from orbit.

Astronaut John W. Young was the commander of the Apollo 16 mission; Thomas K. Mattingly II served as the command module (CM) pilot; and Charles M. Duke, Jr., was the lunar module (LM) pilot. Mattingly was the astronaut originally selected to serve as the CM pilot for the Apollo 13 mission but was removed from that troubled mission just days before launch because of his inadvertent exposure to German measles. The crew named their command and service module (CSM) spacecraft *Casper* and their LM *Orion*.

NASA launched the Apollo 16 mission using a Saturn V rocket (AS-511) on April 16, 1972. Like the four successful Apollo landing missions before them, the astronauts achieved orbit around Earth, traveled along a translunar trajectory, experienced insertion into lunar orbit, and then (for Young and Duke) rode the LM *Orion* to the Moon's surface.

On April 20 at 15:24 (universal time), Young and Duke entered the lunar module and separated from the CSM *Casper* piloted by Mattingly. But the two Moonwalkers had to delay their descent to the lunar surface for six hours because of a malfunction in the yaw gimbal servo loop on the CSM, which was causing oscillations in that spacecraft's service propulsion system (SPS). Ever mindful of the near disaster the service module had caused with the Apollo 13 mission, NASA managers wanted to resolve any lingering issues before two astronauts left for the Moon's surface and the CSM mother spacecraft developed a serious, mission-aborting problem with just one astronaut on board. The engineers at the Johnson Space Center studied the anomaly and determined that the problem would not seriously affect CSM steering. So, NASA mission control gave Young and Duke the go-ahead to descend to the surface in the *Orion* LM.

On April 21, 1972, the *Orion* lunar module landed in the Descartes highland region just north of crater Dolland at 9.0 degrees S latitude and 15.5 degrees E longitude. During their scientific expedition on the lunar surface, Young and Duke performed three Moonwalking extravehicular activities (EVAs), totaling 20 hours and 14 minutes outside the LM. Young became the ninth human being to walk on the Moon and Duke became the 10th. During the three lunar surface EVAs, the *Apollo 16* Moonwalkers covered 16.8 miles (27 km) and collected 208 pounds (94.7 kg) of lunar rocks and soil samples.

The first lunar surface EVA took place on April 21. Young and Duke set up the *Apollo 16* ALSEP, deployed the LRV, and explored some of the nearby craters. While positioning the ALSEP instruments, Young tripped on and broke the cable that ran from the heat flow experiment to the ALSEP's central station, rendering that particular scientific instrument inoperable. On April 22, the two astronauts performed their second lunar surface EVA. They explored a ridge and mountain slope. During the third and final lunar surface EVA, Young and Duke drove the LRV to the North Ray crater. The Moonwalking astronauts collected soil and rock samples as part of each of the three EVAs. After spending a total of 71 hours and two minutes on the Moon's surface, Young and Duke fired the LM's ascent engine on April 24, and the upper portion of *Orion* went into lunar orbit to rendezvous and dock with *Casper.*

After docking with the CSM (*Casper*), Young and Duke transferred the lunar rock samples and some equipment to the command module spacecraft from the LM and then rejoined Mattingly, who had traveled in the CSM around the Moon while they explored its surface. At 20:54 (universal time) on April 24, the astronauts jettisoned the ascent stage of *Orion.* The abandoned spacecraft was supposed to crash into the Moon near the *Apollo 16* landing site, but *Orion* suffered a failure in its attitude control system, which caused the spacecraft to start tumbling and remain in lunar orbit. NASA scientists estimated that the *Orion* ascent stage traveled around the Moon for about a year before decaying from orbit and crashing into the surface at some unknown site.

Because of earlier problems with the CSM's service propulsion system (namely, a misbehaving yaw gimbal servo loop), NASA flight controllers at Houston decided to shorten the Apollo 16 mission by one day. An orbital-shaping maneuver was canceled, and a small scientific subsatellite was spring-launched into a less desirable (but fuctional) elliptical orbit around the Moon. The tiny scientific satellite, mechanically ejected from the scientific instrument module (SIM) bay of the CSM, now traveled in a one-month lifetime, elliptical orbit around the Moon rather than in the planned one-year lifetime, circular orbit. After deploying the tiny satellite, the astronauts prepared *Casper* for the trip back to Earth.

At 2:15 (universal time) on April 25, the *Apollo 16* astronauts successfully fired *Casper*'s SPS rocket as part of the all-important transearth injection maneuver. Later that day, Mattingly performed a deep-space EVA while the CSM traveled through cislunar space on its return journey to Earth. During this 84-minute EVA, Mattingly retrieved camera film from the service module's SIM bay and also inspected instruments and equipment. As the *Casper* approached Earth on April 27, the command module automatically separated from the service module just prior to reentry. The *Apollo 16* command module successfully splashed down in the Pacific Ocean about 215 miles (346 km) southeast of Christmas Island and just three miles (5 km) away from the recovery ship USS *Ticonderoga.* During this scientific expedition to the Moon, the three astronauts had logged a total 265 hours and 51 minutes in space, including a little more than 71 hours on the Moon (Young and Duke). The Apollo 16 mission also accomplished 64 revolutions of the Moon (Mattingly). This was the first mission in which NASA used a television camera (remotely controlled from Earth) to record liftoff of the ascent stage of the lunar module from the Moon's surface. The *Apollo 16* command module *Casper* is now on display at the U.S. Space and Rocket Center in Huntsville, Alabama.

APOLLO 17

The Apollo 17 mission was the final in a series of three J-type missions and also the last human-landing mission on the Moon in the 20th century. The primary mission goals included investigating the lunar surface in the Taurus-Littrow region, placing and activating lunar surface experiments, collecting a wide variety of surface samples, and enhancing space technology capabilities for future human visits to the Moon. NASA mission planners had selected the Taurus-Littrow site for Apollo 17 because the highlands and valley area appeared to contain both older and younger rocks than had been returned to Earth by previous Apollo missions.

Astronaut Eugene A. Cernan was the commander of the mission, astronaut Ronald E. Evans served as the command module (CM) pilot, and astronaut/geologist Harrison H. "Jack" Schmitt was the lunar module (LM) pilot. The astronauts named their command and service module (CSM) *America* and their LM *Challenger*.

The journey to the Moon began with a spectacular early morning launch of a Saturn V rocket (AS-512) from Complex 39-A at the Kennedy Space Center on December 7, 1972. Some 30 minutes after midnight (local time), the ascending rocket lit up the night sky and was visible for hundreds of miles. Except for the unprecedented night launch with a human crew, the flight profile for this essentially flawless trip to the Moon was similar to the one used by the five previous successful lunar-landing missions.

On December 10, the CSM-LM configuration achieved orbit around the Moon. The next day (December 11), Cernan and Schmitt entered the *America* lunar module and prepared for their trip down to the surface. The *America* landed at 19:54 (universal time) on December 11 on the southeastern rim of Mare Serenitatis (the Sea of Serenity) in a valley at Taurus-Littrow, at 20.2 degrees N latitude and 30.8 degrees E longitude. Cernan became the 11th person to walk on the Moon and Schmitt the 12th and last under the Apollo Project. The two astronauts performed three extravehicular activities (EVAs) on the lunar surface, totaling 22 hours and four minutes. During their three EVAs, Cernan and Schmitt traversed a total of 18.6 miles (30 km) and collected 243 pounds (110.5 kg) of lunar rocks and soil. In the first surface EVA (on December 12), the astronauts deployed and drove the lunar roving vehicle (LRV). They also set up the *Apollo 17* ALSEP. As part of the second EVA (on December 13), Schmitt discovered orange soil—a surprising find that created a great deal of excitement in the scientific community. Their third and final EVA began late in the day (universal time) on December 13 and ended some eight hours later on December 14. During this EVA, the astronauts drove the LRV and collected many soil and rock samples. As Cernan ascended the *Challenger*'s ladder, he took the last step on the Moon in the 20th century. On December 14 at 22:54 (universal time), the ascent stage of the lunar module lifted off from the Moon's surface, marking the end to the first era of human visits. As previously mentioned, the ladder on the descent stage of *Challenger* bears a plaque inscribed with this message: "Here man completed his first exploration of the Moon, December 1972 A.D. May the spirit of peace in which we came be reflected in the lives of all mankind."

In the early morning of December 15, the LM *Challenger* docked with the CSM *America,* piloted by Evans. Cernan and Schmitt transferred the large collection of lunar rocks and soil into the command module and then assisted Evans in preparing the spacecraft for the voyage back to Earth. The crew jettisoned the ascent stage of *America,* which then crashed into the Moon about one hour later at 19.96 degrees N latitude and 30.50 degrees E longitude (lunar coordinates). The impact point was about 9.3 miles (15 km) from the *Apollo 17* landing site.

Translunar injection took place on December 16, after the *America* CSM had completed its 75th revolution of the Moon. Evans performed a deep-space EVA on December 17, while the *Apollo 17* spacecraft journeyed through cislunar space toward Earth. The command module separated from the service module at 18:56 (universal time) on December 19. Some 70 minutes later, Cernan, Evans, and Schmitt splashed down in the Pacific Ocean about 400 miles (644 km) southeast of the Samoan Islands and four miles (6.5 km) from the recovery ship USS *Ticonderoga.* During this mission, the crew logged a total of 301 hours and 52 minutes in space. The *Apollo 17* command module *America* is now on exhibit at the Johnson Space Center in Houston, Texas.

APOLLO 18, 19, AND 20—THE CANCELED MISSIONS

After six successful lunar-landing missions, the Apollo Project came to a relatively abrupt conclusion in late 1972. The planned Apollo 18, 19, and 20 missions had been canceled in 1970 because of severe budget limitations within NASA and the growing desire within the civilian space agency to put its dwindling amount of money into a new spaceflight vehicle concept, called the space shuttle. In the original program plan for the Apollo Project, NASA intended to perform three lunar-landing missions beyond Apollo 17. The original plan had recommended the following set of missions after *Apollo 12* (presented here in terms of mission type, mission number, and region of Moon, respectively): H-2 (*Apollo 13*) Fra Mauro; H-3 (*Apollo 14*) Littrow; H-4 (*Apollo 15*) Censorinus; J-1 (*Apollo 16*) Descartes; J-2 (*Apollo 17*) Marius Hills; J-3 (*Apollo 18*) Copernicus; J-4 (*Apollo 19*) Hadley; J-5 (*Apollo 20*) Tycho.

However, after the dramatic failure of the Apollo 13 mission, NASA officials rescheduled Apollo 14 to essentially replace Apollo 13 in exploring the Fra Mauro site. Apollo 13 was not the only "disaster" that befell the Apollo Project in 1970. Because of intense budget constraints and fading political interest in a continued exploration effort by human beings, NASA quietly canceled the Apollo 20 mission in January 1970. Then, in September 1970, further budget pressures caused NASA managers to cancel the original Apollo 15 and Apollo 19 missions. They then reshuffled the lunar exploration program and proceeded as follows: H-4 (*Apollo 14*) Fra Mauro; J-1 (*Apollo 15*) Hadley Rille-Apennine Mountain foothills; J-2 (*Apollo 16*) Descartes region; and J-3 (*Apollo 17*) Taurus-Littrow region.

NASA never officially assigned astronaut crews to the canceled missions, although the agency's crew assignment policy of moving the backup crew for a mission to become the prime flight crew three missions later has led to some interesting speculation about who might have been involved. One NASA report suggests the following selection of mission commanders: *Apollo 18* (Richard Gordon); *Apollo 19* (Fred Haise); and *Apollo 20* (Charles Conrad). But, almost four decades later, there is really no way of knowing what crew assignments might have actually been made.

As discussed in chapter 7, NASA used the remaining Apollo Project rocket and spacecraft hardware to support the Skylab Project and the Apollo-Soyuz Test Project (ASTP). Today many space technology advocates and visionaries view the end of the Apollo Project as simply the "end of the beginning" of human exploration of the Moon, Mars, and beyond.

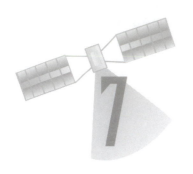

Outposts in the Sky: Early Space Stations

Aspace station is an orbiting space system that is designed to accommodate long-term human habitation in space. The concept of people living and working in artificial habitats in outer space appeared in 19th-century science-fiction literature in stories such as Edward Everett Hale's "Brick Moon" (1869) and Jules Verne's "Off on a Comet" (1878).

At the beginning of the 20th century, Konstantin Tsiolkovsky provided the technical underpinnings for this concept with his visionary writings about the use of orbiting stations as a springboard for exploring the cosmos. Tsiolkovsky, the father of Russian astronautics, provided a more technical introduction to the space station concept in his 1895 work *Dreams of Earth and Heaven, Nature and Man.* He greatly expanded on the idea of a space station in his 1903 work *The Rocket into Cosmic Space.* In this technical classic, Tsiolkovsky described all the essential ingredients needed for a crewed space station, including the use of solar energy, the use of rotation to provide artificial gravity, and the use of a closed ecological system complete with "space greenhouse."

Throughout the first half of the 20th century, the space station concept continued to evolve technically. For example, the German scientist Hermann Oberth described the potential applications of a space station in his 1923 classic treatise *The Rocket to Interplanetary Space* (German title, *Das Rakete zu den Planetenraumen*). The suggested applications included the use of a space station as an astronomical observatory, an Earth-monitoring facility, and a scientific research platform. In 1929, an Austrian named Herman Potocnik (pen name Hermann Noordung) introduced the concept of a rotating, wheel-shaped space station. Noordung called his design "Wohnrad" ("living wheel"). Another Austrian, Guido von Pirquet, wrote many technical papers on spaceflight, including the use of a space station as a refueling node for space tugs. In the late 1920s and early 1930s, von Pirquet also suggested the use of multiple space stations at different locations in cislunar space. After World War II, Wernher von Braun (with the help of space artist Chesley Bonestell) helped popularize the concept of a wheel-shaped space station in the United States.

✧ Defining an American Space Station

Created in 1958, NASA became the forum for the American space station debate. How long should such an orbiting facility last? What was its primary function? How many crewmembers should it accommodate? What orbital altitude and inclination should it be? Should it be built in space or on the ground and then deployed in space? In 1960, space station advocates from every part of the fledgling space industry gathered in Los Angeles for a symposium. They agreed that the space station was a logical goal but disagreed on what it was, where it should be located, or how it should be built.

Then in 1961 President John F. Kennedy decided that the Moon was a worthy target of the American spirit and heritage. A lunar-landing mission had a definite advantage over a space station: Everyone could grasp the concept of landing on the Moon, but few within the aerospace community could agree on a specific concept for the space station. However, this disagreement was actually beneficial. It forced space station designers and advocates to think about what they could do, the cost of design, and what was necessary to make the project a success. What were the true requirements for a space station? How could they best be met? The space station requirements review process started informally within NASA in 1963 and continued up to the end of the 20th century. For over four decades, NASA planners and officials have asked the scientific, engineering, and business communities over and over again: What would you want? What do you need? As answers flowed in, NASA developed a variety of space station concepts to help satisfy these projected requirements. The *International Space Station* (*ISS*) (discussed in chapter 9) is the latest manifestation of this evolving series of space station concepts and hardware approaches.

Even before the Apollo Project had landed men successfully on the Moon, NASA engineers and scientists were busy considering the next giant step in the American human spaceflight program. The leading scenario for that next step involved the simultaneous development of two complementary space technology capabilities. One capability involved a new transportation system that could provide routine access to space. The other capability involved a permanent orbital space station where human beings could live and work in space. The space station might also serve as a base camp from which other, more advanced space technology developments could be initiated, such as a human expedition to Mars or the support of a permanently crewed lunar outpost. As this candidate long-range strategy emerged, it set the stage for the two most significant American human spaceflight activities achieved in the 1970s and 1980s: *Skylab* and the space shuttle—or, as it was officially called, the U.S. Space Transportation System.

By 1968, one of NASA's leading post–Apollo Project candidate missions was an Earth-orbiting space station. Grand ideas for this orbital facility emerged in the brief period of technical and bureaucratic euphoria that accompanied the first human landings on the Moon. Unfortunately, these grand visions were quickly downsized by the reality of severe budget reductions. The end result became a rather hastily contrived project that was primarily forced to use hand-me-down equipment from the abruptly terminated Apollo Project.

In 1969, the year the Apollo 11 mission landed astronauts on the Moon for the first time, strategic planners at the civilian space agency suggested a 100-person permanent orbital facility, with assembly completion scheduled for 1975. Space agency advanced planners called this proposed facility the Space Base. It was to be a large, permanent, Earth-orbiting laboratory for scientific and industrial experiments. The Space Base was also envisioned as home port for a fleet of nuclear-powered space tugs that would carry people and supplies to an outpost on the Moon. Another fleet of nuclear-powered, human-crewed space vehicles might even be outfitted at the Space Base and then sent off on an expedition to explore Mars. Such were the grand space visions in the summer of 1969 as Armstrong and Aldrin left their bootprints on the Moon.

Unfortunately, the most far-reaching components of NASA's post–Apollo mission/project scenario quickly unraveled when the space agency's budget experienced severe restrictions and opposition. First, the last three lunar-landing missions (originally called Apollo 18, 19, and 20) were canceled. This action squashed any further discussion about a permanent lunar outpost. Then, in January 1973, the U.S. government abruptly canceled the joint NASA–Atomic Energy Commission (now Department of Energy) nuclear rocket program—the prime civilian mission rationale for which was the human exploration of Mars. Finally, NASA was given permission by President Richard M. Nixon to pursue the development (beginning in 1972) of a space shuttle. But, there was a catch. In pursuing its new space transportation program, the agency had to agree to defer any plans for a large, permanent space station until after the space shuttle was operational. From one perspective, this approach made little technical sense, because in all the previous concept studies the space station was regarded as the logical orbital destination for any proposed space shuttle–type vehicle intended to ferry people and cargo from Earth's surface into low Earth orbit. Unfortunately, the practice of logical decision making sometimes eludes the bureaucrats in federal agencies, who are forced into making unrealistic fiscal agreements.

NASA's visionary space station advocates were forced to rummage through the leftover hardware of the Apollo Project. Holding firm in their belief that a space station was the next appropriate human spaceflight project in the post-Apollo era, they hastily got together plans for an interim space outpost—the orbital workshop named *Skylab*. (As an historic note, the NASA Project Designation Committee officially approved the name *Skylab* on February 17, 1970.)

There were two competing concepts on how best to use the Apollo Project–era rocket vehicle and spacecraft hardware to construct a human outpost in space. The first concept for a demonstration, manned orbital workshop was called the "wet" workshop configuration. In this concept, a Saturn IB rocket would be launched, and its S-IVB upper stage would then be purged and vented of unused propellants, after which astronauts would refurbish the spent upper stage on orbit and make it fit for human occupancy. The second orbital workshop concept was called the "dry" workshop configuration. In this concept, an empty S-IVB upper stage would be modified on the ground and completely outfitted for human occupancy, prior to launch. Then, a powerful Saturn V rocket would be used to place the massive workshop into orbit around Earth. In the late 1960s, NASA managers selected the "dry" workshop configuration for the *Skylab* space station. They also decided to use three smaller, less powerful Saturn IB boosters to launch surplus Apollo spacecraft to the orbiting workshop. Each spacecraft was a modified Apollo Project command and service module configuration that could carry a crew of three astronauts. As NASA's compromise space station scenario unfolded, three separate teams of *Skylab* astronauts would conduct successive, relatively long-term (28-, 59-, and 84-day) missions in orbit around Earth.

Unfortunately, because of schedule pressures and budget limitations, NASA engineers could not design *Skylab* for a permanent presence in space. For example, they did not design the large, rather comfortable facility to be routinely serviced on orbit—although the *Skylab* crews were able to perform certain repair functions. The first American space station was not designed for evolutionary growth and therefore was subject to rapid technological obsolescence. Finally, *Skylab* was not equipped to maintain its own orbit—a design deficiency that eventually caused its fiery demise on July 11, 1979, over the Indian Ocean and portions of western Australia.

Yet, despite such technical shortcomings, *Skylab* represented an important milestone in the conquest of space and an important chapter in the American program of human spaceflight. *Skylab* demonstrated that people could function in space for periods up to 12 weeks and, with proper exercise, could return to Earth with no ill effects. Specifically, the flight of *Skylab* proved that human beings could operate effectively in a prolonged microgravity environment and that it was not essential to provide artificial gravity for people to live and work in space—at least for periods up to about six months. As discussed later in this chapter and in chapter 9, long-duration flights by Russian cosmonauts and American astronauts on the *Mir* and the *International Space Station* have extended and reinforced these findings, up to a point. However, chronic exposure to microgravity, for periods of a year or longer, appears to cause astronauts and cosmonauts certain undesirable physiological changes (such as bone loss and muscle atrophy)—the consequences of long-duration spaceflight that must be more effectively countered.

The *Skylab* astronauts accomplished a wide range of emergency repairs on station equipment, including freeing a stuck solar panel array (a task that saved the entire mission), replacing rate gyros, and repairing a malfunctioning antenna. On two separate occasions, the crew installed portable Sun shields to replace the original protective equipment that was lost when *Skylab* was launched. These on-orbit activities clearly demonstrated the unique and valuable role people have in space.

✧ *Skylab*—The First American Space Station

On May 14, 1973, the United States launched its first space station, *Skylab*. NASA placed this massive 199,335-pound (90,607-kg) space station into orbit from Complex 39 at the Kennedy Space Center, using the last remaining Saturn V booster (AS-513) from the Apollo Project.

Skylab was composed of five major parts: the Apollo telescope mount (ATM); the multiple docking adapter (MDA); the airlock module; the instrument unit; and the orbital workshop, which included the living and working quarters. The ATM was a solar observatory, and it provided attitude control and experiment pointing for the rest of the cluster. The retrieval and installation of film used in the ATM was accomplished by the astronauts during extravehicular activity. The MDA served as a dock for the modified Apollo spacecraft that taxied the crews to and from the space station. The airlock module was located between the docking port (MDA) and the living and working quarters and contained controls and instrumentation. The instrument unit, which was used only during launch and the initial phases of operation, provided guidance and sequencing functions for the initial deployment of the ATM, its solar arrays, and the like. The orbital workshop was a modified S-IVB stage that had been converted into a two-story space laboratory with living quarters for a crew of three. Although this orbital laboratory was capable of unmanned, in-orbit

A close-up view of NASA's *Skylab* space station, as photographed against an Earth background (Amazon River valley) from the *Skylab 3* command and service module spacecraft during stationkeeping maneuvers prior to docking on July 28, 1973. Due to an accident on launch ascent, there was only one large solar array wing attached to the station's orbital workshop. During *Skylab 2* (the first manned mission to the station), the astronaut crew successfully deployed this solar array wing, which was stuck in its stowed position. The solar shield, which was also deployed by the *Skylab 2* crew, can be seen through the support struts of the Apollo telescope mount. *(NASA)*

storage, reactivation, and reuse, NASA engineers did not design *Skylab* as a permanent orbiting facility.

There were four launches in the *Skylab* project, all taking place from Complex 39 at the Kennedy Space Center. The first launch (called the SL–1 mission) occurred on May 14, 1973. A two-stage Saturn V vehicle (AS-513) placed the unmanned, 90-ton *Skylab* space station into an initial, near-circular 270-mile- (435-km-) altitude orbit around Earth with an inclination of 50 degrees and a period of 93.4 minutes. But this last Saturn V launch did not take place without problems. About 63 seconds after liftoff—as the giant rocket accelerated past 25,000 feet (7,620 m) altitude—atmospheric drag began clawing at *Skylab*'s meteoroid/Sun shield, which had inadvertently deployed.

This cylindrical metal shield was designed to protect the orbital workshop from tiny particles and the Sun's scorching heat. When the shield prematurely deployed as the Saturn V ascended, the atmosphere rushing past ripped the important protective device away from the space station, causing it to trail an aluminum strap that caught on one of the unopened solar wings. As a result, the shield became tethered to the laboratory while at the same time prying the opposite solar wing partly open. Minutes later, as the booster rocket staged, the partially deployed solar wing and meteoroid/Sun shield were flung into space. With the loss of this shield, temperatures inside *Skylab* soared, rendering the space station uninhabitable and threatening the food, medicine, and film stored on board. Despite this malfunction, the Apollo telescope mount—*Skylab*'s major piece of scientific equipment—did deploy properly.

The countdown for the launch of the first *Skylab* crew (called the SL-2 mission) was halted. NASA engineers worked quickly to devise a solar parasol to cover the workshop and to find a way to free the stuck solar wing. On May 25, 1973, astronauts Charles "Pete" Conrad, Jr. (commander); Paul J. Weitz (pilot); and Joseph P. Kerwin, M.D. (scientist pilot); were launched from the Kennedy Space Center by a Saturn IB rocket (AS-206) toward *Skylab*. Their spacecraft, called the *Skylab CSM 1*, was almost identical to the command and service module used in the Apollo Project. However, NASA engineers made some modifications to the spacecraft to allow it to remain semi-dormant while docked to *Skylab*'s multiple docking adapter.

After repairing *Skylab*'s broken docking mechanism, the astronauts entered the overheated space station and erected a sunshade through a space access hatch. The improvised device shaded part of the area where the protective meteoroid/Sun shield had been ripped away. Temperatures within the spacecraft immediately began dropping, and *Skylab* soon became habitable without space suits. But the many experiments stored on board demanded far more electric power than the four ATM solar arrays could generate. *Skylab* could fulfill its scientific mission only if the first crew was able to free the crippled solar wing. Using equipment that resembled long-handled pruning shears and a crowbar, the astronauts, during one of several station-saving extravehicular activities, pulled the stuck solar wing free. The space station was now ready to meet its scientific mission objectives. Before returning to Earth, the first crew conducted Earth resource observation experiments, solar astronomy, medical experiments, and five student-proposed experiments.

On June 22, 1973, the first crew departed from the station and returned to Earth in the command module of the *Skylab CSM 1* spacecraft. The recovery procedure used for the returning *Skylab* crewmen was altered from the ocean recovery procedures used during the Apollo Project. In *Skylab*, after splashdown, the command module and its three astronaut occupants were retrieved simultaneously and lifted directly on board the recovery aircraft carrier. The astronauts then exited from the spacecraft onto a special platform on the ship's hangar deck. The overall process of spacecraft and crew retrieval typically took less than one hour. The SL-2 mission astronauts logged 28 days and 49 minutes in space and performed three EVAs (totaling six hours and 20 minutes). While the human crew was present in the SL-2 mission, *Skylab* made 404 orbits of Earth.

NASA launched the second crewed Skylab mission (designated SL-3) from the

Astronaut Gerald P. Carr, commander for NASA's Skylab 4 mission, jokingly demonstrates weight training in microgravity as he balances astronaut William R. Pogue (*Skylab 4* pilot) upside down on his finger. This clever picture was taken on February 1, 1974, inside *Skylab*'s orbital workshop by the third *Skylab 4* crewmember, astronaut Edward G. Gibson. *(NASA)*

Kennedy Space Center using another Saturn IB rocket (AS-207) on July 28, 1973. This crew consisted of astronauts Alan Bean (commander), Jack Lousma (pilot), and Owen Garriott (scientist pilot). The SL-3 astronauts continued to perform maintenance on the space station and devoted a great deal of their time (about 1,081 hours) to performing a number of scientific and medical experiments. Three EVAs were conducted by the crewmembers, totaling 13 hours and 43 minutes. During one of these spacewalks, the astronauts erected a second Sun shield, a twin-pole device. After spending 59 days and 11 hours in orbit, Bean, Lousma, and Garriott used their Apollo-era spacecraft, called *Skylab CSM 2,* to return to Earth on September 25, 1973. During the SL-3 mission, the space station completed 858 Earth orbits.

The third crewed Skylab mission was called SL-4. NASA used a Saturn IB rocket (AS-208) to send the *Skylab CSM 3* spacecraft and its crew to the orbital workshop from the Kennedy Space Center on November 16, 1973. Astronaut Gerald P. Carr served as the mission commander, William R. Pogue as the pilot, and Edward G. Gibson as the scientist pilot.

Encouraged by the accomplishments of the first two crewed missions to *Skylab*—especially how well the second crew (consisting of astronauts Bean, Lousma, and Garriott) adapted to 59 days of exposure to microgravity after some initial susceptibility to space adaptation syndrome—NASA officials decided to extend the third crewed mission to 84 days and to add more tasks for the crew to perform. To help keep the crew in shape during their prolonged exposure to microgravity, NASA aerospace medicine experts added a treadmill to accompany the onboard ergometer (a bicycle-like device for in-place exercise).

Unfortunately, the excessive workload piled upon this crew caused some psychological problems and tensions during the flight. The primary catalyst was a growing disagreement about work schedules between the crew in orbit and the ground support crew, who kept dictating lengthy work periods. The astronauts felt overwhelmed by these demands, and the ground crew felt that the astronauts were not working hard enough or long enough. Faced with the problem of mistakes being made by the astronauts during rushed experiments and lagging schedules, the ground crew continued to make new demands. Resolution finally occurred after an on-orbit "holiday" followed by an adjustment of the remaining work schedule to give the astronauts more control of their time.

The end result was a marked increase in crew performance and more cooperative communications between the crew and ground control. As an historic note, similar human-factor problems occurred in the Russian space program.

Once the tight work schedules were relaxed, the SL-4 crew actually completed more work than planned. Excellent solar astronomy observations were performed by the astronauts, especially Gibson—a highly trained solar physicist. The SL-4 mission resulted in about 75,000 new telescopic images of the Sun, in the X-ray, ultraviolet, and visible portions of the spectrum. Toward the end of the SL-4 mission, Gibson's patient daily monitoring of the Sun paid off. On January 21, 1974, he filmed the birth of a solar flare—the first unobstructed, space-based, observation of this important phenomenon ever recorded in the history of astronomy.

In mid- to late December 1973, the SL-4 crew was also able to observe and photograph the comet Kohoutek from their unique vantage point above Earth's atmosphere. Like the two crews before them, the third *Skylab* crew spent many hours looking at Earth through the orbital workshop's window and photographing selected surface features. Over the 12-week period, Pogue, Carr, and Gibson were able to watch vegetation change colors and even observe the subtle signatures of ocean currents—such as the warm Gulf Stream flowing northward from the Caribbean Sea up the east coast of the United States and then eastward across the Atlantic Ocean over to Europe. Guided by their own judgment in where to point the handheld cameras, they gathered about 20,000 annotated images from orbit of interesting features on Earth's surface. Their efforts complemented the remote-sensing experiments conducted as part of the SL-4 mission Earth resources experiment.

By the time the crew of the final manned mission splashed down in the Pacific Ocean on February 8, 1974, they had been in space for 84 days and one hour. As part of their long-duration mission, Pogue, Carr, and Gibson had also performed four EVAs (totaling 22 hours and 13 minutes) and orbited Earth 1,214 times.

All three crews demonstrated the technical skills needed for repair and maintenance functions, for performing medical research (related to extended spaceflight), and for conducting scientific experiments (such as the behavior of materials in microgravity). Of perhaps greater importance was the fact that the Skylab missions clearly demonstrated the capability of human beings to perform longer-duration missions in space. Each crew returned in relatively good health and physical condition. NASA also used *Skylab* to prove that space vehicles could be used to rotate crews and resupply an orbiting space station.

After the last astronaut crew departed the space station in February 1974, NASA ground controllers performed some engineering tests on the now vacant *Skylab*. These tests were inappropriate to perform (because of potential risk) when human beings occupied the facility. The results of the tests helped aerospace engineers determine the causes of failures during the overall mission and also to obtain useful data related to the long-term degradation of space systems in orbit. Upon completion of these engineering tests, ground controllers positioned *Skylab* into a stable attitude and then sent commands to the space station, shutting down all remaining active systems. From that point on, the large facility essentially orbited Earth as an abandoned derelict. At the time (circa early 1974), NASA planners thought *Skylab* would remain in a stable orbit long enough, to be re-boosted to a higher altitude by an early flight of the space shuttle (then just entering engineering development). But two factors eventually ruled out the possibility of such a *Skylab* rescue mission. First, the shuttle program experienced a large number of delays and did not fly its first orbital mission until 1981. Second, a greater than predicted amount of solar activity hastened *Skylab*'s decay from orbit.

Unable to maintain its original altitude, the station gradually spiraled toward Earth and finally reentered the atmosphere on July 11, 1979, during orbit 34,981. While most of the large station burned up during reentry, some pieces survived and impacted harmlessly in remote areas of the Indian Ocean and sparsely inhabited portions of Australia.

✧ Apollo–Soyuz Test Project

The Apollo-Soyuz Test Project (ASTP) was a joint United States–Soviet Union space mission that took place in July 1975. The mission involved the central goal of the rendezvous and docking of the *Apollo 18* spacecraft (astronaut crew: Thomas P. Stafford, Vance Brand, and Deke Slayton) and the *Soyuz 19* spacecraft (cosmonaut crew: Alexei Leonov and Valeriy Kubasov). The *Apollo 18* spacecraft was hardware remaining from the Apollo

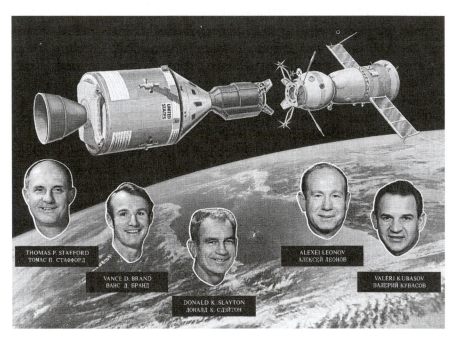

The three American astronauts (left) and two Russian cosmonauts (right) who participated in the first international manned spaceflight, called the Apollo–Soyuz Test Project in July 1975. At the top is an artist's rendering of the American *Apollo 18* spacecraft with its cylindrical docking module about to link up with the Soviet *Soyuz 19* spacecraft. *(NASA)*

Project. For this reason, the American portion of the Apollo-Soyuz Test Project is sometimes referred to as the final mission in the Apollo Project.

Both the *Soyuz 19* and *Apollo 18* spacecraft were launched on July 15, 1975; the American *Apollo 18* spacecraft lifted off from Complex 39 at the Kennedy Space Center approximately seven and one-half hours after the Russians launched the *Soyuz 19* spacecraft from the Baikonur Cosmodrome in Kazakhstan. The Russian cosmonauts maneuvered their *Soyuz 19* spacecraft to the planned orbit for docking at an altitude of 138 miles (222 km) over Europe. The *Apollo 18* astronauts then completed the rendezvous sequence, eventually docking with the Soyuz spacecraft on July 17, 1975, at 2:17 P.M. (U.S. central daylight time). Once the spacecraft were docked, astronaut Thomas Stafford shook hands with Alexei Leonov (b. 1934) in the docking ring of the joined *Apollo 18* and *Soyuz 19* spacecraft.

Stafford, a veteran astronaut, commanded the *Apollo 18* spacecraft. At the time of the mission, he was a major general on loan to NASA from the U.S. Air Force. Stafford had most recently traveled in space in 1969 as commander of the Apollo 10 mission. Leonov, the commander of the *Soyuz 19* spacecraft, was also a veteran cosmonaut. On March 18, 1965, he performed the first tethered extravehicular activity (EVA), when he wore a bulky space suit and exited the orbiting *Voshkod 2* for a 10-minute walk in space. At the time of the ASTP mission, Leonov was major general in the Soviet Air Force. Following this historic international flight, Leonov served as commander of the cosmonaut team (from March 1976 to January 1982) and then as deputy director of the Gagarin Cosmonaut Training Center until his retirement in October 1991.

Their symbolic handshake on orbit ended five years of work and planning for the first international space mission between the United States and the Soviet Union. Up until 1969, the United States and the Soviet Union had been engaged in an unofficial, yet hotly contested space race. The Soviet Union captured an early lead by putting the first man in space and having the first spacewalker (cosmonaut Alexei Leonov), but the United States captured the ultimate prize: the first manned lunar landing. So the ASTP represented a real step forward in the relaxation of cold-war tensions.

Before the two spacecraft could dock, a number of technical problems had to be solved, including the design of a docking module (carried on the front of the *Apollo 18* spacecraft) that allowed crew transfer between two spacecraft that had crew cabins containing different atmospheres. Russian engineers had designed the Soyuz spacecraft with a nitrogen/oxygen-mixed atmosphere that was pressurized at 14 pounds-force per square inch (psi) (96 kilopascals). In contrast, following the tragic *Apollo 1* cabin fire, American aerospace engineers had redesigned the Apollo spacecraft with a pure oxygen atmosphere at 5 psi (34.5 kilopascals). Such technical problems were resolved, as were a number of other issues, including language barriers, different approaches to space crew training, and differences in command and control communications networks. Long before the flight, the astronauts and cosmonauts underwent language training and also trained for the docking mission at space centers in both the United States and the former Soviet Union. ASTP-related interactions broke technical, political, and social barriers and led to the world's first successful international docking mission.

During the next two days, the crews accomplished four transfer operations between the two spacecraft and completed five scheduled experiments. Following the first undocking, a joint solar eclipse experiment was performed. Then, the *Apollo 18* spacecraft accomplished a second successful docking, this time with the *Soyuz 19* apparatus locking the two spacecraft together. The final undocking occurred on July 19. The two spacecraft moved to a stationkeeping distance, and a joint ultraviolet absorption experiment involving a complicated series of orbital maneuvers was performed.

Afterward, the *Apollo 18* spacecraft entered a separate orbit, and both the *Soyuz 19* and *Apollo 18* crews conducted unilateral activities. The *Soyuz 19* landed safely on July 21, after six mission days, and the *Apollo 18* flight successfully concluded on July 24, 1975—nine days after launch. The primary objectives of this first international human-crewed mission were met, including rendezvous, docking, crew transfer, and control center–crew interaction.

✧ Early Russian Space Stations

While the United States was concentrating on the Apollo Project lunar-landing effort, the Soviet Union began embarking on an ambitious space station program. As early as 1962, Russian engineers described a space station composed of modules launched separately and brought together in orbit. Using a Proton booster, the Russians launched the world's first space station, *Salyut 1,* on April 19, 1971. (The Russian word *salyut* means "salute.") The first generation of Russian space stations had one docking port and could not be resupplied or refueled. The stations were launched uncrewed and later occupied by cosmonaut crews. Two types of early Russian space stations existed: Almaz military stations and Salyut civilian stations. In order to confuse Western observers during the cold war, the Soviet officials would refer to both kinds of station as Salyut.

The Almaz military station program was the first approved. When proposed in 1964, it had three parts: the Almaz military surveillance space station, transport logistics spacecraft for delivering military cosmonauts and cargo, and Proton rockets for launching both. All of these spacecraft were built, but none was actually used as originally planned.

Russian engineers completed several Almaz space station hulls by 1970. The Soviet leaders then ordered that the hulls be transferred to a crash program to launch a civilian space station. Work on the transport logistics spacecraft was deferred, and the Soyuz spacecraft originally built for the never-completed Russian manned Moon program was reapplied to ferry crews to the space stations.

Unfortunately, the early first-generation Russian space stations were plagued by failures. For example, the crew of *Soyuz 10,* the first spacecraft sent to *Salyut 1,* was unable to enter the station because of a docking mechanism problem. The *Soyuz 11* crew lived aboard *Salyut 1* for three weeks, but then died during the return to Earth because the air escaped from their spacecraft. Several of the first-generation stations failed to reach orbit or broke up in orbit before the cosmonaut crews could reach them.

Salyut 1 was launched from the Baikonur Cosmodrome into Earth orbit on April 19, 1971. The 40,535-pound (18,425-kg) space station functioned in a civilian research capacity as an observation platform, gathering data in the fields of astronomy, Earth resources monitoring, and meteorology. Soviet sources report the station as being 65.6 feet (20 m) in length and 13.1 feet (4 m) in diameter. The station had four major compartments, three of which were pressurized. The first pressurized compartment served as a transfer compartment and contained a docking cone that allowed Soyuz spacecraft to connect to the station. The second pressurized compartment served as the main habitable living and work volume. The third pressurized compartment on *Salyut 1* housed the station's communications system, life support system, power supply, and other auxiliary. The fourth compartment, which was unpressurized, contained rocket engines and associated control equipment. The *Salyut 1* had two double sets of externally mounted solar cell panels for electric power, which extended like wings from the smaller compartments at each end of the station. There were also chemical batteries, reserve supplies of oxygen and water, and regeneration systems.

The *Soyuz 10* spacecraft, carrying three cosmonauts, took 24 hours to rendezvous with and approach *Salyut 1*. On April 23, 1971, the *Soyuz 10* docked with the station and remained docked for about five and one-half hours, but for unexplained reasons, the cosmonauts did not actually enter the station. The suspected reason appears to have been a problem with the docking mechanism. On June 7, the *Soyuz 11* spacecraft took a little over four hours to dock with *Salyut 1*. This time the crew was able to transfer into the station, where they lived and worked, during 362 "docked" orbits around Earth at an altitude of approximately 131 miles (210 km) and an inclination of 51.6 degrees. But the world's first successful crewed space station mission soon came to a tragic ending.

On June 29, the three cosmonauts, who had just spent 24 days on board the space station, died during reentry operations. Cosmonauts Georgi Dobrovolsky, Vladislav Volkov, and Victor Patseyev suffocated when a vent valve on their *Soyuz 11* spacecraft inadvertently opened and the air rushed out of their crew cabin just as they separated from the space station. After an automatic reentry procedure, the capsule touched down, and startled recovery crews found all three cosmonauts dead in their seats. In July and again in August 1971, Russian ground control personnel fired the station's rocket engines to ensure that *Salyut 1* would not immediately decay out of orbit. Then, after 175 days in orbit, the world's first space station met its demise. Russian ground controllers fired the station's rocket engines for the last time on October 11, 1971, to hasten the orbital decay process, and *Salyut 1* burned up in Earth's atmosphere later that day. Soviet news sources reported several weeks later that the *Soyuz 11* cosmonauts had performed measurements of geological and geophysical objects on Earth's surface in both the visible and infrared portions of the electromagnetic spectrum while working on board *Salyut 1*. The idea of a crewed station in orbit around Earth, first envisioned by Tsiolkovsky some 70 years earlier, had become a reality—although at a tragic cost.

Despite its officially announced civilian station name, Western aerospace analysts regarded *Salyut 2* as the first Almaz military space station. It represented the initial attempt by the Soviet government to operate a military space station—possibly similar in mission to the canceled American *Manned Orbiting Laboratory* (*MOL*). The 40,700-pound (18,500-kg) *Salyut 2* station was launched on April 3, 1973, by a Proton rocket and upper stage vehicle from the Baikonur Cosmodrome. However, before a cosmonaut crew could be sent to the station, it suffered a catastrophic explosion on April 14. The explosion tore away the space station's solar panels, telecommunications equipment, and docking apparatus. The derelict space station then tumbled out of orbit and on about May 28 burned up in Earth's atmosphere.

However, the Russians recovered rapidly from these failures. *Salyut 3, Salyut 4,* and *Salyut 5* supported a total of five crews. In addition to military surveillance and scientific and industrial experiments, the cosmonauts performed engineering tests to help develop the second-generation stations.

Salyut 3, the first operational military Salyut, was launched into a low (136-mile [219-km] by 168-mile [270-km]) orbit on June 25, 1974. Apparently the cosmonauts (Yuri Artukhin and Pavel Popovich) who were launched in the *Soyuz 14* spacecraft to the station on July 3 were able to dock with *Salyut 3* some 32 hours later and then performed military photoreconnaissance operations, although little official information has been released about this spacecraft. The 40,700-pound station had two solar panels that engineers mounted laterally on the facility. *Salyut 3* also had a detachable recovery module designed to return research data, materials, and most likely film capsules back to Earth. The station was only occupied and operated by the cosmonaut crew of *Soyuz 14. Salyut 3*'s second team of cosmonauts (Gennady Sarafanov and Lev Demin), traveling aboard the *Soyuz 15* spacecraft, rendezvoused but was unable to dock successfully with the then vacant station and returned to Earth after the aborted attempt. On September 23, 1974, Russian ground control personnel sent commands to release the unoccupied station's detachable recovery module. The module soon reentered Earth's atmosphere and was recovered successfully by support personnel after landing within the former Soviet Union. In January 1975, the *Salyut 3* station reentered Earth's atmosphere and burned up over the Pacific Ocean.

The Russians launched the 40,700-pound *Salyut 4* on December 26, 1974, from the Baikonur Cosmodrome using a Proton booster with an upper stage rocket. This space station was a civilian station quite similar to *Salyut 1,* except for the fact Russian engineers gave the new spacecraft three large solar panels, which they mounted on the smaller end of the main pressurized compartment for living and working. *Salyut 4* contained scientific instruments, including a solar telescope. The station operated in an approximately 221-mile- (350-km-) by 213-mile- (343-km-) altitude orbit at an inclination of 51.6 degrees. Its cosmonaut crews (from the Soyuz 17 and then Soyuz 18 missions, respectively) performed civilian missions, including astronomy, Earth resources and biomedical observations, and materials processing experiments.

The *Soyuz 17* crew consisted of cosmonauts Alexi Gubarev and Georgi Grechko. After launch from the Baikonur Cosmodrome on January 11, 1975, these cosmonauts docked with and boarded the *Salyut 4* station. Gubarev and Grechko stayed on the station for almost 30 days and then used the docked *Soyuz 17* to return safely to Earth on February 9, 1975. The *Soyuz 18* mission lifted off on May 24, 1975, from the Baikonur Cosmodrome, carrying two cosmonauts (Pytor Klimuk and Vitali Sevastyanov) to the *Salyut 4.* After spending approximately 63 days on the station, the cosmonauts transferred to the docked *Soyuz 18* spacecraft and returned safely to Earth on July 26, 1975. Because the highly publicized Apollo-Soyuz Test Project was taking place during their stay on *Salyut 4,* cosmonauts Klimuk and Sevastyanov became essentially "forgotten" space travelers, even by the government-controlled Soviet press.

There was also an interesting unmanned final mission to *Salyut 4.* An unmanned *Soyuz 20* spacecraft was launched on November 17, 1975, and then docked automatically with the uninhabited space station. The *Soyuz 20* stayed docked with *Salyut 4* until February 16, 1976, when it returned to Earth by remote control under signals sent from ground personnel. It appears that the unmanned Soyuz 20 mission to the space station allowed Russian aerospace engineers to demonstrate the feasibility of automated docking and resupply—a technique now used with the *International Space Station* (see chapter 9). *Salyut 4* reentered Earth's atmosphere and burned up on February 2, 1977.

Salyut 5 was the second successful launch of an Almaz military space station. A Proton rocket placed the 41,800-pound station into a 139-mile- (223-km-) by 162-mile- (260-km-) altitude orbit with an inclination of 81.4 degrees from the Baikonur Cosmodrome on June 22, 1976. The military space station had two solar panels mounted laterally on the center of the spacecraft but otherwise was similar in structure to the *Salyut 3*. Two different teams of cosmonauts inhabited the station. The first cosmonaut team (Boris Volynov and Vitally Zholobov) lifted off from the Baikonur Cosmodrome aboard the *Soyuz 21* spacecraft on July 6, 1976, and docked with *Salyut 5* the next day. Their mission was to last two months on the station, but it was rapidly terminated after just 48 days because of Zholobov's worsening illness. On October 14, 1976, the Soyuz 23 mission from Baikonur carried two cosmonauts (Valery I. Rozhedstvensky and Vyacheslav D. Zudov) to the *Salyut 5* station. However, their attempt to dock with the station failed. Circumstances forced the cosmonauts to make an emergency reentry, and they splashed down in Lake Tengiz at night during a blinding blizzard. Rescue crews could not reach the *Soyuz 23* landing capsule until the next morning. Anticipating the worst, the recovery team was amazed to find the two cosmonauts alive and awaiting rescue.

The final cosmonaut team (Victor Gorbatko and Yuri Glazhkov) to visit *Salyut 5* was launched from the Baikonur Cosmodrome on February 7, 1977, aboard the *Soyuz 24* spacecraft. After docking with the space station, they remained inside the *Soyuz 24* for 24 hours before transferring to the *Salyut 5*. Western observers speculate that the cosmonauts were venting contaminants from the space station's atmosphere and replacing it with a fresh supply of clean air. After entering *Salyut 5*, Gorbatko and Glazhkov participated in a busy but successful mission and returned to Earth in the *Soyuz 24* capsule on February 25. The next day, ground controllers sent signals to the now uninhabited *Salyut 5*. The radio commands separated the station's detachable recovery module and allowed it to be recovered quickly. Although the Soviet government publicly released few details about the mission of *Salyut 5,* Western analysts speculated that, because the detachable module was so rapidly recovered, it probably contained important data and materials (such as reconnaissance imagery) from both the hastily ended *Soyuz 24* crew visit as well as the just concluded *Soyuz 25* crew visit. Whatever the actual case, *Salyut 5* soon depleted its onboard propellant supply for its attitude control rockets, experienced rapid orbital decay, and burned up in Earth's upper atmosphere upon reentry on August

28, 1977. *Salyut 5*'s fiery plunge back to Earth marked the end of the Soviet effort to use space stations for purely military missions. Any future military activities performed by cosmonauts on future space stations would be blended in with so-called civilian research projects.

The second-generation Russian space station was introduced with the launch (on September 29, 1977) and successful operation of the 41,580-pound (18,900-kg) *Salyut 6* station. Several important design improvements appeared on this station, including the addition of a second docking port and the use of an automated *Progress* resupply spacecraft—a space "freighter" derived from the Soyuz spacecraft and demonstrated with the Soyuz 20 mission.

With the second-generation stations, the Russian space station program evolved from short-duration to long-duration stays. Like the first-generation stations, they were launched uncrewed, and their crews arrived later in a Soyuz spacecraft. Second-generation Russian stations had two docking ports. This permitted refueling and resupply by the *Progress* spacecraft, which docked automatically at the aft port. After docking, cosmonauts on the station opened the aft port and unloaded the space freighter. Transfer of fuel to the station was accomplished automatically under supervision from ground controllers.

The availability of a second docking port also meant long-duration resident crews could receive visitors. Visiting crews often included cosmonaut-researchers from the former Soviet bloc countries or countries that were politically sympathetic to the former Soviet Union. For example, the Czech cosmonaut Vladimir Remek visited the *Salyut 6* station in 1978 and became the first space traveler from outside the United States or Russia.

These visiting crews helped relieve the monotony that can accompany a long stay in space. They often traded their Soyuz spacecraft for the one already docked at the station, because the Soyuz spacecraft had only a limited lifetime in orbit. The spacecraft's lifetime was gradually extended from 60 to 90 days for the early *Soyuz Ferry* to more than 180 days for the *Soyuz-TM*. By way of comparison, the *Soyuz TMA* crew transfer (and escape) vehicle used with the *International Space Station* has a lifetime of more than a year.

The *Salyut 6* station received 16 cosmonaut crews, including six long-duration crews. The longest stay time for a *Salyut 6* crew was 185 days. The first *Salyut 6* long-duration crew stayed in orbit for 96 days, surpassing the 84-day space endurance record that had been established in 1974 by the last SL-4 astronaut crew on *Skylab*. The *Salyut 6* hosted cosmonauts from Hungary, Poland, Romania, Cuba, Mongolia, Vietnam, (East) Germany, as well as Czechoslovakia. Twelve *Progress* freighter spacecraft delivered more than 20 tons of equipment, supplies, and fuel. An experimental transport logistics spacecraft called *Cosmos 1267* docked with *Salyut 6* in 1982. The transport logistics spacecraft was originally designed for the Almaz program. *Cosmos 1267* demonstrated that a large module could dock automatically with a space station—a major space technology step toward the multi-modular *Mir* station and the *International Space Station.* The last cosmonaut crew left the *Salyut 6* station on April 25, 1977. The station reentered Earth's atmosphere and was destroyed in July 1982.

The *Salyut 7* space station was launched on April 19, 1982, and was a near twin of the *Salyut 6* station. It was home to 10 cosmonaut crews, including six long-duration crews. The longest crew stay time was 237 days. Guest cosmonauts from France and India worked aboard the station, as did cosmonaut Svetlana Savitskaya, who flew aboard the Soyuz-T-7/ Salyut 7 mission and became the first Russian female space traveler since Valentina Tereshkova in 1963. Savitskaya also became the first woman to walk in space (that is, perform an extravehicular activity) during the Soyuz-T-12/Salyut 7 (in 1984). Unlike the *Salyut 6* station, however, the *Salyut 7* station suffered some major technical problems. In early 1985, for example, Russian ground controllers lost contact with the then unoccupied station. In July 1985, a special crew aboard the *Soyuz-T-13* spacecraft docked with the derelict space station and made emergency repairs that extended its lifetime for another long-duration mission. The *Salyut 7* station finally was abandoned in 1986; it reentered Earth's atmosphere over Argentina in 1991.

During its lifetime on orbit, 13 *Progress* spacecraft delivered more than 25 tons of equipment, supplies, and fuel to *Salyut 7.* Two experimental transport logistics spacecraft, called *Cosmos 1443* and *Cosmos 1686,* docked with the station. *Cosmos 1686* was a transitional vehicle—a transport logistics spacecraft that had been redesigned to serve as an experimental space station module.

During their respective lifetimes, the *Salyut 6* and the *Salyut 7* stations traveled in similar orbital paths around Earth. Each station had a perigee of 136 miles (219 km), an apogee of 171 miles (275 km), and an inclination of 51.6 degrees.

✧ *Mir* Space Station

On February 19, 1986, the Russians introduced a third-generation space station when the core of the *Mir* ("peace") station was placed into a 255-mile (411-km) by 546-mile (878-km) orbit with an inclination of 51.6 degrees by a Proton booster rocket from the Baikonur Cosmodrome.

Mir's design improvements included more extensive automation, more spacious crew accommodations for resident cosmonauts (and later American astronauts), and the addition of a multiport docking adapter at one end of the station. In a very real sense, *Mir* represented the world's first "permanent" space station. When docked with the *Progress-M* and *Soyuz-TM* spacecraft, this station measured more than 107 feet (32.6 m) long and was about 90 feet (27.4 m) wide across its assemblage of modules. The orbital complex consisted of the 44,220-pound (20,100-kg) *Mir* core module and a variety of additional scientific modules, including the Kvant-1 (quantum), Kvant-2, Kristall (crystal), Spektr (spectrum), and Priroda (nature) modules.

The *Mir* core resembled the *Salyut 7* station but had six ports instead of two. The fore and aft ports were used primarily for docking, while the four radial ports that were located in a node at the station's front were used for berthing large modules. When launched in 1986, the core had a mass of about 20 tons, a length of 43 feet (13.1 m), and a diameter of 13.8 feet (4.2 m). *Mir*'s core module consisted of a passage area with five docking ports, a working and living area (containing the command station, hygiene facilities, and eating and sleeping accommodations), and a propulsion section, which included a tunnel that provided access to the Kvant-1 scientific module.

The Kvant-1 module was added to the *Mir* core's aft port on April 9, 1987. Kvant-1 was a scientific module dedicated to astrophysics. This module had an initial mass of approximately 24,310 pounds (11,050 kg), a length of 19 feet (5.8 m), and a diameter of 13.8 feet (4.2 m). In addition to housing astrophysics instruments, the Kvant-1 module also contained life support and attitude control equipment. Although Kvant-1 blocked the core module's aft port, it had its own aft port, which then served as the station's aft port for docking *Progress-M* resupply spacecraft.

The Russians added the 42,900-pound (19,500 kg) Kvant-2 module to the *Mir* complex on December 6, 1989. They based the design of this module on the transport logistics spacecraft originally intended for the Almaz military space station program of the early 1970s. The purpose of Kvant-2 was to provide biological research data and Earth observation data. Kvant-2 carried an EVA airlock, two solar arrays, and life support equipment. The module was 39 feet (11.9 m) long and had a diameter of 14.3 feet (4.35 m). In addition to housing scientific instruments and equipment for technical experiments, the Kvant-2 module contained a shower facility and an airlock that supported extravehicular activity by the crew.

The third expansion module, called Kristall, was added to the *Mir* space station complex on June 10, 1990. Located opposite the Kvant-2 module, the 43,210-pound (19,640-kg) Kristall module provided a more symmetric mass balance for the growing orbital complex. Kristall carried scientific equipment primarily for materials processing and research under microgravity conditions. The 39-foot- (11.9-m-) long and 14.3-foot- (4.35-m-) diameter module also housed Earth observation instruments, had retractable solar arrays, and contained a docking node equipped with a special androgynous interface docking mechanism designed to receive spacecraft with masses of up to 100 tons. This docking unit (originally developed for the former Russian space shuttle *Buran*) was attached to the docking module, which was used by the American space shuttle orbiter vehicles to link with the *Mir* during Phase I of the *International Space Station* (*ISS*) program. The docking module was added to Kristall in November 1995 as part of the STS-74 mission of the space shuttle *Atlantis* to the *Mir*.

In August 1992, the Russians installed a thruster package (called Sofora) at the top of a 45.9-foot- (14-m-) tall mast attached to the Kvant–1 module. This thruster package provided an efficient, propellant-saving way to achieve attitude control of the entire *Mir* complex.

Two other modules, each carrying American equipment, were added to the *Mir* complex as part of the Phase I activities of the *ISS* program. The new scientific modules were called Spektr (added in May 1995) and Priroda (added in April 1996). The 43,210-pound (19,640-kg) Spektr was a habitable science module that carried an international complement of instruments for Earth observation and the study of Earth's atmosphere. This module had a length of 47.2 feet (14.4 m) and a diameter of 14.3 feet (4.35 m). Engineers incorporated four solar arrays capable of generating 6.9 kilowatts of electric power. Like all other *Mir* modules, Spektr had a manipulator arm for repositioning the module to other ports (as necessary) after initial docking operations. The Spektr module was severely damaged on June 25, 1997, when a *Progress* resupply spacecraft collided with it during practice docking operations.

Priroda was primarily a remote-sensing module and carried Earth observation instruments, such as a synthetic aperture radar, a variety of radiometers, and several types of spectrometers. The experiments, contributed by 12 nations, covered the microwave, visible, near-infrared, and thermal infrared portions of the electromagnetic spectrum and used both active and passive remote-sensing techniques. The 43,340-pound (19,700-kg) module had a length of approximately 39.4 feet (12 m) and a diameter of 14.3 feet (4.35 m). Priroda was the last permanent, habitable module added to the *Mir* complex. Unlike the other modules, however, Priroda had none of its own solar power arrays and depended on other portions of the *Mir* complex for electric power.

As previously mentioned, the docking module, constructed in Russia with American cooperation, was delivered by the space shuttle *Atlantis* during the STS-74 mission (November 1995) and berthed at Kristall's androgynous docking port. The docking module became a permanent extension on *Mir* and provided better clearances for space shuttle orbiter vehicles when they docked with *Mir*.

Starting in 1986, the *Mir* space station served as the major element of the Russian human spaceflight program. Early in its orbital operations, *Mir* began to host international crewmembers, or "cosmonaut researchers." By 1995, emphasis on international cooperation in space significantly increased. From 1995 to 1998, *Mir* participated in a series of joint space missions with the United States, which were undertaken as Phase I of the *International Space Station.* The first of these joint missions took place in February 1995 and involved a rendezvous (but not docking) with NASA's space shuttle *Discovery,* during NASA's STS-63 mission. (Chapter 8 discusses the space shuttle and chapter 9 the *International Space Station.*) The STS-63 mission had special importance as a precursor mission and technical dress rehearsal for the series of joint U.S.-Russian missions that would follow, involving the space shuttle rendezvousing and docking with the *Mir* space station. *Discovery* approached within about 36 feet (11 m) of *Mir* and then remained stable for 10 minutes in a position opposite the docking port of the Kristall module. After the close encounter, *Discovery* backed off to about 400 feet (122 m) and performed a fly-around of the Russian space station complex.

This picture shows the space shuttle *Atlantis* connected to Russia's *Mir* space station. The Mir-19 mission crew—cosmonauts Anatoliy Y. Solovyev (commander) and Nikolai M. Budarin (flight engineer)—took the interesting photograph on July 4, 1995. The cosmonauts had temporarily undocked their Soyuz spacecraft from the *Mir* complex and were performing a brief fly-around. They snapped this picture while the STS-71 crew, with the three Mir-18 mission crewmembers aboard, were undocking *Atlantis* from *Mir* for its journey back to Earth. Solovyev and Budarin had been taxied to *Mir* by the STS-71 ascent trip of *Atlantis. (NASA)*

The Mir Principal Expedition 18 involved the participation of American astronaut Norman Thagard, M.D., on board the *Soyuz-TM-21* spacecraft, which was launched from the Baikonur Cosmodrome on March 14, 1995. Thagard served as a cosmonaut-researcher on *Mir* for 115 days, until taken back to Earth by the space shuttle *Atlantis.* On June 29, 1995, at 13:00 (universal time), the shuttle *Atlantis* docked with *Mir* during NASA's STS-71 mission. When the two spacecraft (*Atlantis* and *Mir*) were successfully linked, they were at an altitude of 248 miles (400 km) above the Lake Baikal region of Russia. As part of the STS-71 mission, *Atlantis* also delivered Mir 19 expedition cosmonauts Anatoly Solovyev and Nikolai Budarin to the Russian space station and returned the Mir 18 expedition cosmonauts Vladimir Dezhurov and Gennadiy Strekalov along with their American companion (astronaut Norman Thagard) back to Earth from the *Mir*.

The Mir Principal Expedition 20 (Mir 20) lasted from September 3, 1995, to February 27, 1996, involved use of the *Soyuz-TM-22* spacecraft, and hosted European Space Agency (ESA) astronaut Thomas Reiter under the Euromir 95 project. Reiter, a German, was the first non-Russian *Mir* flight crewmember certified for the position of "flight engineer." The scientific objectives of Euromir 95 were to investigate the effects of microgravity on the human body, to perform experiments involving materials processing in space, and to test new space equipment. Reiter not only devoted about 4.5 hours per day to working on Euromir 95 experiments, but also helped maintain the station's onboard equipment and participated in Russian experiments. During Mir 20, the second shuttle-*Mir* docking experiment took place from November 17–18, when *Atlantis* linked with the Russian space station as part of NASA's STS-74 mission. After spending 179 days in space, Mir 20 cosmonauts Yuri Gidzenko, Sergey Avdeyev, and German (ESA) astronaut Thomas Reiter donned their Sokol launch and entry space suits on Feburary 29, 1996, entered the *Soyuz-TM-22* spacecraft, and returned safely to Earth—landing about 65 miles (105 km) from Arkalyk, Kazakhstan.

The Mir Principal Expedition 21 (Mir 21) began with the launch of the *Soyuz-TM-23* spacecraft on February 21, 1996, from the Baikonur Cosmodrome. The spacecraft carried cosmonauts Yuri Onufrienko (Mir 21 commander) and Yuri Usachev (Mir 21 flight engineer) to the Russian space station. On March 23, the third member of the Mir 21 crew, American astronaut Shannon Lucid, joined them by arriving on board the space shuttle *Atlantis* during the STS-76 mission. *Atlantis*'s flight was also the third shuttle-*Mir* docking mission under Phase I of the *International Space Station*. Lucid transferred to the *Mir* and then ably served in the capacity of cosmonaut-researcher. Among her many duties, Lucid conducted tests and activated U.S. equipment on the Priroda and Spektr modules. She also tried her hand at farming in space, by cultivating dwarf wheat in a special space greenhouse experiment.

On August 17, 1996, the Russians launched the *Soyuz-TM-24* spacecraft from Baikonur to the *Mir*. On board this spacecraft was the new cosmonaut crew for the Mir Principal Expedition 22 (Mir 22) and Claudie Andre-Deshays, a visiting French Space Agency (CNES) cosmonaut-researcher who accompanied the Mir 22 crew into orbit and then performed scientific investigations on the *Mir* station for two weeks. Andre-Deshays would return back to Earth with Onufrienko and Usachev

American astronaut Shannon Lucid exercises on a treadmill set up inside the Russian *Mir* space station as it orbited Earth on March 28, 1996. Regular exercise programs help astronauts and cosmonauts combat some of the undesirable physiological consequences of long-term exposure to the microgravity environment of an orbiting space vehicle. *(NASA)*

when the Mir 21 crew (minus Shannon Lucid) departed *Mir* on September 2 and returned to Earth in the *Soyuz-TM-23* spacecraft. During their participation in the Mir 21 expedition, Onufrienko and Usachev logged 194 days in space. Andre-Deshays spent a total of 17 days in space.

When the Mir 21 cosmonaut team departed the Russian space station, Lucid remained on board the *Mir*. She would return to Earth on the space shuttle *Atlantis* in mid-September. During the intervening time, she continued to monitor experiments she activated during Mir 21. In September 1996, the space shuttle *Atlantis* accomplished the fourth docking mission with the *Mir* station, as part of NASA's STS-79 mission. American astronaut Lucid departed the *Mir* and was replaced by astronaut John E. Blaha, who assumed the position of cosmonaut-researcher. By the time the *Atlantis* landed at the Kennedy Space Center on September 26, Lucid had set the world's record for time in space by a woman on an extended long-duration mission—188 days and five hours. During her approximately six-month stay on *Mir*, Lucid performed numerous experiments, including an interesting study of how plants grew under microgravity conditions.

In late September 1997, *Atlantis* lifted off from the Kennedy Space Center and traveled to *Mir* for the seventh rendezvous and docking mission. As part of the STS-86 mission, the *Atlantis* also performed a fly-around of the *Mir* complex to help the cosmonauts on board *Mir* locate more precisely where the Spektr had been damaged during a previous accident that caused a serious air leak in that module. (On June 25, 1997, a *Progress* resupply spacecraft collided with the Spektr module during practice docking operations.) Astronaut C. Michael Foale completed his stay on the *Mir* as a cosmonaut-researcher and returned to Earth along with the flight crew of *Atlantis*. When the shuttle *Atlantis* landed at the Kennedy Space Center on October 6, 1997, Foale completed a 145-day journey in space, which included the 134 days he spent aboard *Mir*. Astronaut David A. Wolf, who rode *Atlantis* into orbit, replaced Foale as the visiting American cosmonaut-researcher on the *Mir*.

During the STS-89 mission in January 1998, the shuttle *Endeavour* accomplished the eighth docking mission with the Russian space station as part of the overall *Mir*-shuttle linkup program. While the two spacecraft were docked, a fifth crew exchange occurred in which astronaut David A. Wolf departed the *Mir* for Earth via the *Endeavour* and astronaut Andrew S. W. Thomas joined *Mir* as the visiting American cosmonaut-researcher. When the *Endeavour* landed at the Kennedy Space Center on January 31, 1998, Wolf had logged a total of 128 days in space, including the 119 days he spent aboard the Russian space station. Thomas was the last American astronaut to complete a lengthy stay on *Mir*.

In June 1998, the space shuttle *Discovery* performed the ninth and final docking with *Mir,* as part of the STS-91 mission. This linkup in space completed Phase I of the *International Space Station* program and allowed astronaut Andrew S. W. Thomas to depart *Mir* and return to Earth. When *Discovery* opened its hatch and Thomas transferred from *Mir* to the shuttle, the astronaut had completed 130 days of living and working on the Russian space station. Thomas's transfer concluded a total of 907 days spent by a total of seven U.S. astronauts aboard *Mir* as long-duration crewmembers. The *Discovery* landed at the Kennedy Space Center on June 12 to successfully conclude the STS-91 mission. The seven American astronauts who participated in long-duration missions on *Mir* as cosmonaut-researchers are (in chronological order): Norman E. Thagard (Mir 18 expedition), Shannon W. Lucid (Mir 21), John E. Blaha (Mir 22), Jerry M. Linenger (Mir 22/23), C. Michael Foale (Mir 23/24), David A. Wolf (Mir 24), and Andrew S. W. Thomas (Mir 24/25).

As noted previously, the first shuttle-*Mir* crew transfer took place during the STS-71 mission of *Atlantis* in the summer of 1995. The STS-71 mission was also the first shuttle-*Mir* docking mission. As part of these international rendezvous and docking operations, *Atlantis* carried the

Mir 19 expedition crew (cosmonauts Anatoly Y. Solovyev and Nikolai M. Budarin) to the Russian space station and then returned the entire Mir 18 expedition crew (cosmonauts Vladimir N. Dezhurov and Gennady M. Strekalov, along with astronaut Norman E, Thagard) back to Earth—landing at the Kennedy Space Center on July 7, 1995. The Mir Principal Expedition 19 was an all-Russian crew activity, with Solovyev and Budarin staying in space aboard Mir for 75 days, from June 27 to September 11, 1995. Before departing the station on September 11 in the Soyuz-TM-21 spacecraft, the two cosmonauts greeted the international crew of the Mir Principal Expedition 20, who arrived at the station aboard the Soyuz-TM-22 spacecraft.

Once the shuttle-Mir program ended, the Russian government became fiscally challenged to both maintain the aging Mir and actively participate in the follow-on phases of the International Space Station. So the Russians decommissioned Mir and abandoned the station in 1999. For safety reasons, Russian spacecraft controllers successfully deorbited the large space station in March 2001 and intentionally crashed any surviving remnants in a remote area of the Pacific Ocean. They used a Progress M1-5 automatic cargo ship (nicknamed "the Hearse"), which was loaded with attitude-control/orbit-maneuver propellant, to dock with the abandoned space station and nudge it out of orbit in a reasonably controlled fashion. As Mir made its fiery lethal plunge into Earth's upper atmosphere on March 23 (at 06:45 UTC) somewhere over the Pacific Ocean, a somber official at the Russian mission control center outside Moscow publicly announced: "Orbital space station Mir has completed its triumphant flight, which has been unprecedented in the history of manned space exploration and which humankind has yet to fully appreciate."

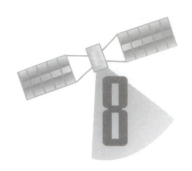

Space Shuttle

NASA's official name for the space shuttle is the *U.S. Space Transportation System* (STS)—a term that encompasses the space shuttle program itself, as well as intergovernmental agency requirements and joint and international projects such as *Spacelab.* The major components of the space shuttle system are the winged orbiter vehicle (often referred to as the "shuttle" or the "space shuttle"); the three space shuttle main engines; the giant external tank, which feeds liquid hydrogen fuel and liquid oxygen (oxidizer) to the shuttle's three main engines; and the two solid rocket boosters.

According to aerospace history reports prepared by NASA, the name *space shuttle* evolved from an accumulation of descriptive references in the press, federal government, and aerospace industry that related to the fundamental concept of a reusable space transportation system. As early concept studies gradually developed into a full-scale NASA program, the name was officially adopted for use by the agency in the late 1960s.

One of the first published uses of the term *shuttle* in the aerospace field appeared in the early 1950s, when Wernher von Braun wrote an article entitled "Crossing the Last Frontier." His article appeared in the March 22, 1952, issue of *Collier's* magazine and discussed the future of rocketry and space travel. Von Braun envisioned space stations in orbit around Earth supplied by large, winged rocket ships that would enter orbit and then return to Earth by "landing like a normal airplane." The German-American space-travel advocate also mentioned the use of a small, rocket-powered "shuttle craft" or "space taxi" to carry men and material between the space station and the larger, airplane-like, rocket ships that took off from Earth.

From its creation as the American civilian space agency in 1958, NASA has supported studies about reusable launch vehicles that could fly into space and then return to Earth for refurbishment and reuse. In the 1950s, NASA (and its predecessor, the National Advisory Committee for Aeronautics) cooperated with the U.S. Air Force on both the X–15 reusable rocket plane program (which actually carried human beings to the threshold of space between 1959 and 1968) and conceptual space plane projects like the X-20 Dyna-Soar ("Dynamic-Soaring") hypersonic boost-glide vehicle program (which ran from 1958 to 1963 but was canceled without ever flying).

Starting in about 1962, the advanced studies and future projects group at NASA's Marshall Space Flight Center in Huntsville, Alabama, began examining the possibility of recovering and reusing the Saturn V launch vehicle. Some engineers at the flight center studied the concept of a winged, flyback Saturn V vehicle, while other engineers explored the idea of specially designed, reusable space vehicle systems intended primarily to support space logistics operations. Then, as the Apollo Project matured, these NASA-sponsored future launch vehicle studies widened to embrace the concept of a fully reusable, economical space transportation system for both manned and unmanned missions.

In 1968, advanced planners at

This dramatic overhead view shows the space shuttle *Atlantis* stacked for flight and resting atop the mobile launcher platform (MLP) at the Kennedy Space Center (ca. August 1996). The orbiter vehicle is mated to the external tank and two solid rocket boosters in preparation for the STS-79 mission. The MLP crawler-transporter takes the stacked shuttle flight vehicle from the 525-foot- (160-m-) tall vehicle assembly building to launch pad 39A. *(NASA/KSC)*

NASA headquarters started using the term *shuttle* for the concept of a reusable space transportation system. By 1969, the term *space shuttle* had become a standardized designation throughout all of NASA and was no longer just found in studies at the headquarters in Washington, D.C. In September of that year, the Space Task Group appointed by President Richard M. Nixon to help define the post-Apollo space objectives for the United States recommended that the nation develop a reusable, economical space transportation system. The group's recommendation reinforced the notion of a shuttle vehicle. Over the next two years (1970 and 1971), aerospace engineers and mission planners participated in a number of intensive design, technology, and cost studies that would shape the final space shuttle program pursued by NASA. On January 5, 1972, President Nixon publicly announced that the United States would develop the space shuttle.

The space shuttle would dominate NASA activities for the next three decades. As initially planned, the space shuttle was to be a delta-winged, aircraft-like aerospace vehicle about as large as a medium-size commercial jet liner, such as the DC-9. Engineers referred to the orbiter vehicle as an *aerospace* vehicle because it would operate in both the atmosphere as well as in outer space. The shuttle orbiter would be launched vertically, while mounted on a large, expendable, liquid-propellant tank, flanked by two recoverable and reusable solid-propellant rocket boosters. NASA planners expected that the shuttle's cargo bay would carry most of the nation's civilian and military payloads. To make the program appear cost-effective, planners put forward optimistic predictions of 60 or 70 flights a year—with each shuttle orbiter having a lifetime of 100 or so space missions. Rapid refurbishment and turnaround of the orbiter vehicle and the recovered solid rocket boosters was also projected. Only the giant external tank would be discarded each mission.

At the start of the shuttle program in the early 1970s, NASA planned for the new space vehicle to be flown by a three-man crew, carry satellites to orbit, repair them on orbit, and later return them to Earth for repair, refurbishment, and reuse. The shuttle vehicle would also carry up to four scientists or engineers into space, possibly in a specially designed pressurized laboratory like *Spacelab*. Following a seven- to 30-day mission in space, the orbiter vehicle would reenter the atmosphere, land like an airplane, and begin preparations for the next flight. These were some of the dominant performance characteristics and operational projections that defined and shaped NASA's space shuttle system in the 1970s.

By the end of 1974, the American aerospace industry, suffering from massive layoffs due to the abrupt conclusion of the Apollo Project, eagerly started "bending metal"—that is, fabricating, assembling, and testing components—for the space shuttle. NASA planners had scheduled the beginning of orbital testing for March 1979 and a completely operational system sometime in 1980.

As will become apparent in the remainder of this chapter, some of these plans and projections came true, but others were wildly off the mark. For example, the estimated turnaround time between flights for an orbiter vehicle and the overall costs incurred during each refurbishment proved excessively optimistic. And the first orbital test flight (called the STS-1 mission) would not occur until April 1981. Technical delays or missed schedule milestones are not uncommon in any new and ambitious aerospace program that attempts to expand the boundaries of space technology along several frontiers at once. NASA engineers encountered many unanticipated difficulties in developing the shuttle—some great and difficult to resolve; others troublesome but relatively straightforward in their solutions. Collectively, these technical challenges caused schedule delays and forced engineers to design compromises that lessened or diluted some of the envisioned performance characteristics and operational features of the new space shuttle. Despite these technical setbacks and increased expenses, what resulted was a marvel of modern aerospace engineering technology and a space vehicle that shaped the practice of human spaceflight by the United States for more than 30 years.

✧ Basic Features of the Space Shuttle System

NASA's space shuttle is a complex, amazing machine with more than 2.5 million parts, including almost 230 miles (370 km) of wire, over 1,440 circuit breakers, some 1,060 plumbing valves and connections, and over 27,000 heat-insulating tiles and thermal blankets. Temperatures experienced by the orbiter vehicle range from –250°F (–157°C) to as high as 3,000°F (1,650°C) as the shuttle reenters Earth's atmosphere.

According to NASA's collection of interesting facts about the space shuttle, at liftoff the vehicle typically has a mass of more than 4.5 million pounds (2 million kg). Over 3.5 million pounds (1.6 million kg) of that liftoff mass belongs to propellants, which are entirely consumed in the next eight and one-half minutes, as the vehicle rises up through the atmosphere and achieves orbit. When ignited at launch, the shuttle's two solid rocket boosters consume more than 10 tons of fuel each second and produce 44 million horsepower (32,824 megawatts)—a quantity equal to the power of 14,700 modern locomotives. When operating at maximum thrust, the three liquid propellant–fueled shuttle main engines produce power at a rate equivalent to 23 times that produced by the Hoover Dam. If the shuttle's main engines were pumping water instead of propellants, they would drain an average-size swimming pool every 25 seconds. The temperatures inside the shuttle's main engines and solid rockets reach more than 6,000°F (3,315°C), which is higher than the boiling point of iron (Fe). Yet the main engine's fuel, liquid hydrogen (LH_2), is the second-coldest liquid on Earth (next to liquid helium) and must be kept at a temperature of –423°F (–253°C). Finally, in about eight and one-half minutes after launch, the orbiter vehicle reaches orbital velocity by having accelerated from zero to 17,400 miles per hour (28,000 km/h)—or about nine times as fast as a rifle bullet. NASA's space shuttle orbiter is definitely an amazing machine and one of the most complex devices ever conceived, constructed, and operated by human beings.

The orbiter is the only component of the space shuttle system that has a name in addition to a part number. The first orbiter built was the

The space shuttle *Endeavour* thunders into space from the Kennedy Space Center (in 1993), powered by its three main engines and two solid rocket boosters (SRBs). The trip into low Earth orbit takes about eight and one-half minutes, during which time the orbiter vehicle jettisons first the SRBs (at an altitude of about 28 miles [45 km]) and then the huge external tank just before entering orbit around Earth. *(NASA/SSC)*

Enterprise (OV–101), which was designed for flight tests in Earth's atmo-sphere rather than operations in space. It is now at the Smithsonian Air and Space Museum located at Dulles Airport outside Washington, D.C. Five operational orbiters were constructed (listed in order of completion): *Columbia* (OV–102), *Challenger* (OV-99), *Discovery* (OV–103), *Atlantis* (OV–104), and *Endeavour* (OV–105). The *Challenger* and its crew were lost in a launch accident on January 28, 1986, and the *Columbia* and its crew were lost in a reentry accident on February 1, 2003.

Shuttles are launched from either Pad 39A or 39B at the Kennedy Space Center, Florida. Depending on the requirements of a particular mission, a space shuttle can carry about 49,900 pounds (22,680 kg) of payload into low Earth orbit. An assembled shuttle vehicle has a mass of about 4.5 mil-lion pounds (2 million kg) at liftoff.

The two solid rocket boosters are each 149 feet (45.4 m) high and 12.1 feet (3.7 m) in diameter. Each has a mass of about 1,298,000 pounds (590,000 kg). Their solid propellant consists of a mixture of powdered aluminum (fuel), ammonium perchlorate (oxidizer), and a trace of iron oxide to control the burning rate. The solid mixture is held together with a polymer binder. Each booster produces a thrust of approximately 3.1 mil-lion pounds-force (13.8 million N) for the first few seconds after ignition. The thrust then gradually declines for the remainder of the two-minute burn to avoid overstressing the flight vehicle. Together with the three main liquid-propellant engines on the orbiter, the shuttle vehicle produces a total thrust of 7.3 million pounds-force (32.5 million N) at liftoff.

Typically, the solid rocket boosters burn until the shuttle flight vehicle reaches an altitude of about 28 miles (45 km) and a speed of 3,090 miles per hour (4,970 km/h). Then they separate and fall back into the Atlantic Ocean to be retrieved, refurbished, and prepared for another flight. After the solid rocket boosters are jettisoned, the orbiter's three main engines, fed by the huge external tank, continue to burn and provide thrust for another six minutes before they, too, are shut down at MECO (main engine cutoff). At this point the external tank is jettisoned and falls back to Earth, disintegrating in the atmosphere with any surviving pieces falling into remote ocean waters.

The huge external tank is 154 feet (47 m) long and 27.6 feet (8.4 m) in diameter. At launch, it has a total mass of about 1,672,550 pounds (760,250 kg). The two inner propellant tanks contain a maximum of 385,000 gallons (1,458,400 L) of liquid hydrogen and 143,400 gallons (542,650 L) of liq-uid oxygen (LO$_2$), respectively The external tank is the only major shuttle flight vehicle component that is expended on each launch. Following the loss of the *Columbia* in February 2003, the external tank underwent major design changes to minimize the generation of launch debris (especially shedded foam and/or ice) that could damage the orbiter vehicle.

The winged orbiter vehicle is both the heart and the brains of America's Space Transportation System. About the same size and mass as a DC-9 commercial jet aircraft, the orbiter contains the pressurized crew com-

REDUCING THE RISK FROM EXTERNAL TANK INSULATION DEBRIS

In 1983, NASA flew a redesigned lightweight external tank, some 10,000 pounds (4,545 kg) lighter than the original tank design, on the STS-6 shuttle mission. This engineering change increased the shuttle's cargo capacity by approximately the same amount. Then, in 1998, a super-lightweight external tank flew on the STS-91 shuttle mission. This further reduced the tank's mass by 7,500 pounds and again increased the shuttle's cargo carrying capacity by an equivalent amount.

The new, super-lightweight external tank was manufactured from a Lockheed-Martin-developed aluminum-lithium alloy that is not only lighter but also 30 percent stronger than the previous tank design. The external tank's aluminum alloy skin is a 10th of an inch (0.25 cm) thick in most places and is covered with polyurethane-like foam that is typically an inch (2.54 cm) or so thick. The foam insulation not only gives the huge, 15-story-tall fuel tank its bright orange appearance but also insulates the liquid hydrogen and liquid oxygen propellants—preventing ice formation on the tank's exterior and protecting the tank's skin from aerodynamic heating during the ascent flight to orbit. About 90 percent of the foam is applied by automatic systems, while the remainder is applied manually.

The tank's improved designs saved liftoff mass and thus improved cargo-carrying performance, but the new tank also had some unforeseen difficulties that led to the loss of Columbia upon its return to Earth at the conclusion of the STS-107 orbital mission. After a successful 16-day mission in space, the seven astronauts aboard Columbia

were in the process of reentry and only about 16 minutes from touchdown at the Kennedy Space Center. Suddenly, the vehicle disintegrated in the skies over East Texas, claiming the lives of all seven crewmembers. Intensive investigations after the tragic accident isolated the cause to debris from the external tank striking the thermal insulation on the leading edge of Columbia's left wing. The debris strike was apparently caused by a large piece of the tank's insulating foam (specifically, the left bipod foam ramp) and inflicted serious damage to the wing's reinforced carbon-carbon (RCC) thermal protection tile. The astronauts were unaware of the serious nature of the wing damage and proceeded with their 16-day mission in space. Upon reentry, the damaged part of the wing provided a ready passageway for intensely hot, atmospheric gases to penetrate and destroy the interior structure of the wing. With its left wing destroyed, the Columbia quickly lost aerodynamic stability and disintegrated.

NASA grounded the shuttle fleet for more than two years while engineers and managers tried to solve the problem of foam and ice debris from the external tank striking and causing potentially lethal damage to the orbiter vehicle on a future shuttle mission. Anxiously, a hundred cameras and a thousands sets of eyes trained on the Discovery as the STS-114 "Return to Flight" mission was launched at the Kennedy Space Center on July 26, 2005. After the Discovery, commanded by astronaut Eileen Collins, was safely in orbit, video analysis revealed that a piece of foam—approximately 36 inches (91.4 cm) long and 11 inches (28 cm) across partment (which can normally carry up to eight crewmembers), the huge cargo bay (which is 60 feet [18.3 m] long and 15 feet [4.57 m] in diameter), and the three main engines mounted on its aft end. The orbiter vehicle

at its widest—came off the external tank. Despite attempts at fixing the problem, the latest foam-shedding event had occurred at 127 seconds into the flight. While NASA engineers on the ground scrutinized every frame of high-resolution imagery available, the astronauts on board the *Discovery* used the laser scanner in the new orbiter boom sensor system (OBSS) to inspect the condition of the orbiter's thermal protection tiles. Everyone wanted to make sure no serious damage had been done to the *Discovery*. If potentially lethal damage was detected, the astronauts were to take refuge in the *International Space Station* (*ISS*) while rescue missions were mounted.

Before docking with the *International Space Station*, shuttle commander Collins performed the first ever rendezvous pitch maneuver when *Discovery* was about 600 feet (183 m) below the station. Thanks to her skillful maneuvering, the *Discovery* gently flipped end over end at a rate of just 0.75 degree per second. As the shuttle slow flipped in front of them, the *Expedition 11* crewmembers took a large number of high-resolution digital photographs of *Discovery*'s underside—paying special attention to detect and record any obvious damage in the vehicle heat-resistant tiles. The space station crew then downlinked these images to the engineers on the ground who were closely monitoring the situation.

The large collection of imagery data available to the ground support team revealed some small dings in the tiles and two areas where the gap fillers between the tiles were protruding. Fortunately, they did not observe any tiles with serious damage. Engineers and managers deliberated for the next two days while *Discovery* docked with

the *ISS* and off-loaded supplies and equipment. As a special precaution, shuttle mission specialist Stephen K. Robinson performed an extravehicular activity while attached to the foot restraint in the space station's robot arm (called the Canadarm2). While spacewalking, he gently pulled the two protruding gap fillers from between the thermal protection tiles. *Discovery* was then considered ready for reentry. The STS-114 mission came to a successful conclusion, as *Discovery* touched down at Edwards Air Force Base in California on August 9, 2005.

But the problem with the external tank persisted, and no more shuttle launches would take place until there was a more satisfactory solution to the foam-shedding problem. NASA engineers revisited this problem and came up with another modification of the external tank (a design configuration called ET–119) that completely did away with the protuberance air-load ramps. Other engineering modifications were incorporated into ET–119, making the tank safer for use on the next shuttle flight, called the STS-121 mission or the "Second Return to Flight Mission." On July 4, 2006, *Discovery* roared off its pad at the Kennedy Space Center, rendezvoused and docked with the space station, and then safely returned to Earth, touching down at the Kennedy Space Center on July 17. Imagery and videos collected during launch, on-orbit inspections, and postflight visual inspections of *Discovery*'s thermal tile system suggested that the engineering changes made to external tank–119 may have been sufficient to mitigate the problem of shedding foam that could damage the orbiter vehicle on its ascent to space.

itself is 121 feet (37 m) long, 56 feet (17 m) high, and has a wingspan of 79 feet (24 m). Since each of the operational vehicles varies slightly in construction, an orbiter generally has an empty mass of between 167,200 pounds (76,000 kg) to 173,800 pounds (79,000 kg).

Each of the three main engines on an orbiter vehicle is capable of producing a thrust of 375,300 pounds-force (1,668,000 N) at sea level and 470,250 pounds-force (2,090,000 N) in the vacuum of space. These engines burn for approximately eight minutes during launch ascent and together consume about 64,000 gallons (242,250 L) of cryogenic propellants each minute, when all three operate at full power.

An orbiter vehicle also has two smaller orbital maneuvering system (OMS) engines that operate only in space. These engines burn nitrogen tetroxide as the oxidizer and monomethyl hydrazine as the fuel. The propellants are supplied from onboard tanks carried in the two pods at the upper rear portion of the vehicle. The OMS engines are used for major maneuvers in orbit and to slow the orbiter vehicle for reentry at the end of its mission in space. On most missions, the orbiter enters an elliptical orbit, then coasts around Earth to the opposite side. The OMS engines then fire just long enough to stabilize and circularize the orbit. On some missions, the OMS engines also are fired soon after the external tank separates, to place the orbiter vehicle at a desired altitude for the second OMS burn that then circularizes the orbit. Later OMS engine burns can raise or adjust the orbit to satisfy the needs of a particular mission. A shuttle flight can last from a few days to more than a week or two.

After deploying the payload spacecraft (some of which can have attached upper stages to take them to higher-altitude operational orbits, such as a geostationary orbit), operating the onboard scientific instruments, making scientific observations of Earth or the heavens, or performing other aerospace activities (such as rendezvous and docking with the space station), the orbiter vehicle reenters Earth's atmosphere and lands. This landing usually occurs at either the Kennedy Space Center in Florida (primary site) or the Edwards Air Force Base in California (first alternate site)—depending on weather conditions at the primary landing site. Unlike prior manned spacecraft, which followed a ballistic trajectory, the orbiter (now operating like an unpowered glider) has a cross-range capability of about 1,240 miles (2,000 km)—that is, it can move to the right or left off the straight line of its reentry path. The landing speed is between 210 miles per hour (340 km/h) and 225 miles per hour (365 km/h). After touchdown and rollout, the orbiter vehicle is made safe by a ground crew with special equipment. This safing operation is also the first step in preparing the orbiter for its next mission in space.

The orbiter's crew cabin has three levels. The uppermost is the flight deck, where the commander and pilot control the mission. The mid-deck is where the galley, toilet, sleep stations, and storage and experiment lockers are found. Also located in the mid-deck are the side hatch for passage to and from the orbiter vehicle before launch and after landing, and the airlock hatch into the cargo bay and to outer space to support on-orbit extravehicular activities. Below the mid-deck floor is a utility area for air and water tanks.

The orbiter's large cargo bay is adaptable to numerous tasks. It can carry satellites, large space platforms such as the *Long-Duration Exposure*

SPACELAB

Spacelab was an orbiting laboratory facility delivered into space and sustained while in orbit within the huge cargo bay of the space shuttle orbiter. Developed by the European Space Agency in cooperation with NASA, *Spacelab* featured several interchangeable elements that were arranged in various configurations to meet the particular needs of a given flight. The major elements were a habitable module (short or long configuration) and pallets. Inside the pressurized habitable research module, astronaut scientists (payload specialists) worked in a relatively comfortable, shirtsleeve environment and performed a variety of experiments under microgravity conditions.

Several platforms (called pallets) could also be placed in the orbiter's cargo bay behind the habitable module. Any instruments and experiments mounted on these pallets were exposed directly to the space environment when the shuttle's cargo bay doors were opened after the aerospace vehicle achieved orbit around Earth. A train of pallets could also be flown without the concurrent use of the habitable module.

Various configurations of these *Spacelab* elements were located within the orbiter's cargo bay to support the scientific objectives of a particular mission. The habitat module was designed to be carried alone or with one or more pallets. The pallets themselves were designed to be carried into space on missions that did not use the habitable module. Despite the absence of the habitable module, a "pallet-only" configuration was still designated as a Spacelab mission.

The *Spacelab* habitable module came in two 13.1-foot- (4-m-) diameter segments. The core segment housed data-processing equipment and utilities for both the pressurized module and pallets when flown together. It also had laboratory fixtures such as air-cooled experiment racks and a workbench. The second section, called the experiment segment, provided more pressurized workspace and additional experiment racks. The core segment could be flown by itself (the "short-module" configuration) or coupled in tandem with the experiment segment (the "long-module" configuration). The short-module configuration consisted of the core segment and two cone-shaped end sections and measured about 14 feet (4.26 m) in length. The long-module configuration had a maximum outside length, including end cones, of 23 feet (7 m).

The *Spacelab* pallets were uniform. Each pallet was a U-shaped aluminum frame and

(continues)

(continued)

panel platform 13.1 feet wide and 10 feet (3 m) long. Scientists and engineers connected the experiment equipment to a series of "hard points" on the main structure of the pallet. Up to five pallets could be flown on a single mission in the pallet-only configuration of *Spacelab*. When the pallets were flown without a habitable module, subsystems needed for equipment operations (which normally would be housed in the core segment of the habitable module) were placed in a pressurized cylinder mounted to the front frame of the first pallet. Engineers called this cylinder the *igloo*.

When a habitable module was flown as part of a Spacelab mission, an access tunnel connected the module with the mid-deck level of the orbiter cabin. This access funnel was about 3.3 feet (1 m) in diameter. The hatch between the orbiter cabin and the access tunnel was left open during a mission, so the orbiter cabin, tunnel, and *Spacelab* habitable module all shared the same pressure and common air. The tunnel had lighting and handrails to allow easy passage (under microgravity conditions) between *Spacelab* and the orbiter mid-deck. The length of the access tunnel varied with the configuration of *Spacelab* in the cargo bay. A tunnel 8.8 feet (2.7 m) in length was used for missions during which the *Spacelab* habitable module was carried in the forward portion of the orbiter's cargo bay. A longer tunnel (with a length of about 19 feet [5.8 m]) was available for use on missions in which the module was carried in the aft portion of the orbiter's cargo bay.

The first Spacelab mission (called STS-9/*Spacelab 1*) was launched in November 1983. It was a highly successful joint NASA and European Space Agency mission consisting of both the habitable module and an exposed instrument platform. The final Spacelab mission (called STS-55/*Spacelab D-2*) was launched in April 1993. It was the second flight of the German (Deutsche) *Spacelab* configuration and continued microgravity research that had started with the first German Spacelab mission (STS-61A/*Spacelab D-1*) flown in October 1985. NASA, other European Space Agency countries, and Japan contributed some of the 90 experiments conducted during the Spacelab D-2 mission.

Facility, and even an entire scientific laboratory, such as the European Space Agency's *Spacelab* to and from low Earth orbit. It also serves as a workstation for astronauts to repair satellites, a foundation from which to erect space structures, and a place to store and hold spacecraft that have been retrieved from orbit for return to Earth.

Mounted on the port (left) side of the orbiter's cargo bay behind the crew quarters is the remote manipulator system (RMS), which was developed and funded by the Canadian government. The RMS is a robot arm and hand with three joints similar to those found in a human being's shoulder, elbow, and wrist. There are two television cameras mounted on the RMS near the "elbow" and "wrist." These cameras provide visual information for the astronauts who are operating the RMS from the aft station on the orbiter's flight deck. The RMS is about 49 feet (15 m) in length and can move anything, from astronauts to satellites, to and from the cargo bay as well as to different points in nearby outer space.

✧ Space Shuttle Missions

This section provides brief summaries of selected space shuttle missions. The missions presented here span the period from the STS-1 mission (first shuttle launch) on April 12, 1981, to the STS-121 mission (launched on July 4, 2006). The carefully chosen selection provides a comprehensive view of the many great triumphs in human spaceflight that have accompanied use of NASA's space shuttle system. Two very painful tragedies—the *Challenger* accident (1986) and the *Columbia* accident (2003)—are also discussed.

STS-1 MISSION

The maiden spaceflight of the shuttle program began on April 12, 1981, when *Columbia* lifted off from Pad 39A at the Kennedy Space Center and carried astronauts John W. Young (commander) and Robert L. Crippen (pilot) into orbit. This first flight was successful, and the astronauts were able to test all the major components of the space transportation system, including the orbiter's reaction control system and orbital maneuvering system. After 36 orbits of Earth, Young and Crippen glided the aerospace plane back through the atmosphere and made a safe landing at Edwards Air Force Base in California. Edwards AFB served as the primary shuttle-landing site for most of the early shuttle missions. Postflight inspection of *Columbia* showed that the orbiter vehicle had sustained significant thermal protection system tile damage on launch, including 16 tiles lost and 148 damaged. Engineers attributed the tile loss to an overpressure wave created at launch by the powerful solid rocket boosters. Subsequent modifications to the water sound-suppression system at the launch pad eliminated this problem. The STS-1 mission was a major milestone in human spaceflight and space technology, since it represented the first time a human crewed space vehicle traveled into space, completed its mission, and returned to Earth, landing much like a modern jet aircraft (although at a slightly steeper glide slope and velocity) so the aerospace vehicle could be refurbished then sent on another trip into space.

STS-3 MISSION

NASA's STS-3 mission was both the third flight of *Columbia* and the third space shuttle mission. Astronaut Jack R. Lousma served as commander, and astronaut C. Gordon Fullerton was the pilot. *Columbia* lifted off from Pad 39A at the Kennedy Space Center on March 22, 1982. Among the many vehicle qualification activities performed during this third orbital flight test, the astronauts extensively tested the remote manipulator

SUMMARY OF SPACE SHUTTLE MISSIONS

The table below provides a chronological summary of all the space shuttle flights from 1981 up to December 31, 2006.

NASA SPACE SHUTTLE LAUNCHES (1981–2006)

YEAR	LAUNCHES
1981	STS-1, STS-2
1982	STS-3, STS-4, STS-5
1983	STS-6, STS-7, STS-8, STS-9
1984	STS 41–B, STS 41–C, STS 41–D, STS 41–G, STS 51–A
1985	STS 51–C, STS 51–D, STS 51–B, STS 51–G, STS 51–F, STS 51–I, STS 51–J, STS 61–A, STS 61–B
1986	STS 61–C, STS 51–L (*Challenger* accident)
1987	No launches
1988	STS-26, STS-27
1989	STS-29, STS-30, STS-28, STS-34, STS-33
1990	STS-32, STS-36, STS-31, STS-41, STS-38, STS-35
1991	STS-37, STS-39, STS-40, STS-43, STS-48, STS-44
1992	STS-42, STS-45, STS-49, STS-50, STS-46, STS-47, STS-52, STS-53
1993	STS-54, STS-56, STS-55, STS-57, STS-51, STS-58, STS-61
1994	STS-60, STS-62, STS-59, STS-65, STS-64, STS-68, STS-66
1995	STS-63, STS-67, STS-71, STS-70, STS-69, STS-73, STS-74
1996	STS-72, STS-75, STS-76, STS-77, STS-78, STS-79, STS-80
1997	STS-81, STS-82, STS-83, STS-84, STS-94, STS-85, STS-86, STS-87
1998	STS-89, STS-90, STS-91, STS-95, STS-88
1999	STS-96, STS-93, STS-103
2000	STS-99, STS-101, STS-106, STS-92, STS-97
2001	STS-98, STS-102, STS-100, STS-104, STS-105, STS-108
2002	STS-109, STS-110, STS-111, STS-112, STS-113
2003	STS-107 (*Columbia* accident)
2004	No launches
2005	STS-114
2006	STS-121, STS-115, STS-116

Source: NASA (as of December 31, 2006)

system—the orbiter's versatile robot arm. Even though this mission was primarily a test qualification flight, NASA's Office of Space Science (OSS) took advantage of the ride into space to fly a collection of experiments and scientific instruments (collectively called OSS–1 payload), which were mounted on a *Spacelab* pallet in the orbiter bay.

Space sickness (space adaptation sickness) and a malfunctioning toilet were some of the difficulties encountered by the astronauts on this eight-day mission. On March 30, the shuttle reentered the atmosphere after making 130 revolutions of Earth and landed at Runway 17, Northrup Strip, White Sands, New Mexico—the alternate landing site that was chosen due to wet runway conditions at Edwards AFB, California. This was the first and only landing of the shuttle (to date) in New Mexico. The orbiter experienced some brake damage upon landing at White Sands, and a dust storm caused extensive contamination of the vehicle while it awaited a ferry flight (on top of a modified 747 commercial jetliner) back to the Kennedy Space Center in Florida.

STS–7 MISSION

NASA launched the STS-7 mission on June 18, 1983, from Pad 39 at the Kennedy Space Center. This was the second flight of the space shuttle *Challenger* and the first flight in the American space program to carry a woman into space—astronaut Sally K. Ride, who served as a mission specialist. The other STS-7 crewmembers were Robert L. Crippen (commander), Frederick H. Hauck (pilot), John M. Fabian (mission specialist), and Norman E. Thagard (mission specialist). The mission specialist is the space shuttle crew member and NASA career astronaut responsible for coordinating payload/space shuttle vehicle interaction. During the payload operation phase of a shuttle flight, the mission specialist is the crewmember who directs the allocation of orbiter vehicle and crew resources to accomplish payload-related mission objectives. STS-7 set a briefly held world spaceflight record of five people aboard a single space vehicle. The crew deployed two commercial communications satellites: *Anik C-2* for Telsat Canada and *Palapa-B1* for Indonesia. Once a safe distance from *Challenger,* each communications satellite was carried to its higher operational orbit by firing an attached upper stage solid rocket motor, called the payload assist module-D (PAM-D). The astronauts deployed, rendezvoused with, and retrieved the German-built *Shuttle Pallet Satellite,* which took the first panoramic images of the entire orbiter in space. The planned mission was extended for two additional revolutions of Earth because of poor weather condition at the primary landing site in Florida. On June 24, during revolution 98, *Challenger* and its crew landed safely at Edwards AFB in California. The crew had spent six days, two hours, and 24 minutes in space.

This photograph shows astronaut Sally K. Ride talking to ground controllers from the flight deck of the space shuttle *Challenger*, as the space vehicle orbited Earth during the STS–7 mission in June 1983. Ride, serving as a mission specialist on this shuttle flight, became the first American woman to travel in space. *(NASA)*

STS–9 MISSION

When NASA successfully launched *Columbia* on November 28, 1983, the space shuttle was carrying the European Space Agency's *Spacelab* (habitable module) in its huge cargo bay on the orbital laboratory's first mission into space. Throughout the mission, *Spacelab* remained cradled within the orbiter vehicle's cargo bay. Once on orbit, the shuttle astronauts opened the cargo bay's doors, exposing the habitable module and any companion pallets to outer space. The crew of the STS-9 (Spacelab-1) mission consisted of John W. Young (commander); Brewster H. Shaw, Jr. (pilot); Owen K. Garriott (mission specialist); Robert A. R. Parker (mission specialist); Bryon K. Lichtenberg (payload specialist); and Ulf Merbold (payload specialist). The German Merbold was the first astronaut to represent the European Space Agency and the first non-American to fly on a United States space vehicle. Within NASA's space shuttle program, the payload specialist is the noncareer astronaut who flies as a passenger aboard the orbiter vehicle and is responsible for achieving the payload/experiment objectives. He/she is the onboard scientific expert in charge of the operation of a particular payload or collection of experiments. Altogether 73 separate investigations were carried out during the STS-9/SL-1 mission. These experiments included the fields of astronomy, physics, Earth observations, atmospheric and space physics, life sciences, and materials sciences. The mission was the first time a single vehicle carried six persons into space at the same time. The STS-9/SL-1 mission ended on December 8, when *Columbia* landed at Edwards Air Force Base in California.

STS 41-B MISSION

On February 3, 1984, NASA successfully launched *Challenger* on the STS 41-B mission. With this flight, the space agency began using a rather unusual alphanumeric designation for shuttle missions. For this particular mission, the "4" designated the originally scheduled year of launch (in this case 1984). The second numerical digit, "1," meant the launch took place from the Kennedy Space Center. NASA had reserved the number "2" for any shuttle launches that took place from Vandenberg Air Force Base in California—but, while such shuttle launches into polar orbit were considered, none ever occurred. Finally, the "B" indicates that the STS 41-B mission was the second shuttle launch within that fiscal year. The fiscal year is an arbitrary calendar of budgetary convenience used by the U.S. government. At the time, the fiscal year (FY 84) started on October 1, 1983, making the STS-9 mission that launched on November 28, 1983, the first shuttle launch in FY-84.

Confusing? Quite, even for experienced aerospace industry personnel. When NASA launched *Discovery* for the STS-26 mission on September 29, 1988, agency officials returned to the original, less complicated designation system. In this original flight designation system, each shuttle flight was listed in an ascending numerical order, based on when the flight was *originally planned* to launch in a multiyear projected schedule. Equipment problems with a particular orbiter vehicle, schedule slips in the delivery of the payload, or the availability of launch windows for a particular mission sometimes caused NASA to launch a higher-numbered mission before it launched a lower-numbered mission. For example, in 1999, NASA launched the following missions: STS-96 (May 27), STS-93 (July 23), and STS-103 (December 19).

The crew STS 41-B mission included Vance D. Brand (commander), Robert L. Gibson (pilot), Bruce McCandless II (mission specialist), Ronald E. McNair (mission specialist), and Robert L. Stewart (mission specialist). One of the highlights of this mission was the untethered space walks performed by McCandless and Stewart, using NASA's new manned maneuvering unit (MMU). The MMU is a self-propelled backpack that allowed the spacewalking astronaut to travel in free flight almost 320 feet (98 m) away from the orbiter.

The STS 41-B crew also deployed two communications satellites, *Westar-VI* and *Palapa B-2,* during this mission. However, the two satellites failed to reach proper operational orbits when each satellite's upper stage rocket engine (a payload assist module) did not ignite. This "temporary failure" had a happy ending, because both satellites were later retrieved and brought back to Earth by the crew of the STS 51-A mission (November 1984). On February 11, 1984, the STS 41-B mission ended successfully as *Challenger* became the first shuttle to return from space and land at the Kennedy Space Center.

STS 41-C MISSION

NASA launched the space shuttle *Challenger* on the STS 41-C on April 13, 1984. The astronaut crew included Robert L. Crippen (commander), Francis R. Scobee (pilot), George D. Nelson (mission specialist), James D. A. van Hoften (mission specialist), and Terry J. Hart (mission specialist). This successful launch involved the first direct ascent trajectory for the space shuttle. The astronauts performed an extravehicular activity (EVA) and used the manned maneuvering unit to make the first satellite service call on orbit. The *Challenger* rendezvoused with and retrieved the *Solar Maximum Mission* (*Solar Max*) spacecraft, which had failed after four years in space. With the *Solar Max* satellite securely anchored in the shuttle's cargo bay, EVA astronauts Nelson and van Hoften replaced a faulty attitude control system and one science instrument on the satellite. Following the in-space repair, the *Challenger*'s crew once again released the *Solar Max* satellite into orbit around Earth.

The other major highlight of the STS 41-C mission involved the deployment of the *Long Duration Exposure Facility* (*LDEF*). NASA's *LDEF* was a large (about the size of a bus), free-flying, passive spacecraft that exposed numerous trays of experiments to the space environment during an extended 69-month mission in orbit around Earth from April 1984 to January 1990. *LDEF* gathered information on the possible consequences of the space radiation environment, atomic oxygen, meteoroids, spacecraft contamination, and space debris on aerospace hardware and spacecraft components. The *LDEF* was deployed on orbit with the intention of retrieving it after about 10 months in space. However, it was not captured and returned to Earth until the STS-32 mission in January 1990 by the crew of *Columbia.*

STS 41-D MISSION

The first orbital mission of the shuttle *Discovery* was a lesson in patience and perseverance. The crew consisted of Henry W. Hartsfield, Jr. (commander); Michael L. Coats (pilot); Judith A. Resnik (mission specialist); Richard M. Mullane (mission specialist); Steven A. Hawley (mission specialist); and Charles D. Walker (payload specialist).

The first attempt to launch *Discovery* took place on June 25, 1984, and was scrubbed during the T-9 minute hold before scheduled liftoff due to failure of the orbiter's backup general purpose computer (GPC). The second launch attempt on June 26 aborted on the pad at T-4 seconds before main engine ignition and just a split second before the solid rocket motors were to be ignited (an irreversible action). The last-minute abort occurred because *Discovery*'s GPC detected an anomaly in the orbiter vehicle's number 3 main engine. NASA returned *Discovery* to the orbital processing facility and replaced the faulty main engine. The third attempt to launch *Discovery* took place on August 29, but liftoff was delayed when launch team personnel noticed a discrepancy in flight software. *Discovery* successfully lifted off and ascended into orbit on August 30—but not before it experienced an unplanned launch delay of almost seven minutes when a private aircraft intruded into the restricted warning area off Cape Canaveral.

Once on orbit, the crew of *Discovery* successfully deployed three commercial communications satellites: *Satellite Business System* (SBS-4), *Telstar 3-C,* and *Syncom IV-2* (also called *LEASAT-2*). The astronauts also deployed an experimental, 102-foot- (31.1-m-) tall, 13-foot- (4-m-) wide solar cell array from the cargo bay. In the crew cabin, payload specialist Walker (a McDonnell Douglas engineer) tended the company's commercial experiment, called the Continuous Flow Electrophoresis System, which was designed to separate materials under microgravity conditions.

STS 51–C MISSION

On January 24, 1985, the shuttle *Discovery* lifted off from Pad 39A at the Kennedy Space Center to start the STS 51-C mission—the first shuttle mission dedicated exclusively to the Department of Defense (DOD). Only a few details have been publicly released about this mission. The crew consisted of Thomas K. Mattingly II (commander), Loren J. Shriver (pilot), James F. Buchli (mission specialist), Ellison S. Onizuka (mission specialist), and Gary E. Payton (payload specialist). During the STS 51-C mission, the crew deployed the inertial upper stage (IUS) booster (developed by the U.S. Air Force) along with a classified DOD payload, and the mission met its objectives. This was the first mission for the versatile IUS propulsion system. Any further details about the STS 51-C mission remain classified. The *Discovery* landed on January 27 at the Kennedy Space Center.

STS 51–J MISSION

The STS 51-J mission was the second shuttle mission dedicated to the Department of Defense (DOD) and also the maiden flight of the space shuttle *Atlantis.* The crew included Karol J. Bobko (commander), Ronald J. Grabe (pilot), Robert L. Stewart (mission specialist), David C. Hilmers

The crew members of *Challenger*'s tragic STS–51L mission as they stood in the White Room at Pad 39B of the Kennedy Space Center during a preflight training session (January 1986). The deceased astronauts are (from left to right): Sharon "Christa" McAuliffe, Gregory Jarvis, Judith A. Resnik, Francis R. "Dick" Scobee, Ronald E. McNair, Michael J. Smith, and Ellison S. Onizuka. *(NASA/KSC)*

(mission specialist), and William A. Pailes (payload specialist). *Atlantis* was launched from Pad 39A at the Kennedy Space Center on October 3, 1985, and landed four days later (October 7) at Edwards Air Force Base in California. During the mission, *Atlantis* made 64 revolutions of Earth. All other details of this DOD-dedicated shuttle mission remain classified.

STS 61–A MISSION

The STS 61-A mission involved a dedicated flight of the first German (Deutsche) *Spacelab* (D-1), conducted in the long-habitation configuration of the orbital laboratory. *Challenger* carried *Spacelab D-1* into orbit on October 30, 1985, from Pad 39A at the Kennedy Space Center. The crew included Henry W. Hartsfield (commander); Steven R. Nagel (pilot); James F. Buchli (mission specialist); Guion S. Bluford, Jr. (mission specialist); Reinhard Furrer (payload specialist from Germany); Ernst Messerschmid (payload specialist from Germany); and Wubbo J. Ockels (payload specialist from the European Space Agency). The *Spacelab D-1* encompassed 75 numbered experiments, involving materials science in microgravity, life sciences, space technology, and communications technology. After seven days in space, the *Challenger* touched down at Edwards Air Force Base in California. The STS 61-A/Spacelab D-1 mission was the first American human spaceflight mission in which the primary payload was sponsored by another country, namely (West) Germany. The eight-person crew was the largest shuttle crew up to that date.

STS 51-L MISSION—*CHALLENGER* ACCIDENT

The space shuttle *Challenger* lifted off Pad B, Complex 39, at the Kennedy Space Center in Florida at 11:38 A.M. (EST) on January 28, 1986, on the STS 51-L mission. At just under 74 seconds into the flight, an explosion occurred, causing the loss of the vehicle and its entire crew, consisting of astronauts Francis R. (Dick) Scobee (commander), Michael John Smith (pilot), Ellison S. Onizuka (mission specialist), Judith Arlene Resnik (mission specialist), Ronald Erwin McNair (mission specialist), S. Christa McAuliffe (payload specialist), and Gregory Bruce Jarvis (payload specialist). Christa McAuliffe was a schoolteacher from New Hampshire who was flying on board *Challenger* as part of NASA's Teacher-in-Space Project, and Gregory Jarvis was an engineer representing the Hughes Aircraft Company. The other five were members of NASA's career astronaut corps.

In response to this tragic event, President Ronald Reagan appointed an independent commission, the Presidential Commission on the Space Shuttle *Challenger* Accident. The commission was composed of people not connected with the STS 51-L mission and was charged to investigate the accident fully and to report their findings and recommendations back to the president.

The consensus of the presidential commission and participating investigative agencies was that the loss of the space shuttle *Challenger* and its crew was caused by a failure in the joint between the two lower segments of the right solid rocket booster (SRB) motor. The specific failure was the destruction of the seals (O-rings) that were intended to prevent hot gases from leaking through the joint during the propellant burn of the SRB. The commission further suggested that this joint failure was due to a faulty design that was unacceptably sensitive to a number of factors. These factors included the effects of temperature, physical dimensions, the character of materials, the effects of reusability, processing, and the reaction of the joint to dynamic loading.

The commission also found that the decision to launch the *Challenger* on that particular day was flawed and that this represented a contributing cause of the accident. (Launch day for the STS 51-L mission was an unseasonably cold day in Florida.) Those who made the decision to launch were unaware of the recent history of problems concerning the O-rings and the joint. They were also unaware of the initial written recommendation of the contractor advising against launch at temperatures below 53°F (11.7°C) and of the continuing opposition of the engineers at Thiokol (the manufacturer of the solid rocket motors) after the management reversed its position. Nor did the decision makers have a clear understanding of the concern at Rockwell (the main NASA shuttle contractor, builder of the orbiter vehicle) that it was not safe to launch because of the ice on the launch pad. The commission concluded that if the decision makers had known all of these facts, it is highly unlikely that they would have decided to launch the STS 51-L mission on January 28, 1986.

From an historic perspective, the primary goal planned for the STS 51-L mission was the deployment of NASA's *Tracking Data Relay Satellite 2.* The schoolteacher-astronaut, Christa McAuliffe, was to conduct a set of lessons from orbit as part of the Teacher-in-Space Project. The other payload specialist, astronaut-engineer Gregory Jarvis, was to perform microgravity experiments of potential commercial value.

The *Challenger* explosion burst the apparent level of complacency that seemed to have crept into the NASA flight-management infrastructure. The large number of previous, successful shuttle flights made spaceflight look almost routine—a task to be carried out with almost "airline-like" regularity. Nothing could be further from the truth! Riding a powerful, chemically fueled rocket into space was then and remains today a hazardous and dangerous undertaking. As a result of this tragic accident, NASA suspended all shuttle flights until detailed investigations were concluded and safety improvements made. The STS 51-L disaster also caused a fundamental change in NASA's space shuttle program. From that day forward, the shuttle would no longer be used to place commercial satellites into orbit.

STS-26 MISSION

When *Discovery* returned the shuttle fleet to space on September 26, 1988—following the *Challenger* accident—more than 200 safety improvements and modifications were ushered in. The improvements in the space shuttle included a major redesign of the solid rocket boosters, the addition of a crew escape and bailout system, stronger landing gear, more powerful flight control computers, updated inertial navigation equipment, and several updated avionics units.

The STS-26 mission had the following crew: Frederick H. Hauck (commander), Richard O. Covey (pilot), John M. Lounge (mission specialist), David C. Hilmers (mission specialist), and George D. Nelson (mission specialist). This conservative four-day mission with an all-veteran astronaut crew marked the return of the shuttle fleet to spaceflight following the STS 51-L disaster in January 1986. On the first day in orbit, the crew successfully deployed the primary payload, NASA's *Tracking and Data Relay Satellite 3* (*TDRS-3*), with its attached inertial upper stage (IUS) rocket vehicle. After deployment in low Earth orbit, the IUS propelled the *TDRS-3* spacecraft to its operational, geostationary orbit. During launch and entry, the astronauts wore new partial-pressure flight suits. On October 3, during its

64th revolution of Earth, *Discovery* reentered the atmosphere and touched down safely at Edwards Air Force Base in California.

STS–31 MISSION

The STS-31 mission involved the deployment of the first of NASA's four "Great Observatories," the *Hubble Space Telescope* (*HST*)—a space-based instrument that helped revolutionize modern optical astronomy. The astronaut crew included Loren J. Shriver (commander); Charles F. Bolden, Jr. (pilot); Steven A. Hawley (mission specialist); Bruce McCandless II (mission specialist); and Kathryn D. Sullivan (mission specialist). *Discovery* lifted off Launch Pad 39A at the Kennedy Space Center on April 24, 1990, and deployed the *HST* into a 380-mile- (611-km-) altitude orbit around Earth.

The 24,200-pound (11,000-kg) free-flying astronomical observatory is 43 feet (13.1 m) long and has a diameter of 14 feet (4.27 m). Soon after the *HST* began operating in space, astronomers on Earth noticed a spherical aberration in the telescope's optical system. A practical solution was found, and the STS-61 mission of *Endeavour* (December 1993) became the first on-orbit servicing and repair mission involving this important telescope.

One of the interesting secondary payloads of the STS-31 mission was an IMAX Cargo Bay camera, which was used to document operations outside the crew cabin. There was also a handheld IMAX camera for use inside the crew cabin. The crew measured the radiation level inside the shuttle's pressurized cabin using the radiation monitoring equipment instrument. This set of data was especially interesting because the *HST* deployment mission involved the highest orbital altitude used by a shuttle vehicle to date.

On April 29, *Discovery* touched down safely at Edwards Air Force Base in California. By the end of the mission, the orbiter had completed 80 revolutions of Earth, and the crew had logged five days, one hour, and 16 minutes of space travel. Finally, this mission demonstrated the first use of carbon brakes at landing. After touchdown, the orbiter vehicle rolled out a distance of 8,889 feet (2,710 m) before stopping.

STS–49 MISSION

The STS-49 mission witnessed the maiden flight of the shuttle *Endeavour*. Launched on May 7, 1992, from Pad 39B at the Kennedy Space Center, *Endeavour*'s first flight marked the debut of many shuttle improvements, including a drag chute to assist braking during landing, improved nose-wheel steering, lighter and more reliable hydraulic power units, and updates to a variety of avionics equipment.

The crew consisted of Daniel C. Brandenstein (commander), Kevin P. Chilton (pilot), Bruce E. Melnick (mission specialist), Thomas D. Akers (mission specialist), Richard J. Hieb (mission specialist), Kathryn C. Thornton (mission specialist), and Pierre J. Thuot (mission specialist). This mission was the first ever American spaceflight that featured four extravehicular activities (EVAs)—one of which involved three space suited crew members simultaneously working outside the shuttle's pressurized cabin. The crew successfully captured and redeployed the *INTELSAT VI* (*F-3*) spacecraft, which had been stranded in an unusable orbit since its March 1990 launch by a Titan III expendable rocket vehicle from Cape Canaveral. Satellite capture and rescue operations required several days and three separate EVAs. The third EVA was an unprecedented three-person operation, in which astronauts Hieb, Thuot, and Akers spent about eight hours and 29 minutes grappling with the large 38.4-foot- (11.7-m-) tall and 11.8-foot- (3.6-m-) diameter communications satellite. With the help of their fellow astronauts, who operated the shuttle's remote manipulator system, the three astronauts placed the 5,630-pound (2,560-kg) *INTELSAT VI* (*F-3*) atop an awaiting perigee kick motor (PKM) in the open cargo bay. The crew's collective efforts marked the first time astronauts attached a live rocket to an orbiting satellite.

On the eighth day of the STS-49 mission, the astronauts used the *Endeavour*'s remote manipulator system to deploy the communications satellite and its newly attached rocket motor from the cargo bay. When *Endeavour* was a safe distance away, ground controllers for the International Telecommunications Satellite (INTELSAT) Corporation sent a signal for the PKM to fire. The PKM burn sent the rescued communications satellite to geostationary orbit. Hard work by the STS-49 crew salvaged a valuable communications satellite that could accommodate 120,000 phone calls and three television channels simultaneously. *INTELSAT VI* (*F-3*) was built by Hughes Space and Communications Group (now part of the Boeing Company) for INTELSAT and represented a new series of spacecraft that were the largest communications satellites flown to date.

On May 16, during its 141st revolution of Earth, *Endeavour* reentered the atmosphere and landed at Edwards Air Force Base in California. The seven astronauts of the STS-49 mission had been in space for eight days and 21 hours.

STS-61 MISSION

NASA launched *Endeavour* on December 2, 1993, on the first *Hubble Space Telescope* (*HST*) servicing mission. The crew included Richard O. Covey (commander), Kenneth D. Bowersox (pilot), F. Story Musgrave (mission specialist), Jeffrey A. Hoffman (mission specialist), Kathryn C. Thornton (mission specialist), Tom Akers (mission specialist), and Claude Nicollier (a Swiss mission specialist representing the European Space Agency). The STS-61 was one of the most complex and challenging human space missions ever attempted by the United States. During a record-setting five back-to-back extravehicular activities (EVAs), totaling 35 hours and 28 minutes, two teams of spacewalking astronauts completed the first servicing of *HST*, and (on the fourth EVA) astronauts Thornton and Akers installed the corrective optics space telescope axial replacement (COSTAR) unit, which was designed to redirect incoming light to three of four remaining *HST* instruments to compensate for a flaw in the primary mirror of the telescope. The crew redeployed the *HST* on flight day nine, allowing the refurbished optical telescope to continue its astronomical mission in a vastly improved way.

After 10 days, 19 hours, and almost 59 minutes in space, *Endeavour* and its crew returned safely to Earth during orbital revolution 163—touching down at the Kennedy Space Center on December 13, 1993.

STS-71 MISSION

NASA's STS-71 mission marked a number of historic firsts in human spaceflight. First, the liftoff of *Atlantis* on June 27, 1995, from Pad 39A at the Kennedy Space Center was the 100th manned spaceflight conducted by the United States from Cape Canaveral. Next, this mission was the first U.S. space shuttle–Russian *Mir* space station docking and joint on-orbit operations mission—performed as Phase I of the *International Space Station* program. Third, when *Atlantis* docked with *Mir*, the joined spacecraft formed the largest human-made object ever flown in orbit (up to that time). Finally, the STS-71 mission supported the first on-orbit change-out of shuttle crewmembers and passengers.

When *Atlantis* ascended into orbit, the space vehicle carried the following NASA crewmembers: Robert L. "Hoot" Gibson (STS-71 mission commander), Charles J. Precourt (pilot), Ellen S. Baker (payload commander), Gregory J. Harbaugh (mission specialist), and Bonnie J. Dunbar (mission specialist). In addition, two Russian cosmonauts also rode up to the *Mir* space station aboard *Atlantis*. They were Anatoliy Y. Solovyev (Mir Expedition 19 commander) and Nikolai M. Budarin. On its return trip to Earth, *Atlantis* carried the members of Mir Expedition 18—namely, cosmonaut Vladimir N. Dezhurov (Mir 18 commander), cosmonaut Gennady M. Strekalov (Mir 18 flight engineer), and astronaut Norman E. Thagard (Mir 18 cosmonaut-researcher).

Atlantis docked with *Mir* at 9:00 A.M. (EST) on June 29 about 249 miles (400 km) above the Lake Baikal region of the Russian Federation. After hatches on each side of the docking mechanisms opened, the STS-71 crew passed into *Mir* for the welcoming ceremony. The physical link between the two spacecraft consisted of the orbiter docking system (ODS) on *Atlantis* and the androgynous peripheral docking system on *Mir*.

On the same day, the Mir 18 crew officially transferred responsibility for the space station to the shuttle-delivered Mir 19 crew, and the two *Mir* crews subsequently transferred spacecraft. For the next five days, joint U.S.-Russian space operations were performed, including biomedical studies. The astronauts and cosmonauts also transferred equipment to and from *Mir*. The two spacecraft undocked on July 4. At the end of a nine-day, 19-hour, and 22-minute mission in space, *Atlantis* returned to Earth during revolution 153, landing at the Kennedy Space Center on July 7. To ease their reentry into Earth's gravitational environment after spending more than 100 days on *Mir*, the Mir 18 crew (Thagard, Dezhurov, and Strekalov) were able to lay supine in custom-made Russian seats that were installed on *Atlantis* in the orbiter mid-deck area just prior to the initiation of the reentry sequence.

STS-88 MISSION

On December 4, 1998, NASA launched *Endeavour* from Pad 39A to start the STS-88 mission—the first on-orbit assembly mission of the *International Space Station* (*ISS*). The crew consisted of Robert D. Cabana (commander), Frederick W. "Rick" Sturckow (pilot), Nancy J. Currie (mission specialist), Jerry L. Ross (mission specialist), James H. Newman (mission

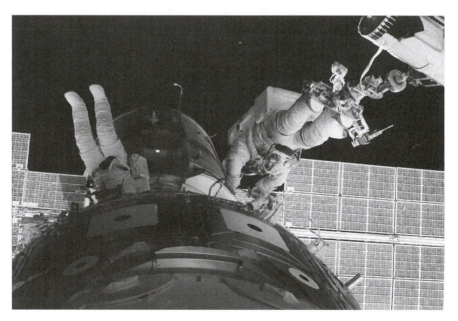

NASA astronauts James H. Newman (left) and Jerry L. Ross (right) work between the *Zarya* and *Unity* (foreground) modules during the first of three space walks on the STS-88 mission of the space shuttle *Endeavour* (December 1998). Newman is tethered to the module, while Ross is anchored at the feet to a mobile foot restraint mounted on the end of the shuttle's remote manipulator system arm. This was the first assembly mission for the *International Space Station*. (*NASA/JSC*)

specialist), and Sergei Konstantinovich Krikalev (Russian cosmonaut and mission specialist).

The primary goal of the 12-day STS-88 mission was to begin assembly of the *ISS*, and the crew met all mission objectives. On December 5, the crew joined the 12.8-ton American-built Unity connecting module to *Endeavour*'s docking system. The following day (December 6), the astronauts used *Endeavour*'s remote manipulator system to capture the Russian-built *Zarya* control module from its co-orbital location and carefully mated the module to *Unity*. Then, astronauts Ross and Newman conducted three space walks to attach cable, connectors, and handrails. In addition to connecting the first two components of the *ISS*, Ross and Newman also used the extravehicular activities (EVAs) to test the new simplified aid for EVA rescue (SAFER) unit—a self-rescue device to be used if a spacewalker became separated from the spacecraft during an EVA.

On December 10, astronaut Cabana and cosmonaut Krikalev floated together from *Endeavour* through the docking mechanism into the *Unity* module portion of the embryonic station. The rest of the crew soon followed. The historic event marked the first human occupancy of the *ISS*. Almost three hours later, Cabana and Krikalev opened the hatch to *Zarya* (now attached to *Unity*) and entered to inspect the interior of the Russian-built module. They powered-up various systems and made sure that everything was as it should be to support future human visits. On December 11, Cabana and Krikalev closed the hatches to *Zarya* and *Unity*, respectively, as they returned to *Endeavour*. As shuttle pilot Rick Sturckow separated *Endeavour* from the station on December 13, the *ISS* began to fly free in space—marking the start of a new era in human spaceflight.

During the STS-88 mission, *Endeavour* completed 185 revolutions of Earth. When the orbiter vehicle touched down at the Kennedy Space Center on December 15, the crew had traveled 11 days, 19 hours, and almost 19 minutes in space.

STS-107 MISSION—*COLUMBIA* ACCIDENT

On January 16, 2003, the space shuttle *Columbia* took off from Pad 39A at the Kennedy Space Center at the start of the STS-107 mission. The primary objective of this 16-day mission was to have the crew perform a variety of science experiments on an essentially 24-hour-a-day basis. Many of the experiments were located in the SPACEHAB commercial experiment module carried in *Columbia*'s cargo bay.

The STS-107 crewmembers included Rick Husband (commander), Willie McCool (pilot), Michael Anderson (payload commander), Kalpana Chawla (mission specialist), David Brown (mission specialist), Laurel Clark (mission specialist), and Ilan Ramon (payload specialist from Israel). Things went exceptionally well during the orbital flight portion

The crew of the space shuttle *Columbia* who perished on February 1, 2003, when the orbiter vehicle disintegrated during reentry operations at the end of the STS–107 mission. They are (left to right): front row, American astronauts Rick Husband, Kalpana Chawla, and William McCool; back row, David Brown, Laurel Clark, Michael Anderson, and Israeli astronaut Ilan Ramon. *(NASA)*

of this mission, and a great deal of useful scientific experimentation was performed. Part of the international dimension of this successful scientific mission was the presence of payload specialist Ramon—the first Israeli citizen to fly in orbit around Earth.

Then, while the orbiter vehicle was gliding back to Earth on February 1, 2003, disaster struck. When *Columbia* was just 16 minutes from its scheduled landing site at the Kennedy Space Center, the orbiter disintegrated in flight, taking the lives of all seven crew persons. Postaccident investigations indicate that a severe heating problem occurred in *Columbia*'s left wing as a result of structural damage from debris impact during launch. Specifically, a debris strike is suspected of having damaged the important reinforced carbon-carbon thermal protection system tiles on the leading edge of the *Columbia*'s left wing. Hot gases generated during reentry penetrated this damaged area into the interior of the wing, destroying it and causing the orbiter vehicle to lose aerodynamic stability and disintegrate during high-speed flight over eastern portions of Texas. Nearly 85,000 pieces of orbiter debris were recovered and shipped to NASA's debris reconstruction team at Kennedy Space Center. In performing a detailed, postaccident analysis of the recovered debris, NASA personnel were able to reconstruct about 38 percent of the *Columbia.* Some of the recovered science experiments could still provide valuable data.

As a result of the threat of debris falling from the external tank during liftoff, NASA officials immediately suspended all shuttle missions and pursued design changes in the external tank, implemented new operational procedures, and instituted on-orbit inspection and repair techniques—all focused on avoiding another *Columbia*-type tragedy during reentry. These changes influenced both the STS-114 mission (July 2005) and the STS-121 mission (July 2006), and will continue to influence all remaining shuttle launches until the current orbiter fleet (*Discovery, Atlantis,* and *Endeavour*) is retired in 2010.

STS-114 MISSION

On July 26, 2005, NASA launched the shuttle *Discovery* on the STS-114 mission, also known as the Return to Flight mission. STS-114 was the first shuttle mission since the loss of the *Columbia* on February 1, 2003. NASA managers and engineers spent two and one-half years researching and implementing safety improvements for the remaining three orbiters. The efforts included greater in-depth examination of the reinforced carbon-carbon thermal protection system panels used on the leading edges of the orbiter vehicle's wings, plus replacing bolts and instituting new foam application procedures for the tanks. A great deal of concern surrounded this launch, and it became the most documented liftoff in the history of the shuttle program. *Discovery*'s climb to orbit was extensively documented through a system of new and upgraded ground-based cameras, radar systems, and airborne cameras flown on high-altitude aircraft.

The crew for the STS-114 mission consisted of Eileen Collins (commander), James Kelly (pilot), Charles Camarda (mission specialist), Wendy Lawrence (mission specialist), Stephen Robinson (mission specialist), Andrew Thomas (mission specialist), and Soichi Noguchi (mission specialist from the Japan Aerospace Exploration Agency).

The large quantity of imagery collected during *Discovery*'s launch and additional imagery provided by the laser systems on *Discovery*'s new orbiter boom sensor system (OBSS) laser scanner, as well as data from sensors embedded in the shuttle's wings, helped mission managers determine the status and condition of *Discovery*'s thermal protection system. The renewed cause for concern was foam shedding and a possible lethal debris strike. Imagery during launch showed a piece of foam being shed from the external tank, as well as smaller tile and foam dings. Imagery also showed that there were two areas in the thermal protection system on the underside of *Discovery* where gap fillers were protruding. As *Discovery* headed for the planned rendezvous and docking with the *International Space Station* (*ISS*), NASA mission managers were busy assessing whether

The space agency's prolonged flight moratorium after the *Columbia* accident ended on July 26, 2005, when NASA started the STS-114 mission to the *International Space Station* by successfully launching the *Discovery* from the Kennedy Space Center. *(NASA)*

the orbiter vehicle could safely reenter Earth's atmosphere. If not, the alternative was to have the crew seek refuge in the *ISS* while a rescue mission (most likely involving *Atlantis*) was mounted.

Once reaching orbit and prior to arriving at the *ISS*, *Discovery*'s mission specialists used the new Canadian-built OBSS to inspect the vehicle for any tile damage. The OBSS contains two types of lasers and a high-resolution television camera, mounted at the end of the boom. Data from this system provided the astronauts a quick assessment of the thermal protection system state of health. No major problems were initially detected, but two protruding gap fillers and a puffed-out piece of thermal blanket near the cockpit caused some initial concern.

When *Discovery* neared the *ISS*, mission commander Collins performed the first ever rendezvous pitch maneuver about 600 feet (183 m) below the station. As the shuttle slowly flipped end over end (at a gentle rate of just 0.75 degree per second), station commander Sergei Krikalev and *ISS* flight engineer John Phillips used digital cameras with high-powered 800-mm and 400-mm lenses to photograph *Discovery*'s thermal protective tiles and key areas around the vehicle's nose and main landing gear doors. All the digital imagery collected by *ISS* crew were then transmitted to the team of NASA engineers and flight support personnel on the ground.

The *Discovery* then docked with the *ISS*, and the STS-114 crew set about the business of visiting the station and making deliveries of much needed supplies and equipment. Prior to the first of three extravehicular activities (EVAs) by the crew of *Discovery*, mission specialist Lawrence and shuttle pilot Kelly guided the station's robotic arm (called Canadarm2) to lift the multipurpose logistics module (MPLM) Raffaello from *Discovery*'s cargo bay for attachment to the *Unity* module.

The MPLM Raffaello was built by the Italian Space Agency and designed for transport back and forth to the *ISS* by the shuttle. Raffaello was equipped with 11 racks containing supplies, hardware, equipment and the Human Research Facility 2 (HRF-2) rack for transfer to the station. HRF-2 contains a refrigerated centrifuge that can separate biological substances of differing densities. Once Raffaello was unstowed from the shuttle and berthed with the *ISS*, the astronauts transferred its supplies and the HRF-2 into the station. They then placed 7,055 pounds (3,207 kg) of old equipment, unneeded hardware, and trash inside Raffaello, which was returned to Earth stowed inside the *Discovery*'s cargo bay.

To prepare the shuttle for its return trip to Earth, astronaut Robinson performed the STS-114 mission's third EVA—a six-hour space walk. During this daring EVA, he traveled while attached to the space station's Canadarm2 to the site on the underside of *Discovery* where the two gap fillers were protruding from between the shuttle's thermal protection tiles. He gently pulled them out.

Discovery undocked from the space station and headed home. After 13 days, 32 hours, and almost 33 minutes in space, the vehicle and its crew

The seven astronauts who flew on the successful STS–114 mission of the space shuttle *Discovery* in July 2005. They are (in front row, left to right): James M. Kelly (pilot), Wendy B. Lawrence (mission specialist), and Eileen M. Collins (commander). In the back row are (left to right): Stephen K. Robinson, Andrew S. W. Thomas, Charles J. Camarda, and Soichi Noguchi (all mission specialists). Noguchi represents the Japanese Aerospace Exploration Agency. *(NASA)*

safely landed at Edwards Air Force Base in California on August 9, 2005. Despite the safe return of *Discovery,* NASA officials again grounded the shuttle fleet. The overall concern was that this persistent foam-shedding problem of the external tank had to be more effectively resolved before another crew would be placed at risk by launch ascent debris.

STS-121 MISSION

The launch of *Discovery* and its seven-member crew on July 4, 2006, marked the second mission in NASA's Return to Flight sequence following the *Columbia* accident. During this 13-day mission, the crew tested new equipment and procedures that increased the safety of the operational shuttle fleet. The crew included Steve Lindsey (commander), Mark Kelly (pilot), Piers Sellers (mission specialist), Mike Fossum (mission specialist), Lisa Nowak (mission specialist), Stephanie Wilson (mission specialist), and Thomas Reiter (German mission specialist representing the European Space Agency). Astronaut Reiter rode *Discovery* into orbit and, following docking with the *International Space Station* (*ISS*), transferred to the station and became the third member of *ISS* Expedition 13. The STS-121 mission built upon the analyses of the safety improvements debuted in the STS-114 mission (June 2005).

Discovery docked with the *ISS* and delivered more than 28,000 pounds (12,725 kg) of equipment and supplies. When the shuttle landed at the Kennedy Space Center on July 17, the crew (save for Reiter who remained on the station) had logged 12 days, 18 hours, and 38 minutes in space.

✧ Beyond the Space Shuttle

Despite the tragic losses of the *Challenger* and its seven-person crew in 1986 (during the STS 51-L mission) and the *Columbia* and its seven-person crew in 2003 (during the STS-107 mission), the space shuttle program has accumulated more than 100 successful missions and played an important role in human spaceflight by the United States. However, following completion of the *International Space Station,* NASA plans to retire the aging orbiter fleet—*Discovery, Atlantis,* and *Endeavour*—in about 2010. On September 19, 2005, Michael Griffin, the NASA administrator, introduced a new spacecraft, called *Orion,* or the crew excursion vehicle (CEV), which will be designed to carry four astronauts to and from the Moon, to support up to six crewmembers on future missions to Mars, and to deliver crew and supplies to the *International Space Station.*

On June 30, 2006, NASA announced the names of the next generation of launch vehicles capable of returning humans to the Moon and eventually sending the first team of explorers to Mars. The crew launch vehicle is called Ares I, and the cargo launch vehicle is called Ares V. NASA officials selected the name "Ares" because it is the ancient Greek name for Mars. Therefore, the names given to the new launch vehicles also incorporate their long-term, visionary mission. The use of the Roman numerals "I" and "V" pays homage to the Apollo Project's spectacularly successful Saturn I and Saturn V rockets—the first large U.S. launch vehicles conceived and developed to specifically support human spaceflight.

Between 2014 and 2015, the crew exploration vehicle will succeed the shuttle orbiter as NASA's spacecraft for human space exploration. Ares I will be an in-line rocket configuration topped by the CEV, its service module, and a launch-abort system. The launch vehicle's first stage will be a single, five-segment reusable solid rocket booster derived from the space shuttle program's reusable solid rocket motor and will burn a specially formulated and shaped solid propellant called polybutadiene acrylonitride. NASA engineers are now designing a new forward adapter that will mate the Ares I vehicle's first stage to the vehicle's second stage. This forward adapter will be equipped with booster separation motors to disconnect the stages during ascent.

This is an artist's rendering of NASA's new crew launch vehicle, called Ares I, taking off from the Kennedy Space Center in Florida, circa 2015. The two-stage rocket is shown topped by the crew exploration vehicle, its service module, and a launch abort system. *(NASA/MSFC)*

The second or upper stage of the Ares I vehicle is an entirely new element, propelled by a J-2X main engine that is fueled by liquid hydrogen (LH_2) and liquid oxygen (LO_2). The J-2X engine is an evolved variation of two historic rocket engine predecessors: the powerful J-2 upper stage engine of the Saturn IB and Saturn V rockets that propelled the Apollo Project–era spacecraft to the Moon and the J-2S engine. The J-2S engine is a simplified version of the J-2 engine that NASA rocket engineers developed and flight-tested in the early 1970s. But, unlike the highly successful J-2 engine, the J-2S engine was never flown in space as part of a human-crewed mission.

This is an artist's rendering of NASA's new cargo launch vehicle, called the Ares V. Starting in about 2015, the Ares V will serve as NASA's primary launch vehicle for safe, reliable delivery of resources to space—from the hardware and materials for establishing a permanent base on the Moon, to fresh food, water, and other supplies needed to extend a human presence beyond Earth orbit. *(NASA/MSFC)*

The primary mission of the Ares I will be to carry crews of four to six astronauts into orbit around Earth. During the first two and one-half minutes of flight, the first-stage boosters will power the vehicle to an altitude of about 200,000 feet (61,000 m). After all the rocket's solid propellant is consumed, the reusable booster will separate and the liquid-propellant upper-stage J-2X engine will ignite, powering the crew exploration vehicle to an altitude of about 63 miles (101 km). Then, the vehicle's upper stage separates and the CEV's service module propulsion system completes the spacecraft's trip into space—eventually placing the CEV into a 185-mile-

(298-km-) altitude, circular orbit around Earth. Once in orbit, the CEV and its service module can rendezvous and dock with either the space station or with a combined lunar lander and Earth departure stage vehicle that will then take the astronauts to the Moon. NASA plans to send an operational CEV to the space station by no later than 2015. The first human-crewed lunar excursion mission with the CEV is scheduled for around 2020. Since the Ares I will be able to lift more than 55,000 pounds (25,000 kg) into low Earth orbit, NASA may also use the new launch vehicle to deliver priority resources and supplies to the *International Space Station.*

Ares V will be a heavy-lift vehicle, the first stage of which has five RS-68 liquid-hydrogen/liquid-oxygen engines mounted below a large version of the space shuttle's external tank and two five-segment solid-propellant rocket boosters. The RS-68 rocket engine will be an upgraded version of the engines currently used in the Delta IV rocket—a powerful vehicle developed by Boeing Corporation in the 1990s for the U.S. Air Force's evolved expendable launch vehicle program and for commercial applications. The upper stage of the Ares V cargo vehicle will use the same J-2X engine as employed in the Ares I vehicle. The Ares V vehicle will stand approximately 360 feet (110 m) tall on the launch pad and will be capable of placing 286,000 pounds (130,000 kg) into low Earth orbit or sending 144,000 pounds (65,455 kg) to the Moon. NASA plans to use the Ares V as its primary rocket vehicle for the safe, reliable delivery of resources to space—from large-scale hardware and materials for establishing a permanent base on the Moon, to fresh food, water, and other staples needed to extend human presence beyond Earth orbit. Under present plans, the Ares V is not being designed by NASA to transport people into space.

Looking beyond the space shuttle, the Ares I and Ares V vehicles appear on the near-term technology horizon. These launch vehicles are vital components of the space transportation infrastructure of NASA's Constellation Program—a contemporary program to carry human explorers back to the Moon and then onward to Mars and other destinations in the solar system.

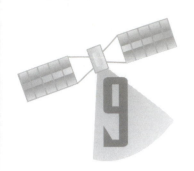

International Space Station

In January 1984, as part of his State of the Union address, President Ronald Reagan called for a space station program that would include participation by countries allied with the United States. With this presidential mandate, NASA established a Space Station Program Office in April 1984 and requested proposals from the American aerospace industry. By March 1986, the baseline design was the dual keel configuration, a rectangular framework with a truss across the middle for holding the station's living and working modules and solar arrays.

Japan, Canada, and the European Space Agency (ESA) each signed a bilateral memorandum of understanding in the spring of 1985 with the United States, agreeing to participate in the space station project. In 1987, the station's dual keel configuration was revised to compensate for a reduced space shuttle flight rate in the wake of the *Challenger* accident on January 28, 1986. (See chapter 8.) The revised baseline had a single truss with the built-in option to upgrade to the dual keel design. The need for a space station "lifeboat," called the *assured crew return vehicle,* was also identified.

In 1988, President Reagan named the proposed space station *Freedom.* With each annual budget cycle, *Freedom*'s design underwent modifications as the United States Congress called for reductions in its cost. The truss was shortened, and the planned U.S.-built habitation module and laboratory module were both reduced in size. The truss was to be launched in sections with subsystems already in place. Despite these redesign efforts, NASA and its contractors were able to produce a substantial amount of hardware.

A major change in the direction of the space station came with the end of the cold war. Quite dramatically, starting in about 1989, cooperation in space exploration between the United States and the new Russian

Federation became the norm rather than the exception, as exemplified by the Apollo-Soyuz Test Project of July 1975.

One important step to international cooperation took place in 1992, when the United States agreed to purchase Russian Soyuz spacecraft to serve as *Freedom*'s lifeboat. This action heralded the greatly increased level of cooperation between the United States and Russia that enabled the development of a truly international space station. Another important activity, the space shuttle–*Mir* program (later called Phase I of the *International Space Station* (*ISS*) program) also began that year. The shuttle-*Mir* program used existing assets (primarily U.S. space shuttle orbiter vehicles and the Russian *Mir* space station) to provide the joint operational experience and to perform the joint research that would eventually lead to the successful construction and operation of the *ISS.* (See chapter 8.)

In 1993, President Bill Clinton called for *Freedom* to be redesigned once again to reduce costs and to include more international involvement. The White House staff selected a design option that was called *Space Station Alpha*—a downsized configuration that would use about 75 percent of

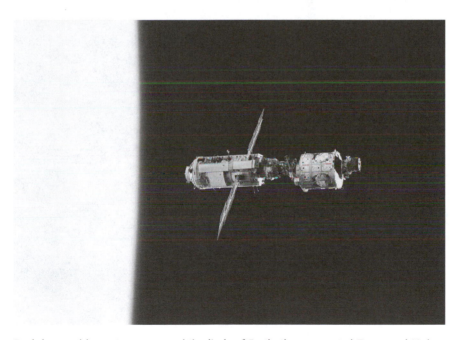

Backdropped by outer space and the limb of Earth, the connected *Zarya* and *Unity* modules float majestically in orbit after having been released from the *Endeavour*'s cargo bay on December 13, 1998. The six-crew members of the STS-88 mission, who had earlier spent the majority of their on-duty mission time working on the tandem of space hardware, watched from *Endeavour* as the shuttle slowly moved away from the embryonic *International Space Station*. *(NASA/JSC)*

the hardware designs originally intended for *Freedom*. After the Russians agreed to supply major hardware elements (many of which were originally intended for a planned *Mir 2* space station), *Space Station Alpha* became officially known as the *International Space Station*.

The *International Space Station* program was divided into three basic phases: Phase I (an expansion of the shuttle-*Mir* docking mission program) provided U.S. and Russian aerospace engineers, flight controllers, and cosmonauts and astronauts the valuable experience needed to cooperatively assemble and build the *ISS*. Phase I officially began in 1995 and involved more than two years of continuous stays by a total of seven American astronauts aboard the Russian *Mir* space station and nine shuttle-*Mir* docking missions. This phase of the program ended in June 1998 with the successful completion of the STS-91 mission—a mission in which the space shuttle *Discovery* docked with *Mir* and "downloaded" (that is, returned to Earth) astronaut Andrew Thomas, the last American occupant of the Russian space station. Phases II and III involve the in-orbit assembly of the station's components—Phase II the core of the *ISS* and Phase III its various scientific modules.

The crew of the space shuttle *Discovery* snapped this panoramic image of the *International Space Station* (*ISS*) on July 15, 2006, just after the two spacecraft had undocked. Following its rendezvous and docking with the *ISS* as part of the STS-121 mission, *Discovery* successfully returned to Earth, landing at the Kennedy Space Center on July 17. *(NASA)*

An historic moment in aerospace history took place near the end of the 20th century. On December 10, 1998, STS-88 shuttle mission commander Robert Cabana and Russian cosmonaut and mission specialist Sergei Krikalev swung open the hatch between the shuttle *Endeavour* and the first element of the *ISS*. With this action, the STS-88 astronauts completed the first steps in the orbital construction of the *ISS*. In late November 1998, a Russian Proton rocket had successfully placed the NASA-owned, Russian-built *Zarya* ("Sunrise") control module into a perfect parking orbit. A few days later, in early December, the shuttle *Endeavour* carried the American-built *Unity* connecting module into orbit for rendezvous with *Zarya*. Astronauts Jerry Ross and James Newman then performed three arduous extravehicular activities (totaling 21 hours and 22 minutes) to complete the initial assembly of the space station. When their spacewalking efforts were complete, Cabana and Krikalev were literally able to open the door of an important new space station era. As predicted by Konstantin Tsiolkovsky at the start of the 20th century, humankind will soon leave the "cradle of Earth" and build permanent outposts in space on the road to the stars.

The *Columbia* accident in February 2003 has caused a significant delay in the completion of the station. Assuming the remaining shuttle orbiter fleet (*Discovery, Atlantis,* and *Endeavour*) can continue to fly orbital missions, NASA officials project the year 2010 for the completion of the station and the retirement of the orbiter fleet. The STS-121 mission of *Discovery* to the *ISS* in July 2006 is considered an important milestone in that overall schedule.

✧ Basic Facts about the *International Space Station*

The *International Space Station* (*ISS*) is a major human spaceflight project headed by NASA. Russia, Canada, Europe, Japan, and Brazil are also contributing key elements to this large, modular space station in low Earth orbit that represents a permanent human outpost in outer space for microgravity research and advanced space technology demonstrations. On-orbit assembly began in December 1998, with completion originally anticipated by 2004.

However, the space shuttle *Columbia* accident that took place on February 1, 2003, killing the seven astronaut crewmembers and destroying the orbiter vehicle, exerted a major impact on the *ISS* schedule. Completion is now anticipated by 2010 to coincide with the retirement of the current fleet of NASA's three orbiter vehicles (*Discovery, Atlantis,* and *Endeavour*). Even though the space station is still a work in progress, the crews for the

This image shows the *International Space Station* as viewed by the crew of *Endeavour* as the orbiter departed from the station and performed a fly-around survey during the STS-100 mission (April 2001). Prominently appearing in the lower-central portion of the picture is Canadarm 2–the station's 55-foot- (16.8-m-) long robotic arm added during the *Endeavour*'s visit. *(NASA)*

first 14 expeditions have performed useful science as well as assisting in the on-orbit assembly and checkout activities.

To compensate for unanticipated delays in the space shuttle missions necessary to complete the *ISS*, Russia has provided launch vehicles and spacecraft (such as the Soyuz-*TMA* vehicle) to ferry crews and essential supplies to the station. Recent shuttle flights like the STS-114 mission (July 2005) and the STS-121 mission (July 2006) have provided much needed relief in terms of large-capacity cargo-hauling up to the station and the removal of expended equipment and accumulated trash from the station, keeping the facility provisioned and habitable.

The 68,420-pound (31,100-kg) Russian-built and American-financed *Zarya* ("Dawn") module was the first of numerous modules that make up the *International Space Station* (*ISS*) to be launched. The *ISS* travels around Earth in an approximately circular orbit at an altitude of 240 miles (386 km) with an inclination of 52.6 degrees. The first assembly step of the *ISS* occurred in late November and early December 1998. During a space shuttle–supported orbital assembly operation, astronauts linked *Zarya*, the initial control module, together with *Unity*, the American six-port habitable connection module. *Zarya* is also known as the *Functional Cargo Block*, or *FGM*, when the Russian equivalent acronym is transliterated. Because *Zarya* was the first-orbited element of the *ISS*, its international spacecraft identification (1998-067A) also serves as the spacecraft identification for the *International Space Station*.

The *Unity* module is the first U.S.-built component of the *International Space Station*. It is a six-sided connecting module and passageway (node). *Unity* was the primary cargo of the space shuttle *Endeavour* during the STS-88 mission in early December 1998. Once delivered into orbit, astronauts mated *Unity* to the Russian-built *Zarya* module—delivered earlier into orbit by a Russian Proton rocket that lifted off from the Baikonour Cosmodrome.

Zvezda ("Star") is the Russian service module for the *International Space Station*. The 20-ton module has three docking hatches and 14 win-

The *Soyuz TMA-1* spacecraft approaches the Pirs docking compartment of the *International Space Station* on November 1, 2002. The Russian spacecraft is carrying the *Soyuz 5* taxi crew: Russian commander Sergei Zalyotin, Belgian flight engineer Frank De Winne (representing the European Space Agency), and Russian flight engineer Yuri V. Lonchakov—the three of whom conducted an eight-day visit to the station. Among other engineered improvements, the *Soyuz TMA-1* spacecraft has upgraded computers, a new cockpit control panel, and improved avionics. *(NASA/JSC)*

dows. Launched by a Proton rocket from the Baikonur Cosmodrome on July 12, 2000, the module automatically docked with the *Zarya* module of the orbiting *ISS* complex on July 26. Prior to docking with the *ISS*, *Zvezda* bore the international spacecraft designation 2000-037A.

Once attached and functional, the module became an integral part of the *ISS* and began to serve as the living quarters for the astronaut and cosmonaut crews during the on-orbit assembly phase. The first *ISS* crew, called the Expedition One crew, began to occupy the *ISS* and live in *Zvezda* on November 2, 2000. American astronaut Bill Shepherd and Russian cosmonauts Yuri Gidzenko and Sergei Krikalev made the *Zvezda* module their extraterrestrial home until March 14, 2001. At that point, the Expedition Two crew—consisting of cosmonaut Yuri Usachev and astronauts Susan Helms and Jim Voss—replaced them as *Zvezda*'s occupants. In addition to supporting human habitation in space, this module also provides electrical power distribution, data processing, flight control, and on-orbit propulsion for the space station complex.

Destiny is the American-built laboratory module that was delivered to the *International Space Station* by the space shuttle *Atlantis* during the STS-98 mission (February 2001). *Destiny* is the primary research laboratory for U.S. payloads. The aluminum module is 28 feet (8.5 m) long and 14.1 feet

(4.3 m) in diameter. It consists of three cylindrical sections and two end cones with hatches that can be mated to other space station components. There is also a 20-inch- (0.51-m-) diameter window located on the side of the center segment. An exterior waffle pattern strengthens the module's hull, and its exterior is covered by a space debris shield blanket made up of material similar to that used in bulletproof vests worn by law enforcement personnel on Earth. A thin aluminum debris shield placed over the blanket provides additional protection against space debris and meteoroids.

NASA engineers designed the laboratory to hold sets of modular racks that can be added, removed, or replaced as necessary. The laboratory module includes a human research facility, a materials science research rack, a microgravity science glove box, a fluids and combustion facility, a window observational research facility, and a fundamental biology habitat holding rack. The laboratory racks contain fluid and electrical connectors, video equipment, sensors, controllers, and motion dampers to support whatever experiments are housed in them. *Destiny*'s window, which takes up the space of one rack, is an optical gem that allows space station crewmembers to shoot high-quality photographs and videos of Earth's ever-changing landscape. Imagery captured from this window gives scientists the opportunity to study features such as floods, glaciers, avalanches, plankton blooms, coral reefs, urban growth, and wild fires.

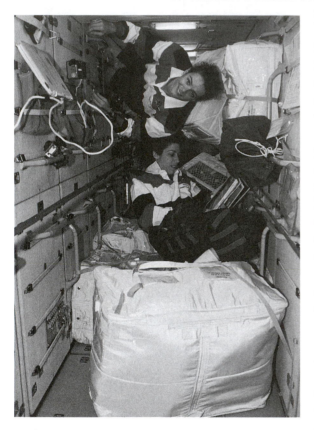

On board the Russian-built *Zarya* module, astronauts Julie Payette (top) and Ellen Ochoa (middle) handle a portion of the supplies that have been moved over from the docked space shuttle *Discovery* during the STS–96 mission (June 1999). Payette represents the Canadian Space Agency, and Ochoa is a career astronaut (mission specialist position) with NASA. *(NASA/JSC)*

As of December 31, 2006, the *ISS* has a habitable volume of 15,000 cubic feet (425 m^3), a mass of 393,700 pounds (178,594 kg), a surface of (solar arrays) 9,600 square feet (892 m^2), a width (across solar arrays) of 240 feet (73 m), a height of 90 feet (27.5 m), and a length of 146 feet (44.5 m) from the *Destiny* laboratory module to the *Zvezda* service module. The length increases to 170 feet (52 m) when a Russian *Progress* spacecraft is docked with the *Zvezda* module. Other modules and pieces of equipment, delivered by the shuttle, are necessary before on-orbit assembly of the station is completed and the system can become fully operational.

NASA plans to retire the space shuttle fleet in 2010, once the *ISS* is completely assembled and operational. From 2010 until about 2014,

This is an artist's rendering of NASA's planned *Orion* crew exploration vehicle arriving at the *International Space Station,* circa 2015. *(NASA/John Frassanito and Associates)*

Russian launch vehicles will continue to ferry crews and bring supplies to the station, by which point NASA's planned Ares I launch vehicle with the *Orion* spacecraft as its payload will begin to fly regular crew rotation and resupply missions to the *ISS.* The European Space Agency may also use its Ariane V launch vehicle to send payloads to the *ISS.*

✧ Space Station Crews

Sustained human exploration of space in this century begins with the *International Space Station.* The station starts a new era of permanent operations in space. NASA's experience during the shuttle-Mir program provided many answers to the question of how to sustain a long-term operation in space. Also, the astronauts and cosmonauts who serve on board the space station will increase humankind's knowledge of what it is like to live in space.

While the *International Space Station* is permanently crewed, the crews rotate during crew exchange flights. As an incoming crew prepares to replace the outgoing crew, there is a handover period. The current space station crewmembers communicate by telecon to the crew on Earth any unique situations they have encountered, new techniques learned, or any topic necessary for life aboard the space station. Once the new crewmembers arrive on board the space station, the outgoing crew briefs them on safety issues, vehicle changes, and payload operations. Listed below in chronological order are the expeditions to the *ISS* as of December 31, 2006.

Expedition One
Launch: 10/31/00
Land: 03/18/01
Time: 140 days, 23 hours, 28 minutes
Crew: Commander William Shepherd, Soyuz Commander Yuri Gidzenko, Flight Engineer Sergei Krikalev

Expedition Two
Launch: 03/08/01
Land: 08/22/01
Time: 167 days, six hours, 41 minutes
Crew: Commander Yury Usachev, Flight Engineer Susan Helms, Flight Engineer James Voss

Expedition Three
Launch: 08/10/01
Land: 12/17/01
Time: 128 days, 20 hours, 45 minutes
Crew: Commander Frank Culbertson, Soyuz Commander Vladimir Dezhurov, Flight Engineer Mikhail Tyurin

Expedition Four
Launch: 12/05/01
Land: 06/19/02
Time: 195 days, 19 hours, 39 minutes
Crew: Commander Yury Onufrienko, Flight Engineer Dan Bursch, Flight Engineer Carl Walz

Expedition Five
Launch: 06/06/02
Land: 12/07/02
Time: 184 days, 22 hours, 14 minutes
Crew: Commander Valery Korzun, Flight Engineer Peggy Whitson, Flight Engineer Sergei Treschev

Expedition Six
Launch: 11/23/02
Land: 05/03/03
Time: 161 days, 19 hours, 17 minutes
Crew: Commander Ken Bowersox, Flight Engineer Nikolai Budarin, Flight Engineer Don Pettit

Expedition Seven
Launch: 04/25/03

Land: 10/27/03
Time: 184 days, 21 hours, 47 minutes
Crew: Commander Yuri Malenchenko, Flight Engineer Ed Lu

Expedition Eight

Launch: 10/18/03

Land: 04/29/04

Time: 194 days, 18 hours, 35 minutes Crew: Commander Michael
   Foale, Flight Engineer Alexander Kaleri, Flight Engineer (ESA)
   Pedro Duque*

* European Space Agency astronaut Duque launched with Expedition 8
   crew on *Soyuz TMA-3* spacecraft and returned with Expedition 7
   crew on *Soyuz TMA-2* spacecraft.

Expedition Nine

Launch: 04/18/04

Land: 10/23/04

Time: 187 days, 21 hours, 17 minutes

Crew: Commander Gennady Padalka, Flight Engineer Mike
Fincke, Flight Engineer André Kuipers*

* European Space Agency astronaut Kuipers launched with Expedition 9
   crew on *Soyuz TMA-4* spacecraft and returned with Expedition 8
   crew on *Soyuz TMA-3* spacecraft.

Expedition Ten

Launch: 10/13/04

Land: 04/24/05

Crew: Commander Leroy Chiao, Flight Engineer Salizhan
Sharipov, Flight Engineer Yuri Shargin*

* Cosmonaut Yuri Shargin launched with the Expedition 10 crew on
   Soyuz TMA-5 spacecraft and returned with Expedition 9 crew on
   Soyuz TMA-4 spacecraft.

Expedition Eleven

Launch: 04/14/05

Land: 10/10/05

Crew: Commander Sergei Krikalev, Flight Engineer John Phillips,
   Flight Engineer (ESA) Roberto Vittori*

* European Space Agency astronaut Vittori launched with Expedition
   11 crew on *Soyuz TMA-6* and returned to Earth with Expedition 10
   crew on *Soyuz TMA-5*.

Expedition Twelve

Launch: 09/30/05

Land: 04/08/06
Crew: Commander William McArthur, Flight Engineer Valery
Tokarev, Space Flight Participant Gregory Olsen*

* Olsen launched with Expedition 12 crew on *Soyuz TMA-7* and then
  returned to Earth with Expedition 11 crew on *Soyuz TMA-6* under
  a commercial contract with the Russian Federation Space Agency.

Expedition Thirteen

Launch: 03/29/06
Land: 09/28/06
Time: 182 days, 23 hours, 44 minutes
Crew: Commander Pavel Vinogradov, Flight Engineer Jeffrey Wil-
liams, Flight Engineer (ESA) Thomas Reiter*, Astronaut Marcos
Pontes (Brazil)**

* Reiter is a European Space Agency astronaut who joined Expedi-
  tion 13 as part of the STS-121 mission of space shuttle *Discovery*
  in July 2006 and returned to Earth as part of STS-116 mission of
  Discovery.
** Brazilian astronaut Pontes launched with Expedition 13 crew on
   Soyuz TMA-8 and then returned to Earth with Expedition 12 crew on
   Soyuz TMA-7 under a commercial contract with the Roscosmos.

Expedition Fourteen

Launch: 09/18/06
Land: 04/18/07 (planned)
Time: On-going mission as of December 31, 2006
Crew: Commander Michael Lopez-Alegria, Flight Engineer Mikhail
Tyurin, Flight Engineer (ESA) Thomas Reiter*, Flight Engineer
(NASA) Astronaut Sunita Williams**, Spaceflight Participant
Anousheh Ansari***.

* Reiter is a European Space Agency astronaut, who joined Expedition 13 as part of STS-121 mission of space shuttle *Discovery* in
  July 2006 and returned to Earth as part of STS-116 mission of *Discovery;* ** NASA astronaut Williams launched with STS-116 crew
  on *Discovery* and then joined Expedition 14 crew; *** Ansari is an
  American-Iranian businesswoman, who launched with Expedition
  14 crew on *Soyuz TMA-9* and returned to Earth with Expedition
  13 crew on *Soyuz TMA-8* under a commercial contract with the
  Roscosmos.

Permanent Moon Bases: Next Stop beyond Low Earth Orbit

On January 14, 2004, President George W. Bush assigned NASA a bold, multi-decade-long "Vision for Space Exploration." His instructions directed officials within the civilian space agency to return the space shuttle to flight (following the *Columbia* accident on February 1, 2003), complete the *International Space Station* (*ISS*), return human beings to the Moon, and then continue on with expeditions to Mars.

The president's sweeping mandate to NASA, if carried to fulfillment, would extend a human presence across the solar system. But fulfillment of this vision requires a sustained and affordable space program that combines the innovative and efficient use of both human explorers and robotic systems. Also required to make this vision a reality is the continued support of the U.S. Congress (primarily by approving the budget to accomplish the multi-decade effort) and of the American people (primarily in the form of social enthusiasm and voter approval). Without these two fundamental building blocks, neither eloquent political rhetoric nor fascinating space technology scenarios will get the job done. The grand vision of human beings creating a permanent presence throughout the inner solar system and beyond has to become an integral part of the proverbial "American dream" in much the same way that landing American astronauts on the Moon during the Apollo Project became an integral part of the national identity during the cold-war era.

In response to the president's directions, during the summer of 2006 NASA introduced a new crew exploration vehicle (CEV), named *Orion*. Starting in about 2015, the *Orion* spacecraft will be launched into space on top of the new Ares I rocket vehicle on missions to the *International Space Station*. By 2020, NASA intends to send astronauts back to the Moon with the *Orion* spacecraft.

This artist's rendering shows NASA's planned *Orion* crew exploration vehicle docked with a lunar lander spacecraft in orbit around the Moon (ca. 2020). *(NASA/John Frassanito and Associates)*

NASA officials named this planned new spacecraft *Orion* after one of the brightest, most familiar, and easily identifiable constellations of stars in the night sky. As currently being developed, *Orion* will be capable of transporting cargo and up to six crewmembers to and from the *ISS*. The new spacecraft will also be able to carry four astronauts on missions back to the Moon. As part of NASA's current strategic plan for space exploration, the *Orion* spacecraft will also support crew transfers for future expeditions to Mars—perhaps starting as early as 2030. (Human exploration of Mars is discussed in the next chapter.)

The *Orion* spacecraft takes its gumdrop-like shape from the successful and familiar command module spacecraft used in the Apollo Project. The space capsule's conical shape is the safest and most reliable design for space vehicles that reenter Earth's atmosphere, especially at the speeds encountered during a direct return from the Moon. But *Orion* is not just a revival of space programs past. The spacecraft will enjoy the latest technology in computers, electronics, life support, propulsion, and thermal protection systems.

As now envisioned by NASA managers and engineers, *Orion* will be 16.5 feet (5 m) in diameter at its base and have a mass of about 55,000 pounds (25,000 kg). The interior volume of the new CEV spacecraft will have approximately 2.5 times the volume of the Apollo Project command module spacecraft. One of the main goals NASA has established for *Orion* is that the spacecraft be capable of supporting the return of human beings to the Moon. This return of astronauts to the Moon is not just for short-term excursions. Rather, *Orion* is intended to allow these future explorers to set up and sustain a permanent base on the lunar surface for both scientific purposes and in preparation for an eventual human mission to Mars.

Why go to Mars by way of the Moon? At first glance, this strategy seems almost like a step backward, since the Apollo Project landed people on the Moon almost four decades ago. In fact, some space exploration advocates inside and outside NASA have been vigorously lobbying for a direct human-crewed mission to Mars—perhaps using an expedition vehicle assembled and outfitted at the *ISS*.

But dashing to Mars with a large, complex, and relatively untested interplanetary space vehicle on a mission taking several hundred days or more is a strategy that invites unwanted technical surprises, incurs extremely high costs, and places enormous stresses upon the human explorers. Any or all of such negative circumstances could spell disaster for not only the human crew but also any future plans to have the human race expand into the solar system. What would the response of the sponsoring governments and their citizens be, if, after spending perhaps $250 billion to $350 billion on a one-shot expedition to Mars, the mission fails and the crew is stranded in interplanetary space or on the Red Planet? Would government officials in these sponsoring nations be eager or even

This artist's rendering depicts the establishment of a human–tended lunar surface base (ca. 2020). *(NASA/JSC)*

allowed by their citizens to quickly invest another $400 billion or so to try again? More likely, many shocked and disappointed citizens might vote down any future missions, as too costly and too risky. This type of adverse social response would basically confine human space travel for the next few centuries or more to low Earth orbit or possibly as far as the Moon. Interplanetary exploration would more than likely be left to a family of progressively more sophisticated space robots.

Perhaps the biggest advantage of the stepwise, conservative approach of going to Mars by way of the Moon is the strategic plan's ability to gradually develop the space technology infrastructure and human-factor experience needed to ensure that any early Mars expedition has a reasonable chance of success.

NASA strategic planners, in responding to the president's "Vision for Space Exploration," have come to more fully appreciate that a return to the Moon is a logical, natural step before sending astronauts to Mars. People can practice living, working, and doing science for extended periods on the lunar surface. A permanent human presence on the Moon, starting in about 2020, will enable astronauts to develop and demonstrate new space exploration technologies and perhaps even harness the Moon's abundant resources—thereby allowing human exploration of more distant locations in the solar system by reducing the overall costs of such interplanetary expeditions. Many NASA planners are hopeful that the establishment of a permanent lunar base and extended human presence on the Moon's surface will ultimately reduce the costs of space exploration activities performed later in this century. For example, lunar-based, human-crewed spacecraft could escape the Moon's lower gravity using much less energy than Earth-based vehicles. The propellants for these interplanetary space vehicles could, in all likelihood, be harvested from native lunar materials.

There is another important dimension to a permanent lunar base that will directly influence detailed exploration of Mars and other distant parts of the solar system. Focusing on its planned goal to have people return to the Moon by 2020, NASA is now increasing the use of robotic spacecraft to enrich the current level of scientific knowledge about solar system bodies and to prepare for more ambitious manned missions. Robot probes, landers, and similar smart-machine vehicles will serve as trailblazers and gather large quantities of important data so that scientists back on Earth can perform detailed studies of the alien worlds and decide what locations are the most suitable candidates for further study and possible human visits. Mars, of course, is at the top of this list. But Mars is a big planet, and mission planners need to know the best location or locations for the initial visit by human explorers. Smart robots will form a strong partnership with human explorers in this century.

The remainder of this chapter discusses some of the properties of the Moon and uses a variety of technically accurate artist renderings to depict some of the events that might take place when human beings return to the Moon and establish a permanent presence there within a decade or so.

✧ The Moon—Earth's Closest Celestial Neighbor

The Moon is Earth's only natural satellite and closest celestial neighbor. While life on Earth is made possible by the Sun, it is regulated by the periodic motions of the Moon. For example, historically the months of the year have been measured by the regular motions of the Moon around Earth, and the tides rise and fall because of the gravitational tug-of-war between Earth and the Moon. Throughout history, the Moon has had a significant influence on human culture, art, and literature. In the Space Age, reaching the Moon with robots and then people proved to be a major stimulus for Earth's contemporary, technology-dependent, global civilization. Starting in 1959 with the U.S. *Pioneer 4* and the Russian *Luna 1* lunar flyby missions, a variety of American and Russian missions have been sent to and around the Moon. The most exciting of these missions were the Apollo Project's human expeditions to the Moon from 1968 to 1972.

In 1994, the *Clementine* spacecraft, which was developed and flown by the Ballistic Missile Defense Organization of the U.S. Department of Defense as a demonstration of certain advanced space technologies, spent 70 days in lunar orbit mapping the Moon's surface. Subsequent analysis of the *Clementine* data offered tantalizing hints that water ice might be present in some of the permanently shadowed regions at the Moon's poles. NASA launched the Lunar Prospector mission in January 1998 to perform a detailed study of the Moon's surface composition and to hunt for signs of the suspected deposits of water ice. Data from the mission strongly suggested the presence of water ice in the Moon's polar regions, although the results require additional confirmation. Water on the Moon (as trapped surface ice in the permanently shadowed regions of the Moon's poles) would be an extremely valuable resource that would open up many exciting possibilities for future lunar base development. In 2008, NASA plans to use another robot spacecraft, called the *Lunar Reconnaissance Orbiter,* to scan for resources in the polar regions, identify candidate landing sites for future robot rovers and human explorers, and measure the Moon's radiation environment.

From evidence gathered by the early robotic lunar missions (such as *Ranger, Surveyor,* and the *Lunar Orbiter* spacecraft) and by the human-crewed Apollo missions, lunar scientists have learned a great deal more about the Moon and have been able to construct a geologic history dating back to its infancy.

Because the Moon does not have any oceans or other free-flowing water and lacks a sensible atmosphere, appreciable erosion, or "weathering," has not occurred there. The primitive materials that lay on its surface for billions of years are still in an excellent state of preservation. Scientists believe that the Moon was formed more than 4 billion years ago and then differentiated quite early, perhaps only 100 million years later. Tectonic activity ceased eons ago on the Moon. The lunar crust and mantle are quite thick, extending inward to more than 497 miles (800 km). However, the deep interior of the Moon is still unknown. It may contain a small iron core at its center, and there is some evidence that the lunar interior may be hot and even partially molten. Moonquakes have been measured within the lithosphere and interior, most being the result of gravitational stresses. Chemically, Earth and the Moon are quite similar, although compared to Earth the Moon is depleted in more easily vaporized materials. The lunar surface consists of highlands composed of alumina-rich rocks that formed from a globe-encircling molten sea and maria made up of volcanic melts that surfaced about 3.5 billion years ago. However, despite all scientists have learned in the past three decades about Earth's nearest celestial neighbor, lunar exploration has only started. Several puzzling mysteries remain, including the origin of the Moon itself.

A new lunar origin theory suggests a cataclysmic birth of the Moon. Scientists supporting this fairly recent theory suggest that near the end of Earth's accretion from the primordial solar nebula materials (that is, after its core was formed but while Earth was still in a molten state), a Mars-size celestial object (called an "impactor") hit Earth at an oblique angle. This ancient explosive collision sent vaporized impactor and molten Earth material into Earth's orbit, and the Moon then formed from these materials.

The surface of the Moon has two major regions with distinctive geologic features and evolutionary histories. First are the relatively smooth, dark areas that Galileo Galilei originally called "maria" (because he thought they were seas or oceans). Second are the densely cratered, rugged highlands (uplands) that Galileo called "terrae." The highlands occupy about 83 percent of the Moon's surface and generally have a higher elevation (as much as three miles [5 km] above the Moon's mean radius). In other places, the maria lie about three miles below the mean radius and are concentrated on the nearside of the Moon—that is, on the side of the Moon always facing Earth.

The main external geologic process modifying the surface of the Moon is meteoroid impact. Craters range in size from tiny pits less than five-millionths of an inch (micrometers) in diameter to gigantic basins hundreds of miles across.

The surface of the Moon is strongly brecciated, or fragmented. This mantle of weakly coherent debris is called regolith. It consists of shocked fragments of rocks, minerals, and special pieces of glass formed by meteoroid impact. Regolith thickness is quite variable and depends on the age of the bedrock beneath and on the proximity of craters and their ejecta blankets. Generally, the maria are covered by 10 feet (3 m) to 52 feet (16 m) of regolith, while the older highlands have developed a lunar "soil" at least 33 feet (10 m) thick.

Advocates of going to Mars by way of the Moon point out some of the interesting physical similarities between the two worlds. The Moon has only one-sixth Earth's gravity, while Mars has just one-third of Earth's gravity. The Moon has no appreciable atmosphere; the Martian atmosphere is highly rarefied. The Moon gets very cold, as low as –400°F (–240°C) in the shadowed regions; the surface temperature on Mars varies between –4°F (–20°C) and –149°F (–100°C).

Perhaps more challenging from a surface operations perspective is that both planetary bodies are covered with silt-fine dust called regolith. The Moon's regolith was created by ceaseless bombardment of micrometeorites, cosmic rays, and energetic particles of the solar wind, which over billions of years have broken down the rocks on the surface. Scientists believe the Martian regolith is the result of impacts of more massive meteorites and even asteroids, in addition to water and wind erosion that has taken place on a daily basis for eons. There are places on both the Moon and Mars where the regolith appears to be more than 33 feet deep.

Operating mechanical equipment in the presence of so much fine dust is a considerable engineering challenge. For one thing, Moon dust is quite coarse and sharp, like fragments of glass. *Apollo 17* astronauts Eugene Cernan and Harrison Schmitt found that these sharp and odd-shaped dust particles would jam the shoulder joints of their space suits, even after short Moonwalks. The troublesome lunar dust penetrated seals, causing the space suits to leak and lose some internal air pressure. Because individual dust particles had the tendency to become electrostatically charged by the Sun's ultraviolet light, when the astronauts walked in sunlit areas fine dust levitated above their knees and sometimes even above their heads. Such electrostatically clinging dust particles are easily tracked back into an astronaut's habitat, where they then become airborne and cause eye and lung irritation. The mitigation of the effect of lunar dust on both robotic systems and astronauts working on the Moon's surface is a major problem.

Scientists anticipate similar problems with dust on Mars, although the Martian dust may not be as sharp as lunar dust because of natural weathering effects on the Red Planet that should have smoothed the edges of the dust particles. However, unlike the windless Moon, dust storms on Mars can whip the particles to erosive velocities (perhaps a 100 miles per

This artist's rendering depicts a lunar surface roving vehicle and the accompanying team of astronauts, who are performing a surface extravehicular activity. The scene uses actual imagery of the lunar surface. *(NASA)*

hour (160 km/h)) or more, causing scouring and premature wearing of every exposed surface. NASA engineers have observed that Martian dust (possibly electrostatically charged) is clinging to the solar panels of the *Spirit* and *Opportunity* surface rovers, blocking sunlight and causing each robot to experience a reduction in the amount of electric power being generated. So dust on both the Moon and Mars will possibly interfere with solar electric generation during surface missions, either robotic or human—especially during operations involving heavy traffic, excavation, construction, and resource harvesting.

Dust mitigation experiments on the Moon, such as the use of thin film coatings to repel dust from space suits and machinery, will help engineers design far more dust-resistant equipment for use on Mars. Testing such technology on the Moon, which is only two or three days' travel from Earth, is far easier and less risky than using Mars, which is six months' travel or more, as a test bed and field laboratory for mission-critical equipment. Even here on Earth, engineers frequently encounter situations in which equipment or some new operating procedure that "worked perfectly" during a computer simulation or in a controlled laboratory test fails mysteriously in a field test or operational application. The cause of the failure or disappointing decrement in performance is often a subtle but disruptive environmental condition or some overlooked real-world synergy between error-compounding events. Such compounded operational circumstances and field problems are quite difficult to model or simulate under controlled laboratory conditions. Conservative aerospace engineers always field-test mission-critical equipment whenever possible. In all likelihood, the success of a future human expedition to Mars will depend to a great extent on the effective use of a permanent lunar base as an "alien-world" test bed.

One strategy that has been popular over the past two decades with Mars mission planners is called the in situ resource utilization (ISRU) scenario. This scenario suggests that future explorers to Mars will take equipment that allows them to "live off the land." The machines would harvest and process local raw materials, such as oxygen (for breathing and rocket fuel), water (for drinking and rocket propellant—when separated into hydrogen and oxygen), and various minerals (for structural materials and radiation shielding). Once again, the permanent lunar base would provide an excellent opportunity to test ISRU equipment from a reliability,

Just three days away from Earth by rocket-propelled space travel, the Moon is a good place to test the hardware and operations for the first human expedition to Mars. This artist's concept shows a simulated Mars mission, including the landing of a lunar environment–adapted Mars excursion vehicle, which could test many relevant Mars expedition systems and technologies. (NASA/JSC; artist, Pat Rawlings)

Scientists currently suspect that the permanently shadowed regions of the lunar poles may contain significant deposits of water ice. If this turns out to be true, harvesting this precious resource would be an important part of any permanent human occupancy of the Moon. This artist's concept shows a solar-powered base in a crater at the Moon's South Pole, harvesting lunar water ice and producing rocket propellant (hydrogen and oxygen) for lunar spacecraft, like the one illustrated. In this depiction, the base's inhabitants are circulating some of the water they have harvested through the dome's cells to provide additional shielding against space radiation. *(NASA/JSC; artist, Pat Rawlings)*

energy consumption, and efficiency perspective.

Perhaps the most exciting extraterrestrial resource "treasure hunt" is that for water. Scientists now speculate that both the Moon and Mars may harbor water deposits frozen in the ground. The current evidence remains indirect, but robotic rover missions to first the Moon and then Mars over the next decade should help identify and quantify any significant deposits. If this water ice could be excavated, thawed out, and separated into hydrogen and oxygen (possibly by electrolysis), "water mining" for life support system consumables and rocket propellants could become a major industry on the Moon and later on Mars. The significance of demonstrating this technology at the lunar base should not be underestimated, because adequate supplies of water harvested in the polar regions of the Moon would dramatically decrease the economic burden of lunar base operations and should also significantly reduce the overall costs of outfitting a major human expedition to Mars. The Moon plays a pivotal role in human expansion into space beyond low Earth orbit.

✦ Lunar Bases and Settlements

There are many factors (some favorable and some unfavorable) and physical resource assessments that will dramatically influence and shift any lunar base development scenario suggested in 2007. Recognizing such limitations in contemporary technical projections, this section provides a generalized overview of what might take place over the remainder of this century if human space exploration activities include the development of a permanent lunar base.

When human beings return to the Moon, it will not be for a brief moment of scientific inquiry, as occurred in NASA's Apollo Project, but rather as permanent inhabitants of a new world. They will build bases from which to explore the lunar surface, establish science and technology laboratories that take advantage of the special properties of the lunar environment, and harvest the Moon's resources (including the suspected deposits of lunar ice in the polar regions) in support of humanity's extra-terrestrial expansion.

A lunar base is a permanently inhabited complex on the surface of the Moon. In the first permanent lunar base camp, a team of from 10 up to perhaps 100 lunar workers will set about the task of fully investigating the Moon. The word permanent here means that human beings will always occupy the facility, but individuals probably will serve tours of from one

This artist's rendering depicts a teleoperated robot rover that has completed its survey of candidate sites for a pending human landing mission (shown in the background during the final stage of its descent). In the space exploration scenario illustrated, the robot rover had been delivered to the Moon's surface by the robotic lunar lander (dubbed "Artemis"), appearing on the left. Controllers on Earth then teleoperated the robot rover and performed site surveys—as suggested by the tracks in the picture. (NASA; artist, Pat Rawlings)

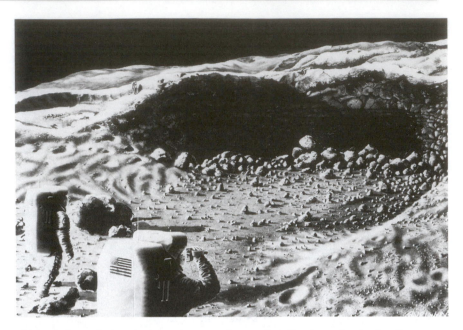

This artist's rendering depicts a lunar surface exploration team beginning its investigation of a small lava tunnel (ca. 2020). Their main objective is to determine if the lava tunnel could serve as a natural shelter (primarily against meteorites and space radiation) for the habitation modules of a lunar base. *(NASA)*

to three years before returning to Earth. Some workers at the base will enjoy being on another world. Some will begin to experience isolation-related psychological problems similar to the difficulties often experienced by members of a scientific team who "winter-over" in Antarctic research stations. Still other workers will experience injuries or even fatal accidents while working at or around the lunar base.

For the most part, however, the pioneering lunar base inhabitants will take advantage of the Moon as a science-in-space platform and perform the fundamental engineering studies needed to confirm and define the specific roles the Moon will play in the full development of space for the remainder of this century and in centuries beyond. For example, the discovery of frozen volatiles (including water) in the perpetually frozen recesses of the Moon's polar regions could change lunar base logistics strategies and accelerate development of a large lunar settlement of up to 10,000 or more inhabitants. Many lunar base applications have been proposed. Some of these concepts include: (1) a lunar scientific laboratory complex, (2) a lunar industrial complex to support space-based manufacturing, (3) an astrophysical observatory for solar system and deep-space surveillance, (4) a fueling station for orbital transfer vehicles that travel through cislunar space, and (5) a training site and assembly point for the first human expedition to Mars.

Social and political scientists suggest that a permanent lunar base could also become the site of innovative political, social, and cultural developments—essentially rejuvenating our concept of who we are as intelligent beings and boldly demonstrating our ability to beneficially apply advanced technology in support of the positive aspects of human destiny. Another interesting suggestion for a permanent lunar base is its use as a field operations center for the rapid-response portion of a planetary defense system that protects Earth from threatening asteroids or comets.

As lunar activities expand, the original lunar base could grow into an early settlement of about 1,000 more or less permanent residents. Then, as the lunar industrial complex develops further and lunar raw materials, food, and manufactured products start to support space commerce throughout cislunar space, the lunar settlement itself will expand to a population of around 10,000. At that point, the original settlement might spawn several new settlements—each taking advantage of some special location or resource deposit elsewhere on the lunar surface.

In the next century, this collection of permanent human settlements on the Moon could continue to grow, reaching a combined population of about 500,000 persons and attaining a social and economic "critical mass" that supports true self-sufficiency from Earth. This moment of self-sufficiency for the lunar civilization will also be an historic moment in human history. For from that time on, the human race will exist in two distinct and separate "biological niches"—people will be either *terran* or *nonterran* (that is, extraterrestrial).

With the rise of a self-sufficient, autonomous lunar civilization, future generations will have a choice of worlds on which to live and prosper. Of course, such a major social development will most likely produce its share of cultural backlash in both worlds. Citizens of the 22nd century may start seeing personal ground vehicles with such bumper-sticker slogans as "This is my world—love it or leave it"; "Terran go home"; or even, "Protect terrestrial jobs—ban lunar outsourcing!"

The vast majority of lunar base development studies includes the use of the Moon as a platform from which to conduct science in space. Scientific facilities on the Moon will take advantage of the unique environment to support platforms for astronomical, solar, and space science (plasma) observations. The unique environmental characteristics of the lunar surface include low gravity (one-sixth that of the Earth), high vacuum, seismic stability, low temperatures (especially in permanently shadowed polar regions), and a low radio noise environment on the Moon's farside.

Astronomy from the lunar surface offers the distinct advantages of a low radio noise environment and a stable platform in a low-gravity

In this artist's rendering, a disoriented lunar worker has tumbled down a 100-foot-(30-m-) high escarpment and fractured his right foot. The crew of an emergency "medivac lunar hopper" provides quality care in the field (note the rescue worker who is reviewing a heads-up display of medical data) prior to removing the accident victim to the medical facilities at the main base. *(NASA; artist, Pat Rawlings)*

environment. The farside of the Moon is permanently shielded from direct terrestrial radio emissions. As future radio telescope designs approach their ultimate (theoretical) performance limits, this uniquely quiet lunar environment may be the only location in all cislunar space where sensitive radio wave–detection instruments can be used to full advantage, both in radio astronomy and in our search for extraterrestrial intelligence (SETI). In fact, radio astronomy, including extensive SETI efforts, may represent one of the main "lunar industries" late this century. In one sense, SETI performed by lunar-based scientists could be viewed as "extraterrestrials" searching for other extraterrestrials.

The Moon also provides a solid, seismically stable, low-gravity, high-vacuum platform for conducting precise interferometric and astrometric observations. For example, the availability of ultrahigh-resolution (micro-arcsecond) optical, infrared, and radio observatories will allow astronomers to carefully search for Earthlike extrasolar planets encircling nearby stars, out to perhaps several hundred light-years of distance.

A lunar scientific base also provides life scientists with a unique opportunity to extensively study biological processes in reduced gravity (one-sixth that of Earth) and in low magnetic fields. Genetic engineers can conduct their experiments in comfortable facilities that are nevertheless physically isolated from the Earth's biosphere. Exobiologists can experiment with new types of plants and microorganisms under a variety of simulated alien-world conditions. Genetically engineered "lunar plants," grown in special greenhouse facilities, could become a major food source, while also supplementing the regeneration of a breathable atmosphere for the various lunar habitats.

The true impetus for large, permanent lunar settlements will most likely arise from the desire for economic gain—a time-honored stimulus that has driven much technical, social, and economic development on Earth. The ability to create useful products from native lunar materials will have a controlling influence on the overall rate of growth of the lunar civilization. Some early lunar products can now be easily identified. Lunar ice, especially when refined into pure water or dissociated into the important chemicals hydrogen (H_2) and oxygen (O_2) represents the Moon's most important resource. Other important early lunar products include (1) oxygen (extracted from lunar soils) for use as a propellant by orbital transfer vehicles traveling throughout cislunar space, (2) raw (i.e., bulk, minimally processed) lunar soil and rock materials for space radiation shielding, and (3) refined ceramic and metal products to support the construction of large structures and habitats in space.

The initial lunar base can be used to demonstrate industrial applications of native Moon resources and to operate small pilot factories that provide selected raw and finished products for use both on the Moon and in Earth orbit. Despite the actual distances involved, the cost of shipping a pound of "stuff" from the surface of the Moon to various locations in cislunar space may prove much cheaper than shipping the same stuff from the surface of Earth.

The Moon has large supplies of silicon, iron, aluminum, calcium, magnesium, titanium, and oxygen. Lunar soil and rock can be melted to make glass—in the form of fibers, slabs, tubes, and rods. Sintering (a process whereby a substance is formed into a coherent mass by heating but without melting) can produce lunar bricks and ceramic products. Iron metal can be melted and cast or converted to specially shaped forms using powder metallurgy. These lunar products would find a ready market as shielding materials, in habitat construction, in the development of large space facilities, and in electric power generation and transmission systems.

Lunar mining operations and factories can be expanded to meet growing demands for lunar products throughout cislunar space. With the rise of lunar agriculture (accomplished in special enclosed facilities), the Moon may even become our "extraterrestrial breadbasket"—providing the majority of all food products consumed by humanity's extraterrestrial citizens.

One interesting space commerce scenario involves an extensive lunar surface mining operation that provides the required quantities of materials in a preprocessed condition to a giant space manufacturing complex located at Lagrangian libration point 4 or 5 (L_4 or L_5). (See chapter 12.) These exported lunar materials would consist primarily of oxygen, silicon, aluminum, iron, magnesium, and calcium locked into a great variety of complex chemical compounds. It has often been suggested by space visionaries, such as Krafft A. Ehricke, that the Moon will become the chief source of materials for space-based industries in the latter part of this century.

Numerous other tangible and intangible advantages of lunar settlements will accrue as a natural part of their creation and evolutionary development. For example, the high-technology discoveries originating in a complex of unique lunar laboratories could be channeled directly into appropriate economic and technical sectors on Earth, as "frontier" ideas,

This artist's concept shows an astronaut unloading a recently delivered habitat module using a lunar crane. The module will be placed on the flatbed part of the open rover vehicle train and then driven to the main base (background) for integration into an evolving human outpost on the Moon, circa 2020. *(NASA/MSFC; artist, Pat Rawlings)*

techniques, products and so on. The permanent presence of people on another world (a world that looms large in the night sky) will continuously suggest an *open-world philosophy* and a sense of cosmic destiny to the vast majority of humans who remain behind on the home planet. The human generation that decides to venture into cislunar space and to create permanent lunar settlements will long be admired, not only for its great technical and intellectual achievements but also for its innovative cultural accomplishments. Finally, it is not too remote to speculate that the descendants of the first lunar settlers will become first the interplanetary, then the interstellar, portion of the human race. The Moon can be viewed as humanity's stepping-stone to the universe.

✧ Krafft A. Ehricke and the Vision of a Lunar Civilization

The German-American Krafft Arnold Ehricke (1917–84) was the talented rocket engineer who conceived advanced propulsion systems for use in the American space program of the late 1950s and 1960s. One of his most important technical achievements was the design and development of the Centaur upper stage rocket vehicle—the first American rocket vehicle to use liquid hydrogen (LH_2) as its propellant. The Centaur rocket vehicle made possible many important military and civilian space missions. As an inspirational space-travel advocate, Ehricke's visionary writings and lectures eloquently expounded upon the positive consequences of space technology. He anchored his far-reaching concept of an extraterrestrial imperative with the permanent human settlement of the Moon playing a central role in the future development of the human race.

Ehricke was born in Berlin, Germany, on March 24, 1917. This was a turbulent time because Imperial Germany was locked in a devastating war with much of Europe and the United States. He grew up in the political and economic chaos of Germany's post–World War I Weimar Republic. Yet, despite the gloomy environment of a defeated and war-devastated Germany, Ehricke was able to develop his lifelong positive vision that space technology would serve as the key to improving the human condition. Following World War I, a major challenge for his parents was that of acquiring schooling of sufficient quality to challenge their son. Unfortunately, Ehricke's frequent intellectual sparring contests with rigid Prussian schoolmasters did not help the situation and earned him a widely varying collection of grades.

By chance, at the age of 12, Ehricke saw Fritz Lang's 1929 motion picture *Die Frau im Mond* (*The Woman in the Moon*). The Austrian filmmaker had hired the German rocket experts Hermann Oberth and Willy Ley to serve as technical advisers during the production of this film. Oberth and Ley gave the film an exceptionally prophetic two-stage rocket design that startled and delighted audiences with its impressive blastoff. Ehricke viewed Lang's film at least a dozen times. Advanced in mathematics and physics for his age, he appreciated the great technical detail that Oberth had provided to make the film realistic. This motion picture served as Ehricke's introduction to the world of rockets and space travel, and he knew immediately what he wanted to do for the rest of his life. He soon discovered Konstantin Tsiolkovsky's theoretical concept of an efficient chemical rocket that used hydrogen and oxygen as its liquid propellants. While a teenager, he also attempted to tackle Oberth's famous 1929 book *Roads to Space Travel,* but he struggled a bit with some of the book's more advanced mathematics.

In the early 1930s, he was still too young to participate in the German Society for Space Travel (Verein für Raumschiffahrt, or VfR), so he experimented in a self-constructed laboratory at home. As Adolf Hitler (1889–1945) rose to power in 1933, Ehricke, like thousands of other young Germans, became swept up in the Nazi youth movement. His free-spirited thinking, however, soon got him into difficulties and earned him an unenviable position as a conscripted laborer for the Third Reich. Just before the outbreak of World War II, he was released from the labor draft so he could attend the Technical University of Berlin. There he majored in aeronautics, the closest academic discipline to space technology. One of his professors was Hans Wilhelm Geiger (1882–1945), the noted German nuclear physicist. Geiger's lectures introduced Ehricke to the world of nuclear energy. Impressed, Ehricke would later recommend the use of nuclear power and propulsion in many of the space-development scenarios he presented in the 1960s and 1970s.

But wartime conditions played havoc with Ehricke's attempt to earn a degree. While enrolled at the Technical University of Berlin, he was drafted for immediate service in the German army and sent to the Western Front. Wounded, he came back to Berlin to recover and resume his studies. In 1942, he obtained a degree in aeronautical engineering from the Technical University of Berlin. But while taking postgraduate courses in orbital mechanics and nuclear physics, he was again drafted into the German army, promoted to the rank of lieutenant, and ordered to serve with a Panzer (tank) Division on the Eastern (Russian) Front. But fortune played a hand, and in June 1942, the young engineer received new orders, this time reassigning him to rocket-development work at Peenemünde. From 1942 to 1945, he worked there on the German army's rocket program under the overall direction of Wernher von Braun.

As a young engineer, Ehricke found himself surrounded by many other skilled engineers and technicians whose collective goal was to produce the world's first modern, liquid-propellant ballistic missile, called the A-4 rocket. This rocket is better known as Hitler's Vengeance Weapon-Two, or simply the V-2 rocket. After World War II, the German V-2 rocket became the ancestor to many of the larger missiles developed by both the United States and the former Soviet Union during the cold war.

Near the end of World War II, Ehricke joined the majority of the German rocket scientists at Peenemünde and fled to Bavaria to escape the advancing Russian army. Swept up in Operation Paperclip along with other key German rocket personnel, Ehricke delayed accepting a contract to work on rockets in the United States by almost a year. He did this in order to locate his wife, Ingebord, who was then somewhere in war-torn Berlin. Following a long search and happy reunion, Ehricke, his wife, and their first child journeyed to the United States in December 1946 to begin a new life.

For the next five years, Ehricke supported the growing United States Army rocket program at White Sands, New Mexico, and Huntsville, Alabama. In the early 1950s, he left his position with the U.S. Army and joined the newly formed Astronautics Division of General Dynamics (formerly called Convair). There he worked as a rocket concept and design specialist and participated in the development of the Atlas—the first American intercontinental ballistic missile. He became an American citizen in 1955.

Ehricke strongly advocated the use of liquid hydrogen as a rocket propellant. While at General Dynamics, he recommended the development of a liquid hydrogen–liquid oxygen propellant upper stage rocket vehicle. His recommendation became the versatile and powerful Centaur upper stage vehicle. In 1965, he completed his work at General Dynamics as the director of the Centaur program and joined the advanced studies group at North American Aviation in Anaheim, California. From 1965 to 1968, this new position allowed him to explore pathways of space-technology development across a wide spectrum of military, scientific, and industrial applications. The excitement of examining future space technologies and their impact on the human race remained with him for the rest of his life.

From 1968 to 1973, Ehricke worked as a senior scientist in the North American Rockwell Space Systems Division in Downey, California. In this capacity, he fully developed his far-ranging concepts concerning the use of space technology for the benefit of humankind. After departing Rockwell International, he continued his visionary space-advocacy efforts through his own consulting company, Space Global, located in La Jolla, California. As the United States government wound down the Apollo Project in the early 1970s, Ehricke continued to champion the use of the Moon and its resources. His far-reaching notion of an extraterrestrial imperative was based upon the creation of a selenospheric (Moon-centered) human civilization in space. Until his death in late 1984, he spoke and wrote tirelessly about how space technology provides the human race the ability to create an unbounded open-world civilization.

Ehricke was a dedicated space visionary who not only designed advanced rocket systems (such as the Atlas-Centaur configuration) that enabled the first golden age of space exploration (1958–87) but also addressed the important yet frequently ignored social and cultural impacts of space technology. He also created original art to communicate many of his ideas. He coined many interesting space technology terms. For example, Ehricke used the term *androsphere* to describe the synthesis of the terrestrial and extraterrestrial environments. The androsphere relates to human integration of Earth's biosphere—the portion of the planet that contains all the major terrestrial environmental regimes, such as the atmosphere, the hydrosphere, and the cryosphere—with the material and energy resources of the solar system. Similarly, the term *astropolis* is his concept for a large, urban-like extraterrestrial facility that orbits in Earth-Moon (cislunar) space and supports the long-term use of the space environment for basic and applied research, as well as for industrial development. Ehricke's *androcell* is an even bolder concept that involves a large, human-made world in space, totally independent of the Earth-Moon system. These extraterrestrial city-states, with human populations of up to 100,000 or more, would offer their inhabitants the excitement of multigravity-level living at locations throughout heliocentric space.

Just weeks before his death on December 11, 1984, Ehricke served as a featured speaker at a national symposium on lunar bases and space activities for the 21st century held in Washington, D.C. Despite being terminally ill, he traveled across the country to give a moving presentation that emphasized the importance of the Moon in creating a multi-world civilization for the human race. He ended his uplifting, visionary discussion by eloquently noting that: "The Creator of our universe wanted human beings to become space travelers. We were given a Moon that was just far enough away to require the development of sophisticated space technologies, yet close enough to allow us to be successful on our first concentrated attempt."

Human Expeditions to Mars

In 1952, *Weltraumfahrt,* the German spaceflight journal, published a special issue containing Wernher von Braun's *Das Marsprojekt* (*The Mars Project*). Von Braun wrote this creative report as an intellectual exercise, while he started working for the U.S. Army and was stationed at Fort Bliss, Texas, in support of V-2 rocket launches from the White Sands Missile Range in New Mexico.

The rocket scientist's *Mars Project* scenario envisioned a flotilla of 10 interplanetary spacecraft that carried a total of no fewer than 70 astronauts. Von Braun suggested that seven of the proposed interplanetary spaceships would function almost exclusively as crew-carrying vessels, while three of the spaceships would serve as cargo carriers on the journey to Mars. The expedition would be assembled in orbit around Earth and then depart on the journey to Mars. Each of the seven passengerspaceships was identical in design and appearance, and all carried sufficient fuel for the return journey to Earth. Each of the three cargo spaceships carried a special winged landing craft (which von Braun called a "landing boat"). Once the flotilla was in orbit around Mars, the astronauts could transfer to the three landing craft and descend to the Martian surface. But the space explorers would have to modify the so-called landing boats in order to return to orbit around Mars.

The proposed Mars landing spacecraft resembled a stubby artillery shell to which enormous wings were attached to let the vehicle glide through the thin Martian atmosphere and then land like a wheeled glider on the surface of the Red Planet. The hulls of these torpedo-like landing spacecraft were about 72 feet (22 m) long, and each had an enormous wingspan of 502 feet (153 m). When it was time to leave the surface of Mars, the astronauts would detach the wings from the three landing craft, raise the stubby, torpedo-like hulls to a vertical position, and then fire each vehicle's rocket engines. Once in orbit, the flotilla would reform as the astronauts transferred back to the seven orbiting passenger spaceships and start their long journey back to Earth. In von Braun's scenario, the three cargo-carrying spaceships would be abandoned in orbit around Mars.

This grand strategy for the exploration of Mars was the first serious technical treatment of interplanetary spaceflight by human beings. Viewed as nothing short of fantastic in the early 1950s, von Braun's modest-size report (published in English in 1953) became the creative basis for the popular space-travel articles he published in *Collier's* magazine and for his cordial business interactions with Walt Disney.

Throughout the Space Age, other space-travel enthusiasts and aerospace engineers have continued to suggest scenarios for the human exploration of Mars. Some of these elaborate scenarios involved the use of nuclear rocket propulsion (either nuclear thermal systems or nuclear electric). Other sce-

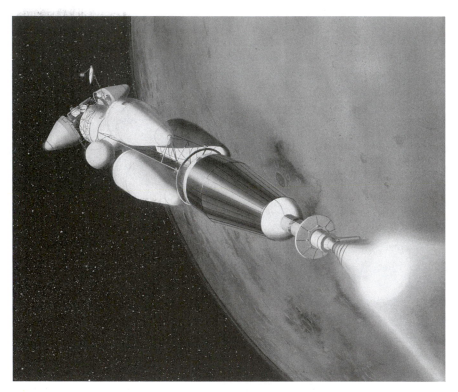

This artist's rendering shows the first human expedition nearing Mars, circa 2030. Upon arrival, the mission's primary propulsion system, a nuclear thermal rocket, fires to insert the space vehicle into the proper parking orbit around the Red Planet. Nuclear propulsion technology can shorten interplanetary trip times and/or deliver more payload mass to the planet for the same initial Earth-orbit (or lunar orbit) departure mass than can chemical propulsion technology. *(NASA; artist, Pat Rawlings)*

narios involved all chemical propulsion systems that would harvest materials on Mars to manufacture the supply of rocket propellants needed to the return journey. Some scenarios involved journeys lasting several hundred days in microgravity conditions, with the Martian explorers being forced to maintain a rigorous exercise and conditioning program while traveling to and from the Red Planet. Other scenarios invoked the use of artificial gravity during the interplanetary journey, to keep the crew prepared for physical activities on the Martian surface at the moment they arrived.

Which scenario will actually unfold in about the year 2030 is open to an enormous amount of technical and political speculation at this time. But no matter which scenario ultimately takes place, several key technology factors must be satisfactorily addressed. The rocket propulsion system used to get the human expedition to Mars must be reliable, efficient, and sufficiently powerful to keep the overall length of the trip as short as possible. A relatively brief journey will minimize stress on life support systems, reduce the crew's exposure to space radiation, and lessen the occurrence of adverse psychological factors that could arise among people confined for

This artist's rendering depicts a Mars transfer vehicle, which provides the onboard astronauts artificial gravity during their long interplanetary journey from Earth orbit to orbit around the Red Planet (ca. 2030). The spacecraft would slowly rotate while the entire vehicle travels forward, producing the equivalent of Mars surface gravity. Under this concept, the Mars expedition astronauts would arrive at the planet in proper physical condition to start exploring the surface. (NASA/JSC)

a long time in tight quarters and under constant physical danger. Other issues, such as the number and composition of the crew, the number of spacecraft committed to the expedition, the objectives of the mission, and the candidate sites to be visited on the Red Planet, must all be resolved.

A human expedition consisting of a single spacecraft represents an all-or-nothing approach, while a multi-spacecraft mission, as originally suggested by von Braun in his *Mars Project*, provides vehicle redundancy and additional safety. Crewmembers from a disabled Mars expedition spacecraft would be able to transfer to other craft in the expedition and continue on with the mission and then return safely to Earth. In planning a multiple-ship Mars expedition, aerospace engineers would have to design each individual spacecraft for possible reconfiguration (during interplanetary flight or once in orbit around Mars) for use as a "lifeboat" to support the rescue of crewmembers from a disabled craft.

One interesting human-factor issue is whether all members of the expedition will be allowed to descend to the surface of Mars. During the lunar-landing missions of NASA's Apollo Project, one of the three crewmembers had to remain in orbit around the Moon, just in case something went wrong on the lunar surface. If one or several crewmembers have to remain in cramped quarters in orbit around Mars while others have the privilege of setting up a base camp and exploring a new world for the first time, some psychological stress factors could arise. Who would make such important choices? Should this momentous selection be made by a team of physically detached mission managers many light-minutes away on Earth or by the Mars expedition crew—perhaps using the time-honored tradition of a lottery or casting dice? Is the decision made before the journey starts or upon arrival at Mars? If mission planners allow all the crewmembers to go down to the surface, who tends the spacecraft waiting to take them home? A decision to allow the crew to go back and forth (in groups) to the surface of Mars adds an enormous complexity to the mission, especially with respect to propellant consumption, transfer vehicle mass, and hardware reliability. For example, how many separate transfer vehicles will the expedition need to ferry the entire expedition team (in small groups) down to the surface of Mars? This prickly question is one of the hundreds that must be carefully addressed and answered over the next two decades, before human beings set out to explore the mysterious red-colored world known as Mars.

This chapter uses a carefully selected collection of artist renderings to describe some of the exciting events and technologies that could surround the first expedition to Mars and then the eventual development of a permanent outpost on the planet's surface. While the illustrations have credible technical content, they also portray contemporary (or past) ideas and concepts. Technical progress over the next two decades could signifi-

This artist's rendering shows a possible surface exploration scenario during the first human expedition to Mars. After driving a short distance from their Ganges Chasma landing site (left, background) on the Red Planet, two explorers stop to perform a surface extravehicular activity and inspect a previously deployed robotic lander and its small rover. The stop also provides the astronauts a convenient opportunity to check the rover's life support subsystems while they are still within walking distance of the base camp. *(NASA/JSC; artist, Pat Rawlings)*

cantly alter the choice of hardware and operational scenario actually used by NASA to take the first human expedition to Mars in about 2030. Therefore, the reader should treat the material presented here as an informative but tentative starting point. The chapter also serves as a creative guide to help a person imagine what wandering across the Red Planet in search of signs of life—extinct or possibly still existing in some protected subsurface biological niche—might be like in the future.

✧ Mars—The Mysterious Red Planet

Mars is the fourth planet in the solar system from the Sun, with an equatorial diameter of 4,222 miles (6,794 km). Throughout human history, Mars, the Red Planet, has been at the center of astronomical thought. The ancient Babylonians, for example, followed the motions of this wandering red light across the night sky and named it after Nergal, their god of war. The ancient Greeks referred to the planet as Ares, their own god of war. In time the Romans, also honoring the mythological god of war, gave the planet its present name.

The presence of an atmosphere, polar caps, and changing patterns of light and dark on the surface caused many pre–Space Age astronomers

This artist's rendering, entitled *20/20 Vision,* depicts a much anticipated exobiology discovery that would have enormous scientific and philosophical significance this century. Shown here is a scientist-astronaut as she examines fossilized evidence of ancient life on Mars, during the first human expedition to the surface of the planet, circa 2030. *(NASA; artist, Pat Rawlings)*

and scientists to consider Mars an "Earthlike planet"—the possible abode of extraterrestrial life. In fact, when actor Orson Welles broadcast a radio drama in 1938 based on H. G. Wells's science-fiction classic *War of the Worlds,* enough people believed the report of invading Martians to create a near panic in some areas of the northeastern United States.

Over the past four decades, however, sophisticated robot spacecraft—flybys, orbiters, and landers—have shattered these romantic myths of a race of ancient Martians struggling to bring water to the more productive regions of a dying world. Spacecraft-derived data have shown that the Red Planet is actually a "halfway" world. Part of the Martian surface is ancient, like the surfaces of the Moon and Mercury, while the other part is more evolved and Earthlike.

In August and September 1975, two Viking spacecraft were launched on a mission to help answer the question, Is there life on Mars? Each Viking spacecraft consisted of an orbiter and a lander. While scientists did not expect these spacecraft to discover Martian cities bustling with intelligent life, the exobiology experiments on the lander were designed to find evidence of primitive life-forms, past or present. Unfortunately, the results sent back by the two robot landers were teasingly inconclusive.

The Viking Project was the first mission to successfully soft-land a robot spacecraft on another planet (excluding the Earth's Moon). All four Viking spacecraft (two orbiters and two landers) exceeded by considerable margins their design goal lifetime of 90 days. The spacecraft were launched in 1975 and began to operate around or on the Red Planet in 1976. When the *Viking 1* lander touched down on the Plain of Chryse on July 20, 1976, it found a bleak landscape. Several weeks later, its twin, the *Viking 2* lander, set down on the Plain of Utopia and discovered a more gentle, rolling landscape. One by one, these robot explorers finished their highly successful visits to Mars. The *Viking 2* orbiter spacecraft ceased operation in July 1978; the *Viking 2* lander fell silent in April 1980; the *Viking 1* orbiter managed at least partial operation until August 1980; the *Viking 1* lander made its final transmission on November 11, 1982. NASA officially ended the Viking mission to Mars on May 21, 1983.

As a result of these interplanetary missions, scientists now know that Martian weather changes very little. For example, the highest atmosphere temperature recorded by either *Viking* lander was -5.8°F (-21°C) (midsummer at the *Viking 1* site), while the lowest recorded temperature was -191°F (-124°C) (at the more northern *Viking 2* site during winter).

The atmosphere of Mars was found to be primarily carbon dioxide (CO_2)—about 95 percent by volume. Nitrogen, argon, and oxygen are present in small percentages, along with trace amounts of neon, xenon, and krypton. The Martian atmosphere contains only a wisp of water (about 1/1000th as much as found in Earth's atmosphere). But even this tiny amount can condense and form clouds that ride high in the Martian atmosphere or form patches of morning fog in valleys. There is also evidence that Mars had a much denser atmosphere in the past—one capable of permitting liquid water to flow on the planet's surface. Physical features resembling riverbeds, canyons and gorges, shorelines, and even islands hint that large rivers and maybe even small seas once existed on the Red Planet.

Mars has two small, irregularly shaped moons, Phobos ("Fear") and Deimos ("Terror"). The longest dimension on irregularly shaped Phobos is 16.8 miles (27 km) and on Deimos 9.3 miles (15 km). The two natural satellites were discovered in 1877 by the American astronomer Asaph Hall (1829–1907). Both tiny moons have ancient, cratered surfaces with some indication of regoliths to depths of possibly 16 feet (5 m) or more. Planetary scientists suggest that the two small moons are actually asteroids that were captured by Mars eons ago.

Scientists also think that several unusual meteorites found on Earth are actually pieces of Mars that were blasted off the Red Planet by ancient meteoroid collisions. One particular Martian meteorite, called ALH84001, has stimulated a great deal of interest in the possibility of life on Mars. In the summer of 1996, a NASA research team at the Johnson Space Center announced that they had found evidence in ALH84001 that "strongly suggests primitive life may have existed on Mars more than 3.6 billion years ago." Inside this ancient Martian rock, the NASA team found the first organic molecules thought to be of Martian origin, several mineral features characteristic of biological activity, and possible microfossils (i.e., very tiny fossils) of primitive, bacteria-like organisms.

Stimulated by the exciting possibility of life on Mars, NASA and other space organizations have launched a variety of robot spacecraft to accomplish more focused scientific investigations of the Red Planet. Starting in 1996, some of these missions have proven highly successful, while others have ended in disappointing failures.

This new wave of exploration started on November 7, 1996, when NASA launched the *Mars Global Surveyor* (*MGS*) from Cape Canaveral Air Force Station in Florida. The spacecraft arrived at Mars on September 12, 1997, an event representing the first successful mission to the Red Planet in two decades. After spending a year and a half carefully trimming its orbit from a looping ellipse to a more useful circular track around the planet, the *MGS* began its mapping mission in March 1999. Using its high-resolution camera, the *MGS* observed the planet from its low-altitude, nearly polar orbit over the course of an entire Martian year—the equivalent of nearly two Earth years. At the conclusion of its primary scientific mission on January 31, 2001, the spacecraft entered an extended mission phase. The *MGS* stopped communicating with scientists on Earth on November 2, 2006. In addition, NASA's Mars Exploration Rover *Opportunity* could no longer detect any signal from the orbiting spacecraft at the end of November. The absence of these signals indicates that the extended, productive mission of the *MGS* has finally ended. The *MGS* has successfully studied the entire Martian surface, atmosphere, and interior, returning an enormous amount of valuable scientific data. Among its most significant scientific contributions so far are high-resolution images of gullies and debris flow features that suggest there may be current sources of liquid water, similar to an aquifer, at or near the surface of the planet. These findings are shaping and guiding upcoming robot missions to Mars.

NASA launched the Mars Pathfinder mission to the Red Planet on December 4, 1996, using a Delta II expendable launch vehicle. The mission, formerly called the Mars Environmental Survey, or MESUR, had as its primary objective the demonstration of innovative, low-cost technology for delivering an instrumented lander and free-ranging robotic rover to the Martian surface. The *Mars Pathfinder* not only accomplished that important objective, but it also returned an unprecedented amount of data and operated well beyond its anticipated design life. From the robot spacecraft's innovative airbag bounce-and-roll landing on July 4, 1997, until its final data transmission on September 27, it returned numerous images of the Ares Vallis landing site and useful chemical analyses of proximate rocks and soil deposits. Data from this successful mission have suggested that ancient Mars was once warm and wet, further stimulating the intriguing question of whether life could have emerged on that planet when liquid water flowed on its surface and its atmosphere was significantly thicker.

However, the exhilaration generated by these two successful missions was quickly dampened by two glaring failures. On December 11, 1998, NASA launched the *Mars Climate Orbiter* (*MCO*). This spacecraft was to serve as both an interplanetary weather satellite and a data relay satellite for another mission, called the Mars Polar Lander (MPL). The *MCO* also carried two science instruments—an atmospheric sounder and a color imager. However, just as the spacecraft arrived at the Red Planet on September 23, 1999, all contact was lost with it. NASA engineers have concluded that because of human error in programming the final trajectory, the spacecraft most probably attempted to enter orbit too deep in the planet's atmosphere and consequently burned up.

NASA used another Delta II expendable launch vehicle to send the *MPL* to the Red Planet on January 3, 1999. The MPL was an ambitious mission to land a robot spacecraft on the frigid Martian terrain near the edge of the planet's southern polar cap. Two small penetrator probes (called *Deep Space 2*) piggybacked with the lander spacecraft on the trip to Mars. After an uneventful interplanetary journey, the *MPL* and its companion *Deep Space 2* experiments were mysteriously lost when the spacecraft arrived at the planet on December 3, 1999.

Undaunted by the disappointing sequential failures, NASA officials sent the 2001 Mars Odyssey mission to the Red Planet on April 7, 2001. The scientific instruments on board the orbiter spacecraft were designed to determine the composition of the planet's surface, detect water and shallow buried ice, and study the ionizing radiation environment in the vicinity of Mars. The spacecraft arrived at the planet on October 24, 2001, and successfully entered orbit around it. After executing a series of aerobrake maneuvers that properly trimmed it into a near-circular polar orbit around Mars, the spacecraft began to make scientific measurements in January 2002. The spacecraft's primary science mission continued through August 2004 and, as of December 2006, *Odyssey* was functioning in an extended mission, which included service as a communications relay for the Mars Exploration Rovers (*Spirit* and *Opportunity*).

In the summer of 2003, NASA launched identical twin *Mars Exploration Rovers* that were to operate on the surface of the Red Planet during 2004. *Spirit* (MER-A) was launched by a Delta II rocket from Cape Canaveral Air Force Station on June 10, 2003, and successfully landed on Mars on January 4, 2004. *Opportunity* (MER-B) was launched from Cape Canaveral Air Force Station on July 7, 2003, by a Delta II rocket and successfully landed on the surface of Mars on January 25, 2004. Both successful landings resembled the airbag bounce-and-roll arrival demonstrated during the Mars Pathfinder mission. Following arrival on the surface of the Red Planet, each rover drove off and began its surface exploration mission in a different location on Mars. As of January 2007, both rovers continue to operate on the surface of Mars. Despite a nonfunctioning right wheel, *Spirit* remains healthy and on the move in the Gusev Crater region. The fully functioning *Opportunity* is providing panoramic (surface-view) images of the Victoria Crater region of the Red Planet.

In 2003, NASA also participated in a mission called Mars Express, sponsored by the European Space Agency (ESA) and the Italian Space Agency. Launched in June 2003, the *Mars Express* spacecraft arrived at the Red Planet in December 2003. Following successful arrival maneuvers, the spacecraft's scientific instruments began to study the atmosphere and surface of Mars from a polar orbit. The main objective of the *Mars Express* is to search from orbit for suspected subsurface water locations. The spacecraft also delivered a small robot lander spacecraft to more closely investigate the most suitable candidate site. This small lander was named *Beagle 2* in honor of the famous ship in which the British naturalist Charles Darwin (1809–82) made his great voyage of scientific discovery. After coming to rest on the surface of Mars, *Beagle 2* was to have performed exobiology and geochemistry research. The *Beagle 2* was scheduled to land on December 25, 2003; however, ESA ground controllers were unable to communicate with the probe, and it was presumed lost. Despite the problems with *Beagle 2,* the *Mars Express* spacecraft has functioned well in orbit around the planet and accomplished its main mission of global high-resolution photogeology and mineralogical mapping of the Martian surface.

On August 12, 2005, NASA successfully launched a powerful new scientific spacecraft called the *Mars Reconnaissance Orbiter (MRO).* The *MRO* arrived at Mars on March 10, 2006, and is now scrutinizing those candidate water-bearing locations previously identified by the *Mars Global Surveyor* and *2001 Mars Odyssey.* The *MRO* is capable of measuring thousands of Martian landscapes with a spatial resolution of between 0.66 foot (0.2 m) and 1.0 foot (0.3 m). By way of comparison, this spacecraft's imaging capability is good enough to detect and identify rocks on the surface of Mars that are as small as a beach ball. The high-resolution imagery data are helping scientists bridge the gap between detailed, localized surface observations accomplished by landers and rovers and the synoptic global measurements made by orbiting spacecraft.

Possibly as early as 2009, NASA plans to develop and launch a long-duration, long-range mobile science laboratory. This effort will demonstrate the efficacy of developing and deploying truly "smart landers"—advanced robotic systems that are capable of autonomous operation, including hazard avoidance and navigation around obstacles to reach promising but difficult-to-reach scientific sites. NASA also proposes to create a new family of small scout missions. These missions would involve airborne robotic vehicles (such as a *Mars airplane*) and special miniature surface landers or penetrator probes, possibly delivered to interesting locations around Mars by the airborne robotic vehicles. These scout missions, beginning in about 2007, would greatly increase the number of interesting sites studied and set the stage for even more sophisticated robotic explorations in the second decade of this century.

As presently planned, NASA intends to launch its first Mars Sample Return Mission in about 2014. Future advances in robot spacecraft technology will enhance and accelerate the search for possible deposits of subsurface water and life (existent or extinct) on the Red Planet. Two decades of intensely focused scientific missions by robot spacecraft will not only increase scientific knowledge about Mars, but the effort will also set the stage for the first Mars expedition by human explorers—now anticipated to take place in about 2030.

✧ Human–Crewed Expedition to Mars

The Mars crewed expedition will be the first visit to another planet by human beings and will most likely occur before the mid-part of this century—possibly as early as 2030. Many current concepts suggest a 600-

to 1,000-day duration mission that most likely will start from Earth orbit, possibly powered by a nuclear thermal rocket. A total crew size of up to 15 astronauts is anticipated. After hundreds of days of travel through interplanetary space, the first Martian explorers will have about 30 days allocated for surface excursion activities on the Red Planet.

The commitment to a human expedition to Mars is an ambitious undertaking that will require strong political and social support that extends over several decades. One nation, or several nations joined in an international, cooperative venture, must be willing to make a lasting statement about the value of human space exploration as an integral part of humans' future civilization. A successful crewed mission to Mars will establish a new frontier that has unusual scientific, social, and philosophical dimensions. If the next generation views the first crewed mission to Mars as a precursor to the permanent settlement of the Red Planet by human beings, then the overall impact of the first expedition on society and civilization will be significantly amplified.

The amount of mass that must be lifted from Earth (or the Moon) can be reduced by as much as 50 percent if a structure called an aerobrake is used. This artist's rendering shows a "molly bolt" lander vehicle design that allows the aerobrake to be deployed in a flat shape for aerodynamic entry and landing, and then retracted to form a smooth conical shape for ascent from the Red Planet. *(NASA/JSC; artist, Pat Rawlings)*

This artist's rendering depicts one crewmember of the first human expedition to Mars (ca. 2030) performing sample collections at an interesting site some distance away from the base camp. *(NASA/JSC)*

Exactly what happens after the first human expedition to Mars is open to wide speculation at present. People here on Earth could simply marvel at another outstanding space exploration first and then settle back to their more pressing terrestrial pursuits. This pattern followed the spectacular Moon-landing missions of NASA's Apollo Project (1969–72). On the other hand, if this first human expedition to Mars is widely recognized and appreciated as the precursor to the permanent occupancy of heliocentric space by the human race, then Mars would truly become the central object of greatly expanded exploration activities and surface operations—perhaps complementing the rise of a self-sufficient lunar civilization.

Outside of the Earth-Moon system, Mars is the most hospitable body in the solar system for humans and is currently the only practical candidate for human exploration and settlement in the mid-decades of this century. Mars also offers the opportunity for in situ resource utilization (ISRU). With ISRU initiatives, the planet can provide air for the astronauts to breathe and fuel for their surface rovers and return vehicle. In fact, ISRU has been regarded as an integral part of the many recent Mars expedition scenarios. In one NASA study, for example, engineers have suggested that a Mars ascent vehicle (for crew departure from surface of planet), critical supplies, an unoccupied habitat, and an ISRU extraction facility, will all be pre-positioned on the surface of the Red Planet *before* the first human explorers ever depart from Earth.

Of course, the logistics of a crewed mission to Mars are complex, and many factors (including ISRU) must be considered before a team of human explorers sets out for the Red Planet with an acceptable level of risk and a reasonable hope of returning safely to Earth. The establishment of a permanent lunar base is now being viewed by NASA long-range planners as a necessary step before human explorers are sent to Mars. (See chapter 10.) The overall hardware performance and human-factor experience gained from extended lunar surface operations should provide Mars mission planners the necessary data and confidence that the expedition hardware ultimately selected significantly reduces both the cost and the risk of the crewed interplanetary journey.

Some of the other important factors that must be carefully considered include the overall objectives of the expedition, the selection of the transit vehicles and their trajectories, the desired stay-time on the surface of Mars, the primary site to be visited, the required resources and equipment, and crew health and safety throughout the extended journey. Due to the nature of interplanetary travel, there will be no quick return to Earth and not even the possibility of supplementary help or rescue from Earth, should the unexpected happen. Once the crew departs from the Earth-Moon system and heads for Mars, they must be totally self-sufficient and flexible enough to adapt to all new situations.

✦ Mars Outpost and Surface Base Concepts

For automated Mars missions, the spacecraft and robotic surface rovers generally will be small and self-contained. For human expeditions to the surface of the Red Planet, however, two major requirements must be satisfied: life support (habitation) and surface transportation (mobility). Habitats, power supplies, and life support systems will tend to be more complex in a permanent Martian surface base that must sustain human beings for years at a time. Surface mobility systems will also grow in complexity and sophistication as early Martian explorers and settlers travel tens to hundreds of miles from their base camp. At a relatively early time in any Martian surface base program, the use of Martian resources to support the base must be tested vigorously and then quickly integrated in the development of an eventually self-sustaining surface infrastructure.

In one candidate scenario, the initial Martian habitats will resemble standardized lunar base (or space station) pressurized modules and would be transported from cislunar space to Mars in prefabricated condition by interplanetary nuclear electric propulsion cargo ships. These modules would then be configured and connected as needed on the surface of Mars

This artist's rendering shows the major components of one possible Mars outpost that could support up to seven astronauts while they explored the surface of the Red Planet. The main components are a habitat module, pressurized rover dock/equipment lock, air locks, and a 52.5-foot- (16-m-) diameter, erectable (inflatable) habitat. Also appearing in the picture are a Mars balloon, an unpressurized rover, a storage work area, a geophysical experiment area, and a local area antenna. In the scenario depicted, many of the elements of this Mars outpost were derived from an earlier lunar test–bed facility. *(NASA/JSC; artist, Mark Dowman of John Frassanito and Associates)*

and covered with about three feet (1 m) or so of Martian soil for protection against the lethal effects of solar flare radiation or continuous exposure to cosmic rays on the planet's surface. Unlike Earth's atmosphere, the thin Martian atmosphere does not shield well against ionizing radiations from space.

Another mid-century Mars base concept involves an elaborate complex of habitation modules, power modules, central base work facilities, a greenhouse, a launch and landing complex, and even a robotic Mars airplane. The greenhouse on Mars would provide astronauts with some much-needed dietary variety. As an early Mars outpost grows into a sufficiently large permanent human settlement, a system of greenhouses will

An artist's rendering of a mid–21st-century Mars base near Pavonis Mons, a large shield volcano on the Martian equator that overlooks the ancient water-eroded canyon in which the base is located. The base infrastructure shown here includes a habitation module, a power module, central base work facilities, a greenhouse, a launch and landing complex, and even a robotic Mars airplane. In the foreground, human explorers have taken their surface rover to an interesting spot, where one of the team members has just made the discovery of the century, a well-preserved fossil of an ancient Martian creature. *(NASA/JSC; artist, Pat Rawlings)*

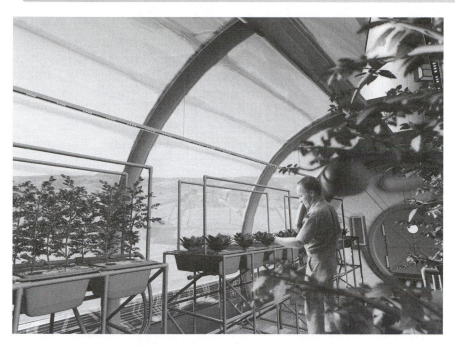

As depicted in this artist's rendering, astronauts exploring Mars will build hydroponic growth labs where vegetables and fruits can be grown. These crops will provide the Martian explorers with dietary variety and additional nutrition. *(NASA/JSC)*

be necessary to establish food self-sufficiency. In time, food grown at the Mars base could be used to supply human space-exploration missions that depart the Red Planet and travel into the asteroid belt and beyond.

✧ Space Policy and the American Presidency

Many otherwise well-informed Americans citizens often forget that a dynamic and ambitious national space program involves several dimensions: an executive vision (usually a clear and focused presidential mandate communicated to NASA and converted into a viable strategic plan), a space technology infrastructure (existing or successfully emerging), a sustained budget (as approved and authorized by the U.S. Congress), and social support (primarily in the form of voter approval and enthusiasm).

In a free and open society, such as the United States, maintaining strong public approval for long-term projects—especially those that bridge several presidential administrations—is a difficult task. The cherished space project of one president's administration often experiences the budget ax by the next presidential administration. Furthermore, Americans quite often get restless with projects that "take too long" to show tangible results. A human expedition to Mars, even if performed in an international effort, will require an unprecedented, sustained commitment by the United States government that spans at least two decades and several presidential administrations. The social and political impediments may prove as challenging—if more so—than the technical hurdles that must be overcome, if human beings are to personally explore the Red Planet this century.

Unfortunately, the gradual movement of the human race off the planet and the diffusion of human beings throughout the solar system is more of an evolutionary process than a revolutionary event. The process will, of course, contain many exciting, clearly discernible milestones (like the first human landing on the Moon on July 20, 1969). However, from the perspective of human history, the emergence of the enabling technical infrastructure (such as a permanent lunar base and an outpost on Mars) will more closely resemble a glacial phenomenon than a volcanic eruption. If history is any indicator of future human behavior, however, the real stimulus for future milestones in off-planet development will most likely be executive decisions and reactionary space agency activities that are triggered by political needs—real or perceived—than the logical societal response to a long-range vision for the maturation of the human species.

President Kennedy's decision in 1961 to send American astronauts to the Moon demonstrates an executive decision about space exploration triggered by dire political necessity. Kennedy's administration faced a serious need to restore the global perception that the United States was still the number-one technical power on the planet. The federal budget was made available for NASA to get the job done. Once accomplished, however, and almost simultaneous with the first two human-crewed Moon landings, another presidential administration (that of Richard M. Nixon) began canceling the remaining Apollo Project missions (Apollo 18, 19 and 20 to be precise) and quickly started downsizing the space agency's relatively large budget in order to pay for a costly war in Southeast Asia.

Any 50-year strategic plan to create a permanent lunar base and then send human beings on to Mars requires a multigenerational sharing of the simple, strategic vision that space exploration is integral to the future well-being of the United States and, by logical extension, to the survival of the human race. In the absence of widespread acceptance of this important vision, space programs like the Apollo Project, the U.S. Space Transportation System, (space shuttle), and the *International Space Station* will remain unpredictably cyclic in nature—with their oscillations in technical capability and achievements often dampening while a new administration pursue the next "politically popular" space project.

The powerful Saturn V rocket is an example of a superbly engineered launch vehicle that essentially became extinct after its last flight on May 14, 1973 (the Skylab 1 mission). NASA is considering resurrecting and improving the Saturn V vehicle's upper stage J-2 rocket engine for use in the upper stages of the new Ares I and Ares V. But most of the fiscal and technical resources the United States invested in this vehicle paid residual dividends. Once the Apollo Project was first downsized and then terminated in the 1970s, surviving aerospace engineers and project managers turned their collective attention to developing a new "reusable" space shuttle vehicle.

Often overlooked was the social cost of this oscillatory approach to space exploration. As the Apollo Project faced severe cutbacks in the early 1970s, thousands of well-trained engineers and technicians suddenly found themselves unemployed or severely underemployed. The race to the Moon of the 1960s had marshaled a great deal of career interest in mathematics, physics, and engineering. But, in the post-Apollo employment "crash" of the 1970s, space exploration no longer served as a bright career beacon. As a result, many of the best and brightest American students began avoiding the "more difficult" science, engineering, physics, and mathematics courses in high school and college. That negative (downward) trend continues to this day in both secondary schools and colleges throughout the United States. Although computer literacy within the general student population remains relatively high, most contemporary students lack an in-depth understanding of the underlying physical principles that define today's technology-based global civilization.

Direct evidence for this oscillatory programmatic behavior in the American space program is easily discovered by reviewing what percentage NASA has received of the total federal budget. Comparison with other national priorities is also helpful. Using 2003 data from the White House Office of Management and Budget, in 1968 (during the peak of the Apollo Project) NASA received 2.7 percent of the total federal budget versus (for example) the 7.3 percent given to health-related federal expenditures. In 1981, NASA received only 0.8 percent of the federal budget (less than one cent of each federal dollar) versus 11.9 percent for health. Finally, in 2003, NASA received 0.7 percent of the federal budget versus 23.5 percent for health. During that same period, military and defense spending also went through a dramatic decline (when expressed as a percent of the total federal budget). Specifically, in 1968, the defense and military expenditures amounted to 45.1 percent of the total budget, 22.7 percent in 1981, and 16.7 percent in 2003. While budget statistics do not represent the entire story, these data do suggest trends in federal support for space exploration.

It is also interesting to review the contributions to human space exploration made by each of the Space Age American presidents. During the administration of Dwight Eisenhower (1953–61), the Space Age dawned, and the American public perceived that the United States was seriously lagging behind the former Soviet Union. Eisenhower had actually approved the development (in secret) of a series of successful military reconnaissance satellites. But his administration was clearly caught off guard by the enormous emotional shock wave and adverse public response to the surprise Soviet *Sputnik 1* launch on October 4, 1957. To rescue national prestige, his administration had to prepare a hasty response by using a modified U.S. Army rocket. The impromptu effort allowed the United States to launch its first satellite (*Explorer 1*) on January 31, 1958. Later that year, NASA was formed on October 1, 1958, and quickly announced the Mercury Project. Civilian space exploration, including the dream of human spaceflight, had become a highly visible instrument of cold-war geopolitics.

President John F. Kennedy's administration (1961–63) witnessed the first American to fly in space (Alan Shepard on May 5, 1961) and the first American to orbit Earth (John Glenn on February 20, 1962). Kennedy's bold and visionary "Urgent Needs" speech to the U.S. Congress on May 25, 1961, launched the Apollo Project and defined American human spaceflight efforts for the next decade.

Following the assassination of Kennedy (November 22, 1963), the administration of Lyndon Baines Johnson (1963–69) pursued the slain president's bold Moon-landing vision and witnessed NASA's steady progress of the Mercury, Gemini, and Apollo Projects. A skilled and forceful politician, Johnson was able to maintain NASA's funding, despite technical setbacks, rising costs, and the *Apollo 1* tragedy (January 27, 1967). At the end of his administration, NASA successfully sent the *Apollo 8* astronauts on the first circumlunar flight—winning (in part) the unofficial cold-war race to send human beings to the Moon.

The administration of Richard M. Nixon (1969–74) harvested the global political benefits of the first lunar-landing mission (*Apollo 11* on July 20, 1969). But, to fund the rising costs of the war in Southeast Asia, Nixon began to slash NASA's budget by canceling the planned Apollo 18, 19 and 20 missions. In January 1972, he approved development of the space shuttle. His administration also witnessed the first American space station project, called *Skylab* (1973–74).

During Gerald Ford's administration (1974–77), NASA astronauts successfully participated in the Apollo-Soyuz Test Project (July 1975)—the first international docking and rendezvous project between the United States and the former Soviet Union. Ford's administration also made major design and performance decisions with respect to the space shuttle. Many of these decisions curtailed anticipated capabilities to accommodate declining space program budgets.

President Jimmy Carter's administration (1977–81) did not include a spaceflight by any American astronaut. Development of the space shuttle continued, but the system again experienced schedule slips due to both technical problems and budget constraints.

During Ronald Reagan's presidency (1981–89), there were many shuttle flights (including STS-1 on April 12, 1981). After the *Challenger* accident (January 28, 1986), Reagan committed the nation to build a replacement shuttle, later called *Endeavour*. But, despite his popularity, Reagan's call for a new space station (named *Freedom*) essentially went unheeded by the U.S. Congress and relatively ignored by the American voters.

The administration of George H. Bush (1989–93) inherited the continually changing (primarily downsizing) concept of space station *Freedom*. Bush's call for human missions to the Moon and Mars went totally unheeded by the U.S. Congress because of the "sticker shock" of the proposed program's estimated total cost (about $450 billion). As the cold war ended, the American public and the U.S. Congress no longer regarded competition in space exploration as an urgent national priority. The first Gulf War also provided American citizens with a strong distraction from thinking about things like space exploration.

During his presidency (1993–2001), Bill Clinton revived and internationalized the space station concept. The first assembly mission of the *International Space Station* occurred during the STS-88 mission of space shuttle *Endeavour* (December 1998). Despite his administration's enthusiastic support for a new, fully reusable launch vehicle, called the X-33 Project, the effort was canceled in March 2001 (by NASA at the direction of the next presidential administration) because of cost overruns and enormous technical roadblocks.

The presidency of George W. Bush (2001–present) witnessed several major, world-changing events, including terrorist attacks on the United States (September 11, 2001), the start of a global war against terrorism, military actions in Afghanistan and Iraq, and the loss of the shuttle *Columbia* (February 1, 2003). On January 14, 2004, Bush presented a bold "Vision for Space Exploration" in which he instructed NASA to return the shuttle to flight, complete the *International Space Station,* and develop the infrastructure to return human beings to the Moon by 2020 and then on to Mars by 2030. NASA subsequently responded to the latest presidential space exploration initiative by introducing the new Ares I and V launch vehicles and the new *Orion* crew exploration spacecraft. NASA officials currently plan to retire the space shuttle in 2010 and begin sending astronauts to the *ISS* using the new Ares I/*Orion* system in about 2015. Yet, how much of President George W. Bush's overall space exploration vision becomes a reality will be determined by many factors, including major political and economic influences that lie well beyond the aerospace arena.

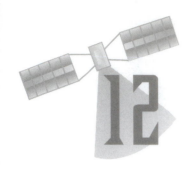

Large Space Settlements— Hallmark of a Solar System Civilization

The large space settlement is the centerpiece of a grand technical vision, involving the construction of human-made mini-worlds that would result in the spread of life and civilization throughout the solar system. Long before the Space Age began, the British physicist and writer John Desmond Bernal (1910–71) speculated in his futuristic 1929 work *The World, the Flesh and the Devil* about the colonization of space and the construction of large, spherical space settlements (now called Bernal spheres). Although Bernal's use of the term *space colony* has yielded to the more politically acceptable term *space settlement,* his basic idea of a large, self-sufficient human habitat in space has stimulated numerous Space Age–era studies. These subsequent studies have spawned other interesting habitat concepts—some engineering extrapolations of Bernal's basic notion and others dramatically different in form or purpose.

This chapter introduces some of the exciting space settlement concepts that emerged as a natural intellectual by-product of the early American human spaceflight programs that culminated in the Apollo Project lunar-landing missions of the 1960s. For example, German-American rocket scientist and space visionary Krafft Arnold Ehricke (1917–84) completed a variety of long-range strategic studies that creatively extrapolated contemporary developments in space technology (such as the lunar-landing missions and NASA's *Skylab* space station) over a century or more. Through his innovative technical concepts involving the androsphere, the androcell, and astropolis, Ehricke attempted to assess and describe the enormous social impacts that the creation of a solar system civilization would have on the human race.

While considering the role of space technology in the synthesis of terrestrial and extraterrestrial environments in the 1960s and 1970s, Ehricke coined the term *androsphere.* Within his far-reaching vision for the human

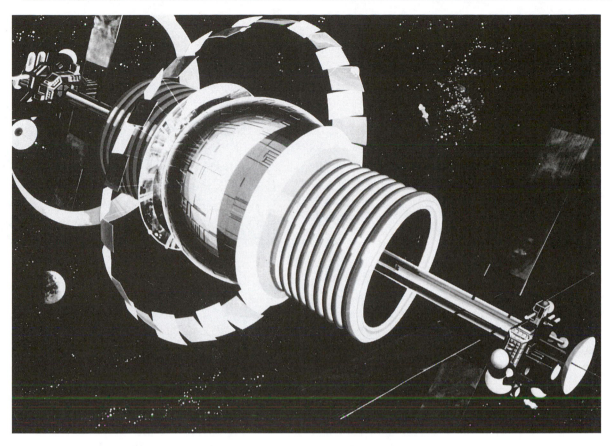

An artist's rendering showing the exterior view of a large space settlement with a spherical (Bernal-type) habitat design. *(NASA/Ames Research Center)*

settlement of space, the androsphere relates to the productive integration of Earth's biosphere (which contains the major terrestrial environmental regimes) and the material and energy resources of the solar system, such as the Sun's radiant energy and the Moon's mineral resources.

✧ Astropolis

Sometime in the late 1960s, Krafft Ehricke began forming his concept for astropolis—an urban facility located in near-Earth space. He envisioned astropolis as a logical growth step beyond the space station. The Earth-orbiting facility would contain several hundred up to perhaps several thousand inhabitants who would live and work in a large human-made world that featured multiple gravity levels. His proposed facility would contain residential sections, a dynarium (a spacious enclosure for human-

SPACE RESOURCES

Generally, when people think about outer space, visions of vast emptiness, devoid of anything useful, come to their minds. However, space is really a new frontier that is rich with resources, including an essentially unlimited supply of (solar) energy, a full range of raw materials, and an environment that is both special (i.e., high vacuum, microgravity, physical isolation from the terrestrial biosphere) and reasonably predictable—although large solar flares represent unpredictable, occasional threats.

Since the start of the Space Age, investigations of meteorites, the Moon, Mars, and several asteroids and comets have provided tantalizing hints about the rich mineral potential of the extraterrestrial environment. For example, NASA's Apollo Project expeditions to the lunar surface established that the average lunar soil contains more than 90 percent of the material needed to construct a complicated space industrial facility. The soil in the lunar highlands is rich in anorthosite, a mineral suitable for the extraction of aluminum, silicon, and oxygen. Other lunar soils have been found to contain ore-bearing granules of ferrous metals, such as iron, nickel, titanium, and chromium. Iron can be concentrated from the lunar soil (called regolith) before the raw material is even refined simply by sweeping magnets over regolith to gather the iron granules scattered within.

Remote-sensing data of the lunar surface obtained in the 1990s by the Department of

In this artist's concept, on the way to the Jovian system, a nuclear thermal rocket-powered interplanetary cargo transfer vehicle refuels in orbit around Mars near the Red Planet's moon Phobos. (NASA; artist, Pat Rawlings, 1996)

Defense's *Clementine* spacecraft and NASA's *Lunar Prospector* spacecraft have encouraged some scientists to suggest that useful quantities of water ice are trapped in the Moon's perpetually shaded polar regions. If this speculation proves true, then "ice mines" on the Moon could provide both oxygen and hydrogen—vital resources for permanent lunar settlements and space industrial facilities. The Moon would be able both to export chemical propellants for propulsion systems and resupply materials for the life support systems of large human habitats constructed in cislunar space.

powered flight and other low-gravity recreational actvities), space industry zones, space agriculture facilities, research laboratories, and exobiology sections called other world enclosures (OWEs).

Ehricke's astropolis would have the ability to recycle air, water, and waste materials. In his basic concept, either nuclear power plants or solar arrays would supply energy. The research section of astropolis would be

Mars's vast mineral-resource potential, frozen volatile reservoirs, and strategic location will make it a critical supply depot for human expansion into the mineral-rich asteroid belt and to the giant outer planets and their fascinating collection of resource-laden moons. Smart robot explorers will assist the first human settlers on Mars, enabling these Martians pioneers to assess quickly and efficiently the full resource potential of their new world. As the early Martian bases mature into large permanent settlements, they will become economically self-sufficient by exporting propellants, life support system consumables, food, raw materials, and manufactured products to feed the next wave of human expansion to the outer regions of the solar system. Cargo spacecraft will routinely travel between cislunar space and Mars, carrying specialty items to eager consumer markets in both extraterrestrial locations.

The asteroids, especially Earth-crossing asteroids, represent another important category of space resources. Recent space missions and analysis of meteorites (many of which scientists believe originate from broken-up asteroids) indicate that carbonaceous (C-type) asteroids may contain up to 10 percent water, 6 percent carbon, significant amounts of sulfur, and useful amounts of nitrogen. S-class asteroids, which are common near the inner edge of the main asteroid belt and among the Earth-crossing asteroids, may contain up to 30 percent free metals (alloys of iron, nickel, and cobalt, along with high concentrations of precious metals). E-class asteroids may be rich sources of titanium, magnesium, manganese, and other metals. Finally, chondrite asteroids, which are found among the Earth-crossing population, are believed to contain accessible amounts of nickel, perhaps more concentrated than the richest deposits found on Earth.

Using smart machines, possibly including self-replicating systems, space settlers in the next century might be able to manipulate large quantities of extraterrestrial matter and move it to wherever it is needed in the solar system. Many of these space resources will be used as the feedstock for the orbiting and planetary surface base industries that will form the basis of interplanetary trade and commerce. For example, atmospheric (aerostat) mining stations could be set up around Jupiter and Saturn, extracting such materials as hydrogen and helium—especially helium 3, an isotope of great potential value in nuclear fusion research and applications. Similarly, Venus could be mined for the carbon dioxide in its atmosphere, Europa for water, and Titan for hydrocarbons. Large fleets of robot spacecraft might even be used to gather chunks of water ice from Saturn's ring system, while a sister fleet of robot vehicles extracts metals from the main asteroid belt. Even the nuclei of selected comets could be intercepted and mined for frozen volatiles, including water ice. Finally, beyond the orbit of Pluto is the Kuiper belt and its population of thousands and thousands of icy planetesimals, which range in size from a few hundred feet in diameter to hundreds of miles in diameter.

dedicated to the long-term use of the space environment for basic and applied research, as well as for the industrial exploitation of the resources of the solar system. The other world enclosures would be located at various distances from the hub of astropolis. Using these special OWE facilities, exobiologists, space scientists, planetary engineers, and interplanetary explorers would be able to simulate the gravitational environment of all major celestial objects in the solar system of interest from the perspective of human visitation and possible settlement. The OWEs would include simulations of the Moon, Mars, Venus, Mercury, selected large asteroids, and many of the major moons of the giant outer planets. Pioneering work performed within these OWE facilities would pave the way for the occupancy of both cislunar space and heliocentric space by human beings.

Ehricke envisioned astropolis as a 4,000- to 15,000-ton-class space complex that would be rotated very slowly at about 92.5 revolutions per Earth day (24 hours). Because of its low angular velocity, the Coriolis force (for example, sideward force felt by an astronaut moving radially in a rotating system, such as a space station with artificial gravity) would cause little disturbance and discomfort even at the greatly reduced artificial gravity levels occurring close to the hub. Research and industrial projects conducted on this type of carefully designed orbiting facility would enjoy excellent, variable gravity-level simulations with minimal Coriolis force disturbance. This beneficial condition would not occur on smaller, more rapidly spinning space stations with mechanically produced artificial gravity.

✧ Androcell

The androcell is a bolder concept that Ehricke proposed in the 1970s. The androcell would be a large, human-made world—an independent, self-contained human biosphere not located on any naturally existing celestial object. In Ehricke's vision, such human-made mini-worlds, or planetellas, would use mass far more efficiently than the natural worlds of the solar system, which formed out of the original solar nebular material some 4 to 5 billion years ago. The naturally formed terrestrial planets (Earth, Mars, Mercury and Venus) and the wide variety of moons found throughout the solar system are essentially "solid" spherical objects of great mass. The surface gravity force on each of these "solid" worlds results from the self-attraction of a large quantity of matter. However, except for the first mile or so, the interior of each of these natural worlds is essentially useless from the perspective of human habitation.

Instead of large quantities of matter, the androcell would use rotation (centrifugal inertia) to provide variable levels of artificial gravity. The unusable solid interior of a natural celestial body is now replaced (through human ingenuity and engineering) with many useful inhabitable layers of airtight, habitable cylinders. In concept, inhabitants of an androcell would be able to enjoy a truly variable lifestyle in a multiple gravity-level miniworld. There would be a maximum gravity level at the outer edges of the androcell, tapering off to essentially zero gravity in the inner cylinder levels closest to the central hub.

One especially important idea contained within Ehricke's overall visionary concept is that the androcell would not be tied to the Earth-Moon system. Rather—with its giant space-based factories, farms, and fleets of merchant spacecraft—the androcell would be free to seek political and economic development throughout heliocentric (Sun-centered) space. Its inhabitants might trade with Earth, the Moon, Mars, or with other androcells. These giant space settlements of 10,000 to perhaps 100,000 or more people represent the Space Age analogy to the city-state of ancient Greece. The multiple gravity-level lifestyle would also encourage migration to and from settlements on other "natural" worlds—perhaps a terraformed Mars, an environmentally subdued Venus, or maybe even one of the larger moons of the giant outer planets. In essence, the androcell represents the cellular division of humanity—since, as residents of autonomous extraterrestrial city-states, their inhabitants could choose to pursue culturally diverse lifestyles throughout the solar system.

Of course, the human race already has the initial, natural androcell—it is called "Spaceship Earth." In time, inhabitants of the humans' parent world would be able to use their technical skills and intelligence to fashion a series of such androcells or other large space settlements throughout the solar system. As the number of these artificial human habitats grows, a swarm of settlements might eventually encircle the Sun, capturing and using its entire energy output. At that point, the solar system–wide civilization of the human race will have created a Dyson sphere, making the next step of cosmic mitosis—migration to the stars—technically, economically, and socially feasible. The Dyson sphere and its implications are discussed in the last section of this chapter.

✦ Space Settlement Concepts Sponsored by NASA

Responding to the energy crisis of the 1970s and the emergence of the satellite power system concept, NASA sponsored a series of university-based concept studies involving space resources and space settlements. These multidisciplinary studies focused on space habitats (originally called space colonies) containing from 1,000 to perhaps 10,000 people who would live, work, and play while supporting space industrialization activity, such as the operation of a large space manufacturing complex or the construction of satellite power systems. While the civilian space agency had no actual plans to develop any of these large space habitats, NASA planners and administrators regarded the studies as an interesting intellectual exercise that would identify various technical, psychological, and social issues that could arise when human beings began to migrate in large numbers beyond the boundaries of Earth.

One popular space settlement design that emerged in the late 1970s out of the NASA-sponsored studies is that of a torus-shaped habitat for about 10,000 people. The space settlement would be located in cislunar space at either Lagrangian libration point four (L_4) or five (L_5). Its inhabitants, all members of a space manufacturing complex workforce (and their families), would return after work to homes on the inner surface of the large torus, which would be nearly one mile (1.6 km) in circumference.

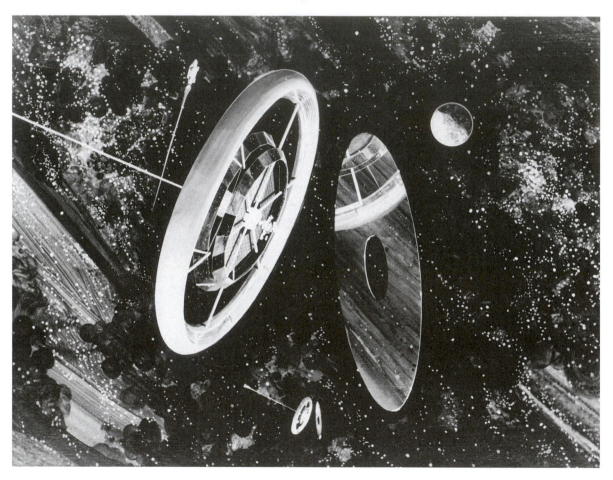

This artist's rendering provides an exterior view of a large space settlement capable of supporting about 10,000 people in cislunar space. As envisioned in various NASA-sponsored studies performed in the late 1970s, the inhabitants of this type of space settlement would harvest materials from the Moon and possibly from near-Earth asteroids to construct large satellite power systems, which would then provide energy to Earth. *(NASA/ Ames Research Center)*

The proposed torus-shaped space settlement would rotate to provide the inhabitants with a gravity level similar to that experienced on the surface of Earth. A nonrotating shell of material would shield the habitat against cosmic rays and solar flare radiation. To minimize cost, this shield could be built up from accumulated slag or waste materials from lunar or asteroid mining operations. On the outside of the shielded habitat living areas, the settlement's inhabitants would grow agricultural crops in special zones that took advantage of the intense continuous stream of sunlight available in space. Docking areas and microgravity industrial zones would be located at each end of the settlement, as well as the large, flat radiator surfaces necessary to reject waste heat away from the facility to outer space.

Another candidate design was that of a large spherical space settlement, based on the original Bernal sphere concept. The giant spherical habitat would be approximately 1.3 miles (2 km) in circumference. Up to 10,000 people would live in residences along the inner surface of the large sphere. Rotation of the settlement at about 1.9 revolutions per minute (rpm) would provide Earthlike gravity levels at the sphere's equator, but there would be essentially microgravity conditions at the poles. Because of the short distances between locations in the equatorial residential zone, passenger vehicles would not be necessary. Instead, the space settlers would travel on foot or perhaps by bicycle. The climb from the residential equatorial area up to the sphere's poles would take about 20 minutes and would lead the hiker past small villages, each at progressively lower levels of artificial gravity. An enclosed corridor at the axis would permit residents to float safely under microgravity conditions out to the settlement's exterior facilities, such as observatories, docking ports, and industrial and agricultural areas. Ringed areas above and below the main sphere in this type of space settlement would be the agricultural toruses.

Another possible space settlement design developed during the NASA-sponsored studies involved a large set of twin 20-mile- (32-km-) long, four-mile- (6.4-km-) diameter cylindrical space settlements. As envisioned, these enormous space settlements could house several hundred thousand people. Each cylinder would rotate around its main axis once every 114 seconds to create an Earthlike level of artificial gravity. Teacup-shaped containers ringing each cylinder would be used as agricultural stations. Each cylinder would also be capped by a space industrial facility and a power station. Large, movable, rectangular mirrors on the sides of each cylinder (hinged at one end to the cylinder) would direct sunlight into the habitat's interior, control the day-night cycles, and even regulate the settlement's seasons. A random number generator somewhere in the mirror's controller loop could be used to provide weather variations that are unpredictable but within certain, previously established, limits. This

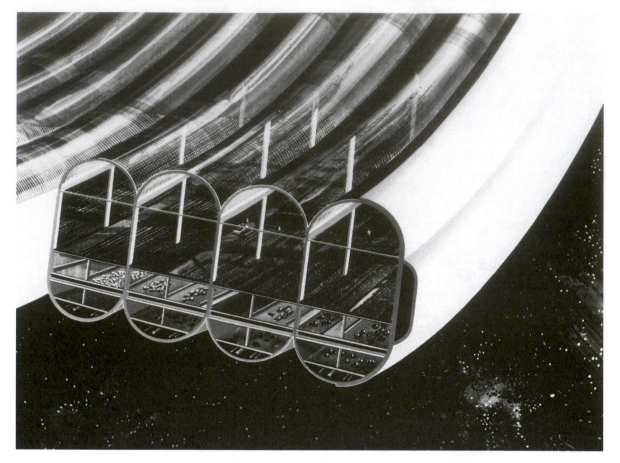

Artist's rendering that provides a cutaway view of the inside of the three-tiered agricultural zone of a large future space settlement located in cislunar space. *(NASA/Ames Research Center)*

type of controlled randomness or programmed chaos might prove necessary in overcoming some of the psychological problems that might arise when human beings live in a totally artificial world.

✧ Design Considerations for Large Space Settlements

The basic space settlement design will have to provide the essentials for life, such as air, food, water, and some level of artificial gravity, necessary to accommodate living in space for extended periods of time. However, the design of the space settlement must not only ensure physiological safety and comfort, but it should also satisfy the psychological and aesthetic needs of the inhabitants.

Just like astronauts and cosmonauts living on the *International Space Station,* human beings living in large space settlements must have an adequate diet. Food in a large space settlement should be nutritious, sufficiently abundant, and even attractive. The space settlers may be able to get their initial food supplies from Earth or (later in this century) from permanent bases on the Moon. But as most initial space settlement designs have assumed, the habitat will include agricultural facilities that accommodate food self-sufficiency.

The NASA-sponsored space settlement studies have suggested that when a food consumption quantity greater than 10,000 person-days is needed in cislunar space, agriculture in space (performed in special greenhouse facilities) becomes economically competitive with the solution of importing food from Earth. It also appears that for a large space settlement a modified form of terrestrial agriculture, based on plants and meat-bearing animals, could solve both nutritional requirements and the need for dietary variety. If properly designed, the habitat's agriculture facility could also serve as a recreational area. Visiting the facility, the inhabitants might enjoy some degree of visual variety and gently unpredictable experiences (encountering a beautiful flower that is just blooming or an exceptionally large tomato on the vine) in an otherwise well-managed, artificially constructed world. Photosynthetic agriculture could be used to help regenerate the space settlement's atmosphere by converting carbon dioxide (CO_2) and generating oxygen (O_2). Space agricultural activities might also serve as a source of pure water from the condensation of humidity produced by transpiration of the plants.

The design of a space settlement should not exert damaging psychological stresses on the inhabitants. A chronic sense of isolation (the shimanagashi syndrome) or an acute sense of artificiality (the solipsism syndrome) must be prevented through variety, diversity, and flexibility of interior designs. The generous use of natural light, sufficient views of outer space, the availability of privacy niches and private places, long lines of sight, and the use of larger overhead clearance (that is, vaulted ceilings where practical) are just some of the suggested environmental design criteria for a large space settlement.

The space settlement must also have a form of government or political organization that permits its inhabitants to enjoy a comfortable lifestyle under conditions that are crowded and physically isolated from other human communities. Because early space settlements may very likely be "company towns" dedicated to some particular space commerce activity, their organizations should support a fairly high level of productivity and should maintain the physical security of the habitat. Without an equitable division of political power and a properly controlled internal security force, the inhabitants of an isolated space community could easily become the victims of despots and self-elected demigods. In contrast, with an appropriate organizational structure, a large space settlement could also emerge as the trailblazing sociopolitical institution that becomes the model for all new human societies in the 22nd century and beyond.

Large space settlements, whatever their final design, population, or political structure, should emerge in the last decades of this century and become the characteristic hallmark and technical centerpiece of humans' solar system civilization. Present-day strategic planners and futurists cannot fully appreciate the impact that (almost) self-sufficient pockets of humanity sprinkled first throughout cislunar space and then across heliocentric space will have on the overall trajectory of civilization. In one exciting extrapolation, as these settlements grow and replicate themselves throughout the solar system, a Dyson sphere—perhaps the first ever constructed in the Milky Way Galaxy—will start to form.

✧ Dyson Sphere

The Dyson sphere would be a huge, artificial biosphere created around a star by an intelligent species as part of its technological growth and expansion within a solar system. This giant structure would most likely be formed by a swarm of artificial habitats and mini-planets capable of intercepting essentially all the radiant energy from the parent star. The captured radiant energy would be converted for use through a variety of techniques such as living plants, direct thermal-to-electric conversion devices, photovoltaic cells, and perhaps other (as yet undiscovered) energy conversion techniques. In response to the second law of thermodynamics, waste heat and unusable radiant energy would be rejected from the cold side of the Dyson sphere to outer space. Based on current knowledge of engineering heat transfer, the heat rejection surfaces of the Dyson sphere might be at temperatures of -100°F (200 K) to 80°F (300 K).

The notion of this gigantic astroengineering project is the brainchild of the British-American theoretical physicist Freeman John Dyson (b. 1923). In essence, what Dyson has proposed is that truly advanced extraterrestrial societies, responding to Malthusian pressures, might eventually expand into their local solar system, ultimately harnessing the full extent of that system's energy and materials resources. Just how much growth does this type of expansion represent?

For now strategic planners can only invoke the principle of mediocrity (namely, that things are pretty much the same throughout the universe) and use humans' home solar system as a model. The energy output from the Sun, which is an average yellow star of G2V spectral type, is approximately 4×10^{26} watts (joules per second). For all practical purposes, scientists treat the Sun as a blackbody radiator at an absolute temperature of approximately 9,980°F (5,800 K). Consequently, the vast majority of its energy output occurs as electromagnetic radiation, predominantly in the wavelength range 0.3 to 0.7 micrometer.

As an upper limit, the available mass in the solar system for such astroengineering construction projects may be taken as the mass of the planet Jupiter, some 4.4×10^{27} pounds (2×10^{27} kg). Contemporary energy consumption now amounts to about 10^{13} watts (joules per second), which corresponds to 10 terawatts. Projecting just a 1 percent growth in terrestrial energy consumption per year, within three millennia humankind's energy consumption needs would amount to the entire energy output of the Sun. Today several billion human beings live in a single biosphere, planet Earth, with a total mass of some 11.0×10^{24} pounds (5×10^{24} kg). A few thousand years from now, the Sun could be surrounded by a swarm of habitats, containing trillions of human beings.

As a brief exercise in the study of how technology induces social change, compare western Europe today with western Europe just two millennia ago—during the peak of the Roman Empire. What has changed, and what remains pretty much the same? Now do the same for the solar system, only this time go forward in time two or three millennia? What will change in a solar system civilization, and what will remain pretty much the same?

NIKOLAI SEMENOVICH KARDASHEV

While considering factors that could influence the search for extraterrestrial intelligence (SETI), the Russian astronomer Nikolai Semenovich Kardashev (b. 1932) suggested that there were three possible types (or levels) of alien civilizations. He based his characterizations on how each type of civilization harnessed energy resources at the planetary, solar system, and galactic scale, respectively. Kardashev reasoned that the more energy a particular civilization controlled, the more powerful the signals they might beam through interstellar space. This interesting line of speculation has become known within the exobiology and SETI communities as the three types of Kardashev civilizations. A Kardashev Type I civilization would be capable of harnessing the total energy capacity of its home planet (a maximum of about 10^{16} watts for an Earthlike planet); a Type II civilization, the energy output of its parent star (a maximum of about about 10^{27} watts for a Sunlike star); while a Type III civilization would be capable of using and manipulating the energy output of their entire galaxy (a maximum of about 10^{38} watts for a galaxy like the Milky Way).

The Dyson sphere is just one way of examining an upper limit for physical growth within this solar system. It is basically "the best humans can do" from an energy and materials point of view in this small corner of the universe. The vast majority of these human-made habitats would probably be located in the ecosphere, or continuously habitable zone around the Sun—that is, at about one astronomical unit away from humans' parent star. This does not preclude the possibility that other habitats, powered by nuclear fusion energy, could be found scattered throughout the outer regions of a somewhat dismantled solar system. Fusion-powered habitats might even serve as the technical precursors for the first interstellar space arks, which could carry a portion of the human race to some of the nearby star systems.

Using the solar system and today's planetary civilization as a reference point, some scientists anticipate that within a few millennia after the start of industrial development an intelligent species might rise from the level of planetary civilization (or Kardashev Type I civilization) and eventually occupy a swarm of artificial habitats that completely surround the parent star, thereby creating a mature Kardashev Type II civilization. Of course, these intelligent creatures might also elect to pursue interstellar travel and galactic migration, as opposed to completing the Dyson sphere within their home star system—initiating a Kardashev Type III civilization.

Freeman Dyson further speculated that such advanced civilizations could be detected by the presence of thermal infrared emission (typically 8.0 to 14.0 micrometers wavelength) from large objects in space that had dimensions of one to two astronomical units in diameter. Although advanced infrared telescopes have not detected such objects, the Dyson sphere is certainly a grand, far-reaching concept. It is also quite interesting to realize that the permanent space stations and space bases constructed in this century, are in a real sense, the first habitats in the swarm of artificial structures that humans could eventually construct as part of a mature solar system civilization. No other period in history provides a generation of people with the unique opportunity of constructing the first artificial habitat in the human race's Dyson sphere.

Conclusions

The ancient Greek philosopher Socrates (ca. 470–399 B.C.E.) noted that: "Man must rise above the Earth—to the top of the atmosphere and beyond—for only thus will he fully understand the world in which he lives." Human spaceflight is the technical embodiment of that special human characteristic to explore and conquer the unknown. To be human is to desire to push the boundaries of knowledge and the frontiers of understanding into previously untraveled regions.

This book has described some of the great triumphs, as well as some of the great tragedies, that have occurred in the past four decades of human spaceflight. Blazing a trail into the unknown often exacts a heavy toll in resources, time, and lives. As history has clearly demonstrated, not all explorers succeed. While some triumph, others fail—perhaps paying dearly for such curiosity with their own lives.

But people will continue to explore the unknown because this trait is so deeply rooted in human nature. Integrated against the backdrop of history, what do the contemporary technical achievements in human spaceflight really mean? Thousands of words can be spent describing how space travel is an elegant, high-technology manifestation of human nature. Or else this important connection between science, technology, and society can be compactly made with a single image.

Once the Apollo astronauts walked on the Moon, the entire human race "came of age" in the universe. Centuries from now, future historians will regard the triumph of the Apollo Project as the most definitive technical milestone in the 20th century, if not in all of human history. Why? Because it is the singularly special event, when intelligent life finally emerged from the cradle of Earth and cautiously first ventured out into the cosmos. The footprints made on the lunar surface by the 12 Apollo

One of the first steps taken on the Moon. This is an image of astronaut Edwin "Buzz" Aldrin's bootprint made during the *Apollo 11* lunar landing on July 20, 1969. *(NASA)*

astronauts now serve as a beacon and a challenge to all future generations. Spaceflight transforms the universe into both a destination and a destiny for the human race.

Chronology

⬥ **ca. 3000 B.C.E. (to perhaps 1000 B.C.E.)**

Stonehenge erected on the Salisbury Plain of Southern England (possible use: ancient astronomical calendar for prediction of summer solstice)

⬥ **ca. 1300 B.C.E.**

Egyptian astronomers recognize all the planets visible to the naked eye (Mercury, Venus, Mars, Jupiter, and Saturn), and they identify over 40 star patterns or constellations

⬥ **ca. 500 B.C.E.**

Babylonians devise zodiac, which is later adopted and embellished by Greeks and used by other early peoples

⬥ **ca. 375 B.C.E.**

The early Greek mathematician and astronomer Eudoxus of Cnidus starts codifying the ancient constellations from tales of Greek mythology

⬥ **ca. 275 B.C.E.**

The Greek astronomer Aristarchus of Samos suggests an astronomical model of the universe (solar system) that anticipates the modern heliocentric theory proposed by Nicolaus Copernicus. However, these ideas, which Aristarchus presents in his work *On the Size and Distances of the Sun and the Moon,* are essentially ignored in favor of the geocentric model of the universe proposed by Eudoxus of Cnidus and endorsed by Aristotle

⬥ **ca. 129 B.C.E.**

The Greek astronomer Hipparchus of Nicaea completes a catalog of 850 stars that remains important until the 17th century

✧ ca. 60 C.E.

The Greek engineer and mathematician Hero of Alexandria creates the aeoliphile, a toylike device that demonstrates the action-reaction principle that is the basis of operation of all rocket engines

✧ ca. 150 C.E.

Greek astronomer Ptolemy writes *Syntaxis* (later called the *Almagest* by Arab astronomers and scholars)—an important book that summarizes all the astronomical knowledge of the ancient astronomers, including the geocentric model of the universe that dominates Western science for more than one and a half millennia

✧ 820

Arab astronomers and mathematicians establish a school of astronomy in Baghdad and translate Ptolemy's work into Arabic, after which it became known as *al-Majisti* (The great work), or the *Almagest,* by medieval scholars

✧ 850

The Chinese begin to use gunpowder for festive fireworks, including a rocketlike device

✧ 1232

The Chinese army uses fire arrows (crude gunpowder rockets on long sticks) to repel Mongol invaders at the battle of Kaifung-fu. This is the first reported use of the rocket in warfare

✧ 1280–90

The Arab historian al-Hasan al-Rammah writes *The Book of Fighting on Horseback and War Strategies,* in which he gives instructions for making both gunpowder and rockets

✧ 1379

Rockets appear in western Europe; they are used in the siege of Chioggia (near Venice), Italy

✧ 1420

The Italian military engineer Joanes de Fontana writes *Book of War Machines,* a speculative work that suggests military applications of gunpowder rockets, including a rocket-propelled battering ram and a rocket-propelled torpedo

✧ 1429

The French army uses gunpowder rockets to defend the city of Orléans. During this period, arsenals throughout Europe begin to test various types of gunpowder rockets as an alternative to early cannons

✧ ca. 1500

According to early rocketry lore, a Chinese official named Wan-Hu attempted to use an innovative rocket-propelled kite assembly to fly through the air. As he sat in the pilot's chair, his servants lit the assembly's 47 gunpowder (black powder) rockets. Unfortunately, this early rocket test pilot disappeared in a bright flash and explosion

✧ 1543

The Polish church official and astronomer Nicolaus Copernicus changes history and initiates the Scientific Revolution with his book *De Revolutionibus Orbium Coelestium* (On the revolutions of the heavenly spheres). This important book, published while Copernicus lay on his deathbed, proposed a Sun-centered (heliocentric) model of the universe in contrast to the longstanding Earth-centered (geocentric) model advocated by Ptolemy and many of the early Greek astronomers

✧ 1608

The Dutch optician Hans Lippershey develops a crude telescope

✧ 1609

The German astronomer Johannes Kepler publishes *New Astronomy,* in which he modifies Nicolaus Copernicus's model of the universe by announcing that the planets have elliptical orbits rather than circular ones. Kepler's laws of planetary motion help put an end to more than 2,000 years of geocentric Greek astronomy

✧ 1610

On January 7, 1610, Galileo Galilei uses his telescope to gaze at Jupiter and discovers the giant planet's four major moons (Callisto, Europa, Io, and Ganymede). He proclaims this and other astronomical observations in his book, *Sidereus Nuncius* (Starry messenger). Discovery of these four Jovian moons encourages Galileo to advocate the heliocentric theory of Nicolaus Copernicus and brings him into direct conflict with church authorities

✧ 1642

Galileo Galilei dies while under house arrest near Florence, Italy, for his clashes with church authorities concerning the heliocentric theory of Nicolaus Copernicus

✦ 1647
The Polish-German astronomer Johannes Hevelius publishes *Seleno-graphia,* in which he provides a detailed description of features on the surface (near side) of the Moon

✦ 1680
Russian czar Peter the Great sets up a facility to manufacture rockets in Moscow. The facility later moves to St. Petersburg and provides the czarist army with a variety of gunpowder rockets for bombardment, signaling, and nocturnal battlefield illumination

✦ 1687
Financed and encouraged by Sir Edmond Halley, Sir Isaac Newton publishes his great work, *Philosophiae Naturalis Principia Mathematica* (Mathematical principles of natural philosophy). This book provides the mathematical foundations for understanding the motion of almost everything in the universe including the orbital motion of planets and the trajectories of rocket-propelled vehicles

✦ 1780s
The Indian ruler Hyder Ally (Ali) of Mysore creates a rocket corps within his army. Hyder's son, Tippo Sultan, successfully uses rockets against the British in a series of battles in India between 1782 and 1799

✦ 1804
Sir William Congreve writes *A Concise Account of the Origin and Progress of the Rocket System* and documents the British military's experience in India. He then starts the development of a series of British military (black-powder) rockets

✦ 1807
The British use about 25,000 of Sir William Congreve's improved military (black-powder) rockets to bombard Copenhagen, Denmark, during the Napoleonic Wars

✦ 1809
The brilliant German mathematician, astronomer, and physicist Carl Friedrich Gauss publishes a major work on celestial mechanics that revolutionizes the calculation of perturbations in planetary orbits. His work paves the way for other 19th-century astronomers to mathematically anticipate and then discover Neptune (in 1846), using perturbations in the orbit of Uranus

✧ **1812**

British forces use Sir William Congreve's military rockets against American troops during the War of 1812. British rocket bombardment of Fort William McHenry inspires Francis Scott Key to add "the rocket's red glare" verse in the "Star Spangled Banner"

✧ **1865**

The French science fiction writer Jules Verne publishes his famous story *De la terre a la lune* (From the Earth to the Moon). This story interests many people in the concept of space travel, including young readers who go on to become the founders of astronautics: Robert Hutchings Goddard, Hermann J. Oberth, and Konstantin Eduardovich Tsiolkovsky

✧ **1869**

American clergyman and writer Edward Everett Hale publishes *The Brick Moon*—a story that is the first fictional account of a human-crewed space station

✧ **1877**

While a staff member at the U.S. Naval Observatory in Washington, D.C., the American astronomer Asaph Hall discovers and names the two tiny Martian moons, Deimos and Phobos

✧ **1897**

British author H. G. Wells writes the science fiction story *War of the Worlds*—the classic tale about extraterrestrial invaders from Mars

✧ **1903**

The Russian technical visionary Konstantin Eduardovich Tsiolkovsky becomes the first person to link the rocket and space travel when he publishes *Exploration of Space with Reactive Devices*

✧ **1918**

American physicist Robert Hutchings Goddard writes *The Ultimate Migration*—a far-reaching technology piece within which he postulates the use of an atomic-powered space ark to carry human beings away from a dying Sun. Fearing ridicule, however, Goddard hides the visionary manuscript; it remains unpublished until November 1972—many years after his death in 1945

✧ **1919**

American rocket pioneer Robert Hutchings Goddard publishes the Smithsonian monograph *A Method of Reaching Extreme Altitudes*. This important work presents all the fundamental principles of modern rocketry. Unfortunately, members of the press completely miss the true significance of his technical contribution and decide to sensationalize his comments about possibly reaching the Moon with a small, rocket-propelled package. For such "wild fantasy," newspaper reporters dubbed Goddard with the unflattering title of "Moon man"

✧ 1923

Independent of Robert Hutchings Goddard and Konstantin Eduardovich Tsiolkovsky, the German space-travel visionary Hermann J. Oberth publishes the inspiring book *Die Rakete zu den Planetenräumen* (The rocket into planetary space)

✧ 1924

The German engineer Walter Hohmann writes *Die Erreichbarkeit der Himmelskörper* (The attainability of celestial bodies)—an important work that details the mathematical principles of rocket and spacecraft motion. He includes a description of the most efficient (that is, minimum energy) orbit transfer path between two coplanar orbits—a frequently used space operations maneuver now called the Hohmann transfer orbit

✧ 1926

On March 16 in a snow-covered farm field in Auburn, Massachusetts, American physicist Robert Hutchings Goddard makes space technology history by successfully firing the world's first liquid-propellant rocket. Although his primitive gasoline (fuel) and liquid oxygen (oxidizer) device burned for only two and one half seconds and landed about 60 meters away, it represents the technical ancestor of all modern liquid-propellant rocket engines.

In April, the first issue of *Amazing Stories* appears. The publication becomes the world's first magazine dedicated exclusively to science fiction. Through science fact and fiction, the modern rocket and space travel become firmly connected. As a result of this union, the visionary dream for many people in the 1930s (and beyond) becomes that of interplanetary travel

✧ 1929

German space-travel visionary Hermann J. Oberth writes the award-winning book *Wege zur Raumschiffahrt* (Roads to space travel) that helps popularize the notion of space travel among nontechnical audiences

✧ 1933

P. E. Cleator founds the British Interplanetary Society (BIS), which becomes one of the world's most respected space-travel advocacy organizations

✧ 1935

Konstantin Tsiolkovsky publishes his last book, *On the Moon,* in which he strongly advocates the spaceship as the means of lunar and interplanetary travel

✧ 1936

P. E. Cleator, founder of the British Interplanetary Society, writes *Rockets through Space,* the first serious treatment of astronautics in the United Kingdom. However, several established British scientific publications ridicule his book as the premature speculation of an unscientific imagination

✧ 1939–1945

Throughout World War II, nations use rockets and guided missiles of all sizes and shapes in combat. Of these, the most significant with respect to space exploration is the development of the liquid propellant V-2 rocket by the German army at Peenemünde under Wernher von Braun

✧ 1942

On October 3, the German A-4 rocket (later renamed Vengeance Weapon Two or V-2 Rocket) completes its first successful flight from the Peenemünde test site on the Baltic Sea. This is the birth date of the modern military ballistic missile

✧ 1944

In September, the German army begins a ballistic missile offensive by launching hundreds of unstoppable V-2 rockets (each carrying a one-ton high explosive warhead) against London and southern England

✧ 1945

Recognizing the war was lost, the German rocket scientist Wernher von Braun and key members of his staff surrender to American forces near Reutte, Germany in early May. Within months, U.S. intelligence teams, under Operation Paperclip, interrogate German rocket personnel and sort through carloads of captured documents and equipment. Many of these German scientists and engineers join von Braun in the United States to continue their rocket work. Hundreds of captured V-2 rockets are also disassembled and shipped back to the United States.

On May 5, the Soviet army captures the German rocket facility at Peenemünde and hauls away any remaining equipment and personnel. In the closing days of the war in Europe, captured German rocket technology and personnel help set the stage for the great missile and space race of the cold war

On July 16, the United States explodes the world's first nuclear weapon. The test shot, code-named Trinity, occurs in a remote portion of southern New Mexico and changes the face of warfare forever. As part of the cold-war confrontation between the United States and the former Soviet Union, the nuclear-armed ballistic missile will become the most powerful weapon ever developed by the human race.

In October, a then-obscure British engineer and writer, Arthur C. Clarke, suggests the use of satellites at geostationary orbit to support global communications. His article in *Wireless World,* "Extra-Terrestrial Relays," represents the birth of the communications satellite concept—an application of space technology that actively supports the information revolution

✧ 1946
On April 16, the U.S. Army launches the first American-adapted, captured German V-2 rocket from the White Sands Proving Ground in southern New Mexico.

Between July and August the Russian rocket engineer Sergei Korolev develops a stretched-out version of the German V-2 rocket. As part of his engineering improvements, Korolev increases the rocket engine's thrust and lengthens the vehicle's propellant tanks

✧ 1947
On October 30, Russian rocket engineers successfully launch a modified German V-2 rocket from a desert launch site near a place called Kapustin Yar. This rocket impacts about 320 kilometers downrange from the launch site

✧ 1948
The September issue of the *Journal of the British Interplanetary Society* publishes the first in a series of four technical papers by L. R. Shepherd and A. V. Cleaver that explores the feasibility of applying nuclear energy to space travel, including the concepts of nuclear-electric propulsion and the nuclear rocket

✧ 1949
On August 29, the Soviet Union detonates its first nuclear weapon at a secret test site in the Kazakh Desert. Code-named First Lightning (Per-vaya Molniya), the successful test breaks the nuclear-weapon monopoly enjoyed by the United States. It plunges the world into a massive nuclear arms race that includes the accelerated development of strategic ballistic missiles capable of traveling thousands of kilometers. Because they are well behind the United States in nuclear weapons technology, the leaders

of the former Soviet Union decide to develop powerful, high-thrust rockets to carry their heavier, more primitive-design nuclear weapons. That decision gives the Soviet Union a major launch vehicle advantage when both superpowers decide to race into outer space (starting in 1957) as part of a global demonstration of national power

✧ 1950

On July 24, the United States successfully launches a modified German V-2 rocket with an American-designed WAC Corporal second-stage rocket from the U.S. Air Force's newly established Long Range Proving Ground at Cape Canaveral, Florida. The hybrid, multistage rocket (called Bumper 8) inaugurates the incredible sequence of military missile and space vehicle launches to take place from Cape Canaveral—the world's most famous launch site.

In November, British technical visionary Arthur C. Clarke publishes "Electromagnetic Launching as a Major Contribution to Space-Flight." Clarke's article suggests mining the Moon and launching the mined-lunar material into outer space with an electromagnetic catapult

✧ 1951

Cinema audiences are shocked by the science fiction movie *The Day the Earth Stood Still.* This classic story involves the arrival of a powerful, humanlike extraterrestrial and his robot companion, who come to warn the governments of the world about the foolish nature of their nuclear arms race. It is the first major science fiction story to portray powerful space aliens as friendly, intelligent creatures who come to help Earth.

Dutch-American astronomer Gerard Peter Kuiper suggests the existence of a large population of small, icy planetesimals beyond the orbit of Pluto—a collection of frozen celestial bodies now known as the Kuiper belt

✧ 1952

Collier's magazine helps stimulate a surge of American interest in space travel by publishing a beautifully illustrated series of technical articles written by space experts such as Wernher von Braun and Willey Ley. The first of the famous eight-part series appears on March 22 and is boldly titled "Man Will Conquer Space Soon." The magazine also hires the most influential space artist, Chesley Bonestell, to provide stunning color illustrations. Subsequent articles in the series introduce millions of American readers to the concept of a space station, a mission to the Moon, and an expedition to Mars

Wernher von Braun publishes *Das Marsprojekt* (The Mars project), the first serious technical study regarding a human-crewed expedition to

Mars. His visionary proposal involves a convoy of 10 spaceships with a total combined crew of 70 astronauts to explore the Red Planet for about one year and then return to Earth

✧ 1953

In August, the Soviet Union detonates its first thermonuclear weapon (a hydrogen bomb). This is a technological feat that intensifies the super-power nuclear arms race and increases emphasis on the emerging role of strategic, nuclear-armed ballistic missiles.

In October, the U.S. Air Force forms a special panel of experts, headed by John von Neumann, to evaluate the American strategic ballistic missile program. In 1954, this panel recommends a major reorganization of the American ballistic missile effort

✧ 1954

Following the recommendations of John von Neumann, President Dwight D. Eisenhower gives strategic ballistic missile development the highest national priority. The cold war missile race explodes on the world stage as the fear of a strategic ballistic missile gap sweeps through the American government. Cape Canaveral becomes the famous proving ground for such important ballistic missiles as the Thor, Atlas, Titan, Minuteman, and Polaris. Once developed, many of these powerful military ballistic missiles also serve the United States as space launch vehicles. U.S. Air Force General Bernard Schriever oversees the time-critical development of the Atlas ballistic missile—an astonishing feat of engineering and technical management

✧ 1955

Walt Disney (the American entertainment visionary) promotes space travel by producing an inspiring three-part television series that includes appearances by noted space experts like Wernher von Braun. The first episode, "Man in Space," airs on March 9 and popularizes the dream of space travel for millions of American television viewers. This show, along with its companion episodes, "Man and the Moon" and "Mars and Beyond," make von Braun and the term *rocket scientist* household words

✧ 1957

On October 4, Russian rocket scientist Sergei Korolev, with permission from Soviet premier Nikita S. Khrushchev, uses a powerful military rocket to successfully place *Sputnik 1* (the world's first artificial satellite) into orbit around Earth. News of the Soviet success sends a political and technical shockwave across the United States. The launch of *Sputnik 1* marks the beginning of the Space Age. It also is the start of the great space race of the cold war—a period when people measure national strength and global prestige by accomplishments (or failures) in outer space.

On November 3, the Soviet Union launches *Sputnik 2*—the world's second artificial satellite. It is a massive spacecraft (for the time) that carries a live dog named Laika, which is euthanized at the end of the mission.

The highly publicized attempt by the United States to launch its first satellite with a newly designed civilian rocket ends in complete disaster on December 6. The Vanguard rocket explodes after rising only a few inches above its launch pad at Cape Canaveral. Soviet successes with *Sputnik 1* and *Sputnik 2* and the dramatic failure of the Vanguard rocket heighten American anxiety. The exploration and use of outer space becomes a highly visible instrument of cold-war politics

✧ 1958

On January 31, the United States successfully launches *Explorer 1*—the first American satellite in orbit around Earth. A hastily formed team from the U.S. Army Ballistic Missile Agency (ABMA) and Caltech's Jet Propulsion Laboratory (JPL), led by Wernher von Braun, accomplishes what amounts to a national prestige rescue mission. The team uses a military ballistic missile as the launch vehicle. With instruments supplied by Dr. James Van Allen of the State University of Iowa, *Explorer 1* discovers Earth's trapped radiation belts—now called the Van Allen radiation belts in his honor.

The National Aeronautics and Space Administration (NASA) becomes the official civilian space agency for the United States government on October 1. On October 7, the newly created NASA announces the start of the Mercury Project—a pioneering program to put the first American astronauts into orbit around Earth.

In mid-December, an entire Atlas rocket lifts off from Cape Canaveral and goes into orbit around Earth. The missile's payload compartment carries Project Score (Signal Communications Orbit Relay Experiment)—a prerecorded Christmas season message from President Dwight D. Eisenhower. This is the first time the human voice is broadcast back to Earth from outer space

✧ 1959

On January 2, the Soviet Union sends a 790-pound- (360-kg-) mass spacecraft, *Lunik 1,* toward the Moon. Although it misses hitting the Moon by between 3,125 and 4,375 miles (5,000 and 7,000 km), it is the first human-made object to escape Earth's gravity and go in orbit around the Sun.

In mid-September, the Soviet Union launches *Lunik 2*. The 860-pound- (390-kg-) mass spacecraft successfully impacts on the Moon and becomes the first human-made object to (crash-) land on another world. *Lunik 2* carries Soviet emblems and banners to the lunar surface.

On October 4, the Soviet Union sends *Lunik 3* on a mission around the Moon. The spacecraft successfully circumnavigates the Moon and takes the first images of the lunar farside. Because of the synchronous rotation of the Moon around Earth, only the near side of the lunar surface is visible to observers on Earth

✦ 1960

The United States launches the *Pioneer 5* spacecraft on March 11 into orbit around the Sun. The modest-sized (92-pound- [42-kg-]) mass spherical American space probe reports conditions in interplanetary space between Earth and Venus over a distance of about 23 million miles (37 million km).

On May 24, the U.S. Air Force launches a MIDAS (Missile Defense Alarm System) satellite from Cape Canaveral. This event inaugurates an important American program of special military surveillance satellites intended to detect enemy missile launches by observing the characteristic infrared (heat) signature of a rocket's exhaust plume. Essentially unknown to the general public for decades because of the classified nature of their mission, the emerging family of missile surveillance satellites provides U.S. government authorities with a reliable early warning system concerning a surprise enemy (Soviet) ICBM attack. Surveillance satellites help support the national policy of strategic nuclear deterrence throughout the cold war and prevent an accidental nuclear conflict.

The U.S. Air Force successfully launches the *Discoverer 13* spacecraft from Vandenberg Air Force Base on August 10. This spacecraft is actually part of a highly classified Air Force and Central Intelligence Agency (CIA) reconnaissance satellite program called Corona. Started under special executive order from President Dwight D. Eisenhower, the joint agency spy satellite program begins to provide important photographic images of denied areas of the world from outer space. On August 18, *Discoverer 14* (also called *Corona XIV*) provides the U.S. intelligence community its first satellite-acquired images of the former Soviet Union. The era of satellite reconnaissance is born. Data collected by the spy satellites of the National Reconnaissance Office (NRO) contribute significantly to U.S. national security and help preserve global stability during many politically troubled times.

On August 12, NASA successfully launches the *Echo 1* experimental spacecraft. This large (100 foot [30.5 m] in diameter) inflatable, metalized balloon becomes the world's first passive communications satellite. At the dawn of space-based telecommunications, engineers bounce radio signals off the large inflated satellite between the United States and the United Kingdom.

The former Soviet Union launches *Sputnik 5* into orbit around Earth. This large spacecraft is actually a test vehicle for the new *Vostok* spacecraft that will soon carry cosmonauts into outer space. *Sputnik 5* carries two dogs, Strelka and Belka. When the spacecraft's recovery capsule functions properly the next day, these two dogs become the first living creatures to return to Earth successfully from an orbital flight

✧ 1961

On January 31, NASA launches a Redstone rocket with a Mercury Project space capsule on a suborbital flight from Cape Canaveral. The passenger astrochimp Ham is safely recovered down range in the Atlantic Ocean after reaching an altitude of 155 miles (250 km). This successful primate space mission is a key step in sending American astronauts safely into outer space.

The Soviet Union achieves a major space exploration milestone by successfully launching the first human being into orbit around Earth. Cosmonaut Yuri Gagarin travels into outer space in the *Vostok 1* spacecraft and becomes the first person to observe Earth directly from an orbiting space vehicle.

On May 5, NASA uses a Redstone rocket to send astronaut Alan B. Shepard, Jr., on his historic 15-minute suborbital flight into outer space from Cape Canaveral. Riding inside the Mercury Project *Freedom 7* space capsule, Shepard reaches an altitude of 115 miles (186 km) and becomes the first American to travel in space.

President John F. Kennedy addresses a joint session of the U.S. Congress on May 25. In an inspiring speech touching on many urgent national needs, the newly elected president creates a major space challenge for the United States when he declares: "I believe that this nation should commit itself to achieving the goal, before this decade is out, of landing a man on the Moon and returning him safely to Earth." Because of his visionary leadership, when American astronauts Neil A. Armstrong and Edwin E. "Buzz" Aldrin, Jr., step onto the lunar surface for the first time on July 20, 1969, the United States is recognized around the world as the undisputed winner of the cold-war space race

✧ 1962

On February 20, astronaut John Herschel Glenn, Jr., becomes the first American to orbit Earth in a spacecraft. An Atlas rocket launches the NASA Mercury Project *Friendship 7* space capsule from Cape Canaveral. After completing three orbits, Glenn's capsule safely splashes down in the Atlantic Ocean.

In late August, NASA sends the *Mariner 2* spacecraft to Venus from Cape Canaveral. *Mariner 2* passes within 21,700 miles (35,000 km) of the planet on December 14, 1962—thereby becoming the world's first successful interplanetary space probe. The spacecraft observes very high surface temperatures (~800°F [430°C]). These data shatter pre–space age visions about Venus being a lush, tropical planetary twin of Earth.

During October, the placement of nuclear-armed Soviet offensive ballistic missiles in Fidel Castro's Cuba precipitates the Cuban Missile Crisis. This dangerous superpower confrontation brings the world perilously close to nuclear warfare. Fortunately, the crisis dissolves when Premier Nikita S. Khrushchev withdraws the Soviet ballistic missiles after much skillful political maneuvering by President John F. Kennedy and his national security advisers

✧ 1964

On November 28, NASA's *Mariner 4* spacecraft departs Cape Canaveral on its historic journey as the first spacecraft from Earth to visit Mars. It successfully encounters the Red Planet on July 14, 1965, at a flyby distance of about 6,100 miles (9,800 km). *Mariner 4*'s close-up images reveal a barren, desertlike world and quickly dispel any pre–space age notions about the existence of ancient Martian cities or a giant network of artificial canals

✧ 1965

A Titan II rocket carries astronauts Virgil "Gus" I. Grissom and John W. Young into orbit on March 23 from Cape Canaveral, inside a two-person Gemini Project spacecraft. NASA's *Gemini 3* flight is the first crewed mission for the new spacecraft and marks the beginning of more sophisticated space activities by American crews in preparation for the Apollo Project lunar missions

✧ 1966

The former Soviet Union sends the *Luna 9* spacecraft to the Moon on January 31. The 220-pound- (100-kg-) mass spherical spacecraft soft lands in the Ocean of Storms region on February 3, rolls to a stop, opens four petal-like covers, and then transmits the first panoramic television images from the Moon's surface.

The former Soviet Union launches the *Luna 10* to the Moon on March 31. This massive (3,300-pound- [1,500-kg-] mass) spacecraft becomes the first human-made object to achieve orbit around the Moon.

On May 30, NASA sends the *Surveyor 1* lander spacecraft to the Moon. The versatile robot spacecraft successfully makes a soft landing (June 1) in the Ocean of Storms. It then transmits over 10,000 images from the lunar surface and performs numerous soil mechanics experiments in preparation for the Apollo Project human landing missions.

In mid-August, NASA sends the *Lunar Orbiter 1* spacecraft to the Moon from Cape Canaveral. It is the first of five successful missions to collect detailed images of the Moon from lunar orbit. At the end of each mapping mission, the orbiter spacecraft is intentionally crashed into the Moon to prevent interference with future orbital activities

✧ 1967

On January 27, disaster strikes NASA's Apollo Project. While inside their *Apollo 1* spacecraft during a training exercise on Launch Pad 34 at Cape Canaveral, astronauts Virgil "Gus" I. Grissom, Edward H. White, Jr., and Roger B. Chaffee are killed when a flash fire sweeps through their spacecraft. The Moon landing program is delayed by 18 months, while major design and safety changes are made in the Apollo Project spacecraft.

On April 23, tragedy also strikes the Russian space program when the Soviets launch cosmonaut Vladimir Komarov in the new *Soyuz* (union) spacecraft. Following an orbital mission plagued with difficulties, Komarov dies (on April 24) during reentry operations, when the spacecraft's parachute fails to deploy properly and the vehicle hits the ground at high speed

✧ 1968

On December 21, NASA's *Apollo 8* spacecraft (command and service modules only) departs Launch Complex 39 at the Kennedy Space Center during the first flight of the mighty Saturn V launch vehicle with a human crew as part of the payload. Astronauts Frank Borman, James Arthur Lovell, Jr., and William A. Anders become the first people to leave Earth's gravitational influence. They go into orbit around the Moon and capture images of an incredibly beautiful Earth "rising" above the starkly barren lunar horizon—pictures that inspire millions and stimulate an emerging environmental movement. After 10 orbits around the Moon, the first lunar astronauts return safely to Earth on December 27

✧ 1969

The entire world watches as NASA's *Apollo 11* mission leaves for the Moon on July 16 from the Kennedy Space Center. Astronauts Neil A. Armstrong, Michael Collins, and Edwin E. "Buzz" Aldrin, Jr., make a long-held dream of humanity a reality. On July 20, American astronaut Neil Armstrong cautiously descends the steps of the lunar excursion module's ladder and steps on the lunar surface, stating, "One small step for a man, one giant leap for mankind!" He and Buzz Aldrin become the first two people to walk on another world. Many people regard the Apollo Project lunar landings as the greatest technical accomplishment in all of human history

◇ 1970

NASA's *Apollo 13* mission leaves for the Moon on April 11. Suddenly, on April 13, a life-threatening explosion occurs in the service module portion of the Apollo spacecraft. Astronauts James A. Lovell, Jr., John Leonard Swigert, and Fred Wallace Haise, Jr., must use their lunar excursion module (LEM) as a lifeboat. While an anxious world waits and listens, the crew skillfully maneuvers their disabled spacecraft around the Moon. With critical supplies running low, they limp back to Earth on a free-return trajectory. At just the right moment on April 17, they abandon the LEM *Aquarius* and board the Apollo Project spacecraft (command module) for a successful atmospheric reentry and recovery in the Pacific Ocean

◇ 1971

On April 19, the former Soviet Union launches the first space station (called *Salyut 1*). It remains initially uncrewed because the three-cosmonaut crew of the *Soyuz 10* mission (launched on April 22) attempts to dock with the station but cannot go on board

◇ 1972

In early January, President Richard M. Nixon approves NASA's space shuttle program. This decision shapes the major portion of NASA's program for the next three decades.

On March 2, an Atlas-Centaur launch vehicle successfully sends NASA's *Pioneer 10* spacecraft from Cape Canaveral on its historic mission. This far-traveling robot spacecraft becomes the first to transit the main-belt asteroids, the first to encounter Jupiter (December 3, 1973) and by crossing the orbit of Neptune on June 13, 1983 (which at the time was the farthest planet from the Sun), the first human-made object ever to leave the planetary boundaries of the solar system. On an interstellar trajectory, *Pioneer 10* (and its twin, *Pioneer 11*) carries a special plaque, greeting any intelligent alien civilization that might find it drifting through interstellar space millions of years from now.

On December 7, NASA's *Apollo 17* mission, the last expedition to the Moon in the 20th century, departs from the Kennedy Space Center, propelled by a mighty Saturn V rocket. While astronaut Ronald E. Evans remains in lunar orbit, fellow astronauts Eugene A. Cernan and Harrison H. Schmitt become the 11th and 12th members of the exclusive Moon walkers club. Using a lunar rover, they explore the Taurus-Littrow region. Their safe return to Earth on December 19 brings to a close one of the epic periods of human exploration

✧ 1973

In early April, while propelled by Atlas-Centaur rocket, NASA's *Pioneer 11* spacecraft departs on an interplanetary journey from Cape Canaveral. The spacecraft encounters Jupiter (December 2, 1974) and then uses a gravity assist maneuver to establish a flyby trajectory to Saturn. It is the first spacecraft to view Saturn at close range (closest encounter on September 1, 1979) and then follows a path into interstellar space.

On May 14, NASA launches *Skylab*—the first American space station. A giant Saturn V rocket is used to place the entire large facility into orbit in a single launch. The first crew of three American astronauts arrives on May 25 and makes the emergency repairs necessary to save the station, which suffered damage during the launch ascent. Astronauts Charles (Pete) Conrad, Jr., Paul J. Weitz, and Joseph P. Kerwin stay onboard for 28 days. They are replaced by astronauts Alan L. Bean, Jack R. Lousma, and Owen K. Garriott, who arrive on July 28 and live in space for about 59 days. The final *Skylab* crew (astronauts Gerald P. Carr, William R. Pogue, and Edward G. Gibson) arrive on November 11 and reside in the station until February 8, 1974—setting a space endurance record (for the time) of 84 days. NASA then abandons *Skylab*.

In early November, NASA launches the *Mariner 10* spacecraft from Cape Canaveral. It encounters Venus (February 5, 1974) and uses a gravity assist maneuver to become the first spacecraft to investigate Mercury at close range

✧ 1975

In late August and early September, NASA launches the twin *Viking 1* (August 20) and *Viking 2* (September 9) orbiter/lander combination spacecraft to the Red Planet from Cape Canaveral. Arriving at Mars in 1976, all Viking Project spacecraft (two landers and two orbiters) perform exceptionally well—but the detailed search for microscopic alien life-forms on Mars remains inconclusive

✧ 1977

On August 20, NASA sends the *Voyager 2* spacecraft from Cape Canaveral on an epic grand tour mission, during which it encounters all four giant planets and then departs the solar system on an interstellar trajectory. Using the gravity assist maneuver, *Voyager 2* visits Jupiter (July 9, 1979), Saturn (August 25, 1981), Uranus (January 24, 1986), and Neptune (August 25, 1989). The resilient, far-traveling robot spacecraft (and its twin *Voyager 1*) also carries a special interstellar message from Earth—a digital record entitled *The Sounds of Earth*.

On September 5, NASA sends the *Voyager 1* spacecraft from Cape Canaveral on its fast trajectory journey to Jupiter (March 5, 1979), Saturn (March 12, 1980), and beyond the solar system

✧ 1978

In May, the British Interplanetary Society releases its Project Daedalus report—a conceptual study about a one-way robot spacecraft mission to Barnard's star at the end of the 21st century

✧ 1979

On December 24, the European Space Agency successfully launches the first Ariane 1 rocket from the Guiana Space Center in Kourou, French Guiana

✧ 1980

India's Space Research Organization successfully places a modest 77-pound-mass (35 kg) test satellite (called *Rohini*) into low Earth orbit on July 1. The launch vehicle is a four-stage, solid propellant rocket manufactured in India. The SLV-3 (Standard Launch Vehicle-3) gives India independent national access to outer space

✧ 1981

On April 12, NASA launches the space shuttle *Columbia* on its maiden orbital flight from Complex 39-A at the Kennedy Space Center. Astronauts John W. Young and Robert L. Crippen thoroughly test the new aerospace vehicle. Upon reentry, it becomes the first spacecraft to return to Earth by gliding through the atmosphere and landing like an airplane. Unlike all previous onetime use space vehicles, *Columbia* is prepared for another mission in outer space

✧ 1986

On January 24, NASA's *Voyager 2* spacecraft encounters Uranus.

On January 28, the space shuttle *Challenger* lifts off from the NASA Kennedy Space Center on its final voyage. At just under 74 seconds into the STS 51-L mission, a deadly explosion occurs, killing the crew and destroying the vehicle. Led by President Ronald Reagan, the United States mourns seven astronauts lost in the *Challenger* accident

✧ 1988

On September 19, the State of Israel uses a Shavit (comet) three-stage rocket to place the country's first satellite (called *Ofeq 1*) into an unusual east-to-west orbit—one that is opposite to the direction of Earth's rotation but necessary because of launch safety restrictions.

As the *Discovery* successfully lifts off on September 29 for the STS-26 mission, NASA returns the space shuttle to service following a 32-month hiatus after the *Challenger* accident

✧ 1989
On August 25, the *Voyager 2* spacecraft encounters Neptune

✧ 1994
In late January, a joint Department of Defense and NASA advanced technology demonstration spacecraft, *Clementine,* lifts off for the Moon from Vandenberg Air Force Base. Some of the spacecraft's data suggest that the Moon may actually possess significant quantities of water ice in its permanently shadowed polar regions

✧ 1995
In February, during NASA's STS-63 mission, the space shuttle *Discovery* approaches (encounters) the Russian *Mir* space station as a prelude to the development of the *International Space Station.* Astronaut Eileen Marie Collins serves as the first female shuttle pilot.

On March 14, the Russians launch the *Soyuz TM-21* spacecraft to the *Mir* space station from the Baikanur Cosmodrome. The crew of three includes American astronaut Norman Thagard—the first American to travel into outer space on a Russian rocket and the first to stay on the *Mir* space station. The *Soyuz TM-21* cosmonauts also relieve the previous *Mir* crew, including cosmonaut Valeri Polyakov, who returns to Earth on March 22 after setting a world record for remaining in space for 438 days.

In late June, NASA's space shuttle *Atlantis* docks with the Russian *Mir* space station for the first time. During this shuttle mission (STS-71), *Atlantis* delivers the *Mir 19* crew (cosmonauts Anatoly Solovyev and Nikolai Budarin) to the Russian space station and then returns the *Mir 18* crew back to Earth—including American astronaut Norman Thagard, who has just spent 115 days in space onboard the *Mir.* The Shuttle-*Mir* docking program is the first phase of the *International Space Station.* A total of nine shuttle-*Mir* docking missions will occur between 1995 and 1998

✧ 1998
In early January, NASA sends the *Lunar Prospector* to the Moon from Cape Canaveral. Data from this orbiter spacecraft reinforce previous hints that the Moon's polar regions may contain large reserves of water ice in a mixture of frozen dust lying at the frigid bottom of some permanently shadowed craters.

In early December, the space shuttle *Endeavour* ascends from the NASA Kennedy Space Center on the first assembly mission of the *International Space Station*. During the STS-88 shuttle mission, *Endeavour* performs a rendezvous with the previously launched Russian-built *Zarya* (sunrise) module. An international crew connects this module with the American-built *Unity* module carried in the shuttle's cargo bay

✧ 1999

In July, astronaut Eileen Marie Collins serves as the first female space shuttle commander (STS-93 mission) as the *Columbia* carries NASA's *Chandra X-ray Observatory* into orbit

✧ 2001

NASA launches the *Mars Odyssey 2001* mission to the Red Planet in early April—the spacecraft successfully orbits the planet in October

✧ 2002

On May 4, NASA successfully launches its *Aqua* satellite from Vandenberg Air Force Base. This sophisticated Earth-observing spacecraft joins the *Terra* spacecraft in performing Earth system science studies.

On October 1, the United States Department of Defense forms the U.S. Strategic Command (USSTRATCOM) as the control center for all American strategic (nuclear) forces. USSTRATCOM also conducts military space operations, strategic warning and intelligence assessment, and global strategic planning

✧ 2003

On February 1, while gliding back to Earth after a successful 16-day scientific research mission (STS-107), the space shuttle *Columbia* experiences a catastrophic reentry accident at an altitude of about 63 km over the Western United States. Traveling at 18 times the speed of sound, the orbiter vehicle disintegrates, taking the lives of all seven crew members: six American astronauts (Rick Husband, William McCool, Michael Anderson, Kalpana Chawla, Laurel Clark, and David Brown) and the first Israeli astronaut (Ilan Ramon).

NASA's Mars Exploration Rover (MER) *Spirit* is launched by a Delta II rocket to the Red Planet on June 10. *Spirit,* also known as MER-A, arrives safely on Mars on January 3, 2004, and begins its teleoperated surface exploration mission under the supervision of mission controllers at the NASA Jet Propulsion Laboratory.

NASA launches the second Mars Exploration Rover, called *Opportunity,* using a Delta II rocket launch, which lifts off from Cape Canaveral Air Force Station on July 7, 2003. *Opportunity,* also called MER-B, successfully lands on Mars on January 24, 2004, and starts its teleoperated surface exploration mission under the supervision of mission controllers at the NASA Jet Propulsion Laboratory

✧ 2004

On July 1, NASA's *Cassini* spacecraft arrives at Saturn and begins its four-year mission of detailed scientific investigation.

In mid-October, the Expedition 10 crew, riding a Russian launch vehicle from Baikonur Cosmodrome, arrives at the *International Space Station* and the Expedition 9 crew returns safely to Earth.

On December 24, the 703-pound- (319-kg-) mass *Huygens* probe successfully separates from the *Cassini* spacecraft and begins its journey to Saturn's moon, Titan

✧ 2005

On January 14, the *Huygens* probe enters the atmosphere of Titan and successfully reaches the surface some 147 minutes later. *Huygens* is the first spacecraft to land on a moon in the outer solar system.

On July 4, NASA's Deep Impact mission successfully encounters Comet Tempel 1.

NASA successfully launches the space shuttle *Discovery* on the STS-114 mission on July 26 from the Kennedy Space Center in Florida. After docking with the *International Space Station*, the *Discovery* returns to Earth and lands at Edwards AFB, California, on August 9.

On August 12, NASA launches the *Mars Reconnaissance Orbiter* from Cape Canaveral AFS, Florida.

On September 19, NASA announces plans for a new spacecraft designed to carry four astronauts to the Moon and to deliver crews and supplies to the *International Space Station*. NASA also introduces two new shuttle-derived launch vehicles: a crew-carrying rocket and a cargo-carrying, heavy-lift rocket.

The *Expedition 12* crew (Commander William McArthur and Flight Engineer Valery Tokarev) arrives at the *International Space Station* on October 3 and replaces the *Expedition 11* crew.

The People's Republic of China successfully launches its second human spaceflight mission, called *Shenzhou 6*, on October 12. Two taikonauts, Fei Junlong and Nie Haisheng, travel in space for almost five days and make 76 orbits of Earth before returning safely to Earth, making a soft, parachute-assisted landing in northern Inner Mongolia

✧ 2006

On January 15, the sample package from NASA's *Stardust* spacecraft, containing comet samples, successfully returns to Earth.

NASA launches the *New Horizons* spacecraft from Cape Canaveral on January 19 and successfully sends this robot probe on its long one-way mission to conduct a scientific encounter with the Pluto system (in 2015) and then to explore portions of the Kuiper belt that lie beyond.

Follow-up observations by NASA's *Hubble Space Telescope*, reported on February 22, confirm the presence of two new moons around the distant planet Pluto. The moons, tentatively called S/2005 P 1 and S/2005 P 2, were first discovered by *Hubble* in May 2005, but the science team wants to examine the Pluto system further to characterize the orbits of the new moons and validate the discovery.

NASA scientists announce on March 9 that the *Cassini* spacecraft may have found evidence of liquid water reservoirs that erupt in Yellowstone Park–like geysers on Saturn's moon Enceladus.

On March 10, NASA's *Mars Reconnaissance Orbiter* successfully arrives at Mars and begins a six-month-long process of adjusting and trimming the shape of its orbit around the Red Planet prior to performing its operational mapping mission.

The Expedition 13 crew (Commander Pavel Vinogradov and Flight Engineer Jeff Williams) arrive at the *International Space Station* on April 1 and replace the Expedition 12 crew. Joining them for several days before returning back to Earth with the Expedition 12 crew is Brazil's first astronaut, Marcos Pontes

On August 24, the members of the International Astronomical Union (IAU) meet for the organization's 2006 General Assembly in Prague, Czech Republic. After much heated debate, the 2,500 assembled professional astronomers decide (by vote) to demote Pluto from its traditional status as one of the nine major planets and place the object into a new class, called a dwarf planet. The IAU decision now leaves the solar system with eight major planets and three dwarf planets: Pluto (which serves as the prototype dwarf planet), Ceres (the largest asteroid), and the large, distant Kuiper belt object indentified as 2003 UB313 (now officially named Eris). Astronomers anticipate the discovery of other dwarf planets in the distant parts of the solar system.

On September 9, NASA launched the space shuttle *Atlanis* on the 12-day duration STS-115 mission to the *International Space Station*.

The Expedition 14 crew (Commander Michael Lopez-Alegria, Flight Engineer Mikhail Tyurin) arrived at the *International Space Station* on September 20 and replaced the Expedition 13 crew.

NASA launched the space shuttle *Discovery* on December 9 on the 12-day duration STS-116 mission to the *International Space Station*. American astronaut Sunita Williams joined the Expedition 14 crew of the *ISS* and the ESA astronaut Thomas Reiter returned to Earth onboard *Discovery*, which landed at Kennedy Space Center on December 22.

Glossary

abort   To cut short or cancel an operation with a rocket, spacecraft, or aerospace vehicle, especially because of equipment failure. NASA's space shuttle system has two types of abort modes during the ascent phase of a flight: the intact abort and the contingency abort. An intact abort is designed to achieve a safe return of the astronaut crew and orbiter vehicle to a planned landing site. A contingency abort involves a ditching operation in which the crew is saved, but the orbiter vehicle is damaged or destroyed.

acceleration of gravity   The local acceleration due to gravity on or near the surface of a planet. On Earth, the acceleration due to gravity (symbol: g) of a free-falling object has the standard value of 32.1740 feet per second per second (9.80665 m/s^2) by international agreement.

acronym   A word formed from the first letters of a name, such as *HST*—which stands for the *Hubble Space Telescope*—or a word formed by combining the initial parts of a series of words, such as lidar—which stands for *l*ight *d*etection *a*nd *r*anging. Acronyms are frequently used in space technology and astronomy.

acute radiation syndrome (ARS)   The acute organic disorder that follows exposure to relatively severe doses of ionizing radiation. A person will initially experience nausea, diarrhea, or blood cell changes. In the later stages, loss of hair, hemorrhaging, and possibly death can take place. Radiation dose equivalent values of about 450 to 500 rem (4.5 to 5 sievert) will prove fatal to 50 percent of the exposed individuals in a large general population. Also called radiation sickness.

aerodynamic heating   Frictional surface heating experienced by an aerospace vehicle or space system as it enters the upper regions of a planetary atmosphere at high velocities. Special thermal protection is needed to prevent structural damage or destruction. For example, NASA's space shuttle orbiter vehicle uses thermal protection tiles to survive the intense aerodynamic heating environment that occurs during reentry and landing.

aerospace   A term, derived from *aero*nautics and *space,* meaning of or pertaining to Earth's atmospheric envelope and outer space beyond it. NASA's space shuttle orbiter vehicle is called an aerospace vehicle because it operates both in the atmosphere and in outer space.

aerospace medicine   The branch of medical science that deals with the effects of flight upon the human body. The treatment of space sickness (space adaptation syndrome) falls within this field.

aerospace vehicle   A vehicle capable of operating both within Earth's sensible (measurable) atmosphere and in outer space. The space shuttle orbiter vehicle is an example.

air   The overall mixture of gases that make up Earth's atmosphere, primarily nitrogen (N_2) at 78 percent (by volume), oxygen (O_2) at 21 percent, argon (Ar) at 0.9 percent, and carbon dioxide (CO_2) at 0.03 percent. Sometimes aerospace engineers use this word for the breathable gaseous mixture found inside the crew compartment of a space vehicle or in the pressurized habitable environment of a space station.

air lock   A small chamber with airtight doors that can be pressurized and depressurized. The air lock serves as a passageway for crew members and equipment between places at different pressure levels—for example, between a spacecraft's pressurized crew cabin and outer space.

alphanumeric (alphabet plus numeric)   Including letters and numerical digits, as for example, the term *JEN75WX11.*

altitude (spacecraft)   In space vehicle navigation, the height above the mean surface of the reference celestial body. Note that the distance of a space vehicle or spacecraft from the reference celestial body is taken as the distance from the center of the object.

androgynous interface   A nonpolar interface; one that physically can join with another of the same design; literally, having both male and female characteristics.

antenna   A device used to detect, collect, or transmit radio waves. A radio telescope is a large receiving antenna, while many spacecraft have both a directional antenna and an omnidirectional antenna to transmit (down-link) telemetry and to receive (uplink) instructions.

apogee The point in the orbit of a spacecraft that is farthest from Earth. The term applies to both the orbit of the Moon as well as to the orbits of artificial satellites around Earth. At apogee, the orbital velocity of a satellite is at a minimum. *Compare with* PERIGEE.

Apollo Lunar Surface Experiments Package (ALSEP) Scientific devices and equipment placed on the Moon by the Apollo Project astronauts and left there to transmit data back to Earth. Experiments included the study of meteorite impacts, lunar surface characteristics, seismic activity on the Moon, solar wind interaction, and analysis of the tenuous lunar atmosphere.

Apollo Project The American effort in the 1960s and early 1970s to place astronauts successfully on the surface of the Moon and return them safely to Earth. President John F. Kennedy (1917–63) initiated the project in May 1961 in response to a growing space technology challenge from the former Soviet Union. Managed by NASA, the Apollo 8 mission sent the first three humans to the vicinity of the Moon in December 1968. The Apollo 11 mission involved the first human landing on another world (July 20, 1969). Apollo 17, the last lunar-landing mission under this project, took place in December 1972. The project is often considered one of the greatest technical accomplishments in all of human history.

Apollo-Soyuz Test Project (ASTP) Joint United States—former Soviet Union space mission (July 1975) that centered on the rendezvous and docking of the *Apollo 18* spacecraft (three astronaut crew) and the *Soyuz 19* spacecraft (two cosmonaut crew).

approach The maneuvers of a spacecraft or aerospace vehicle from its normal orbital position (station-keeping position) toward another orbiting spacecraft for the purpose of conducting rendezvous and docking operations.

Ares I The name given by NASA to the new crew launch vehicle that will start carrying astronauts to the *International Space Station* in about 2015 and back to the Moon in about 2020.

Ares V The name given by NASA to the new heavy-lift launch vehicle that will serve as the agency's primary launch vehicle for the safe, reliable delivery of resources to space, including the hardware and materials needed to establish a permanent base on the Moon in about 2020.

Armstrong, Neil A. (b. 1930)   American astronaut Neil Armstrong is the former X-15 test pilot who served as the commander for NASA's Apollo 11 lunar-landing mission in July 1969. As he became the first human being to set foot on the Moon (July 20, 1969), he uttered these historic words: "That's one small step for (a) man, one giant leap for mankind."

artificial gravity   Simulated gravity conditions established within a spacecraft, space station, or large space settlement. Rotating the human-occupied space system about an axis creates this condition, since the centrifugal force generated by the rotation produces effects similar to the force of gravity within the vehicle.

astro-   A prefix that means star or (by extension) outer space or celestial; for example, astronaut, astronautics, or astrophysics.

astrochimp(s)   Nickname given to the primates used in the early U.S. space program to test space capsule and launch vehicle hardware prior to the commitment of this equipment to human missions. *See also* ENOS; HAM.

astronaut   Within the American space program, a person who travels in outer space; a person who flies in an aerospace vehicle to an altitude of more than 50 miles (80 km). The word comes from a combination of two ancient Greek words that literally mean "star" (*astro*) and "sailor or traveler" (*naut*). *Compare with* COSMONAUT.

atmosphere (cabin)   The breathable environment inside a human-occupied space capsule, aerospace vehicle, spacecraft, or space station.

attitude   The position of an object as defined by the inclination of its axes with respect to a frame of reference. The orientation of a space vehicle that is either in motion or at rest, as established by the relationship between the vehicle's axes and a reference line or plane. Attitude is often expressed in terms of PITCH, ROLL, and YAW.

attitude control system   The onboard system of computers, low-thrust rockets (thrusters), and mechanical devices (such as a momentum wheel) used to keep a spacecraft stabilized during flight and to precisely point its instruments in some desired direction. Stabilization is achieved by spinning the spacecraft or by using a three-axis active approach that maintains the spacecraft in a fixed, reference attitude by firing a selected combination of thrusters when necessary.

auxiliary power unit (APU) A power unit carried on a spacecraft or aerospace vehicle that supplements the main source of electric power on the craft.

backout The process of undoing tasks that have already been completed during the countdown of a launch vehicle, usually in reverse order.

backup crew A crew of astronauts or cosmonauts trained to replace the prime crew, if necessary, on a particular space mission.

Baikonur Cosmodrome The major launch site for the space program of the former Soviet Union and later the Russian Federation. The complex is located just east of the Aral Sea in Kazakhstan (now an independent republic). Also known as the Tyuratam launch site during the cold war, the Soviets launched *Sputnik 1* (1957), the first artificial satellite, and cosmonaut Yuri Gagarin (1934–68), the first human to fly in outer space (1961), from this location.

barbecue mode The slow roll of an orbiting aerospace vehicle or spacecraft to help equalize its external temperature and to promote a more favorable heat (thermal energy) balance. This maneuver is performed during certain missions, because in outer space solar radiation is intense on one side of a space vehicle while the side opposite the Sun can become extremely cold.

Bernal sphere A large, spherically shaped space settlement first proposed in 1929 by the Irish physicist and writer John Desmond Bernal (1910–71).

berthing The joining of two orbiting spacecraft, using a manipulator or other mechanical device, to move one into contact (or close proximity) with the other at a selected interface. For example, NASA astronauts use the space shuttle's remote manipulator system to carefully berth a large free-flying spacecraft (like the *Hubble Space Telescope*) onto a special support fixture located in the orbiter's payload bay during an on-orbit servicing and repair mission. *See also* DOCKING; RENDEZVOUS.

biotelemetry The remote measurement of life functions. Data from biosensors attached to an astronaut or cosmonaut are sent back to Earth (as telemetry) for the purposes of space crew health monitoring and evaluation by medical experts and mission managers. For example, biotelemetry allows NASA medical specialists on Earth to monitor an astronaut's heartbeat and respiration rate during strenuous tasks, like performing an extravehicular activity.

Cape Canaveral   The region on Florida's east-central coast from which the United States Air Force and NASA have launched more than 3,000 rockets since 1950. Cape Canaveral Air Force Station is the major east coast launch site for the Department of Defense, while the adjacent NASA Kennedy Space Center is the spaceport for the fleet of space shuttle vehicles.

cargo bay   The unpressurized middle portion of NASA's space shuttle orbiter vehicle. *See* PAYLOAD BAY.

***Challenger* accident**   NASA's space shuttle *Challenger* was launched from Complex 39-B at the Kennedy Space Center on January 28, 1986, as part of the STS 51-L mission. At approximately 74 seconds into the flight, an explosion occurred that caused the loss of the aerospace vehicle and its entire crew, including astronauts Francis R. Scobee, Michael J. Smith, Ellison S. Onizuka, Judith A. Resnik, Ronald E. McNair, S. Christa Corrigan McAuliffe, and Gregory B. Jarvis.

chaser spacecraft   The spacecraft or aerospace vehicle that actively performs the key maneuvers during orbital rendezvous and docking/berthing operations. The other space vehicle serves as the target and remains essentially passive during the encounter.

circadian rhythms   A biological organism's day/night cycle of living; a regular change in physiological function occurring in approximately 24-hour cycles.

clean room   A controlled work environment for spacecraft and aerospace systems in which dust, temperature, and humidity are carefully controlled during the fabrication, assembly, and/or testing of critical components.

closed ecological life support system (CELSS)   A system that can provide for the maintenance of life in an isolated living chamber or facility through complete reuse of the materials available within the chamber or facility.

cold war   The ideological conflict between the United States and the former Soviet Union from approximately 1946 to 1989, involving rivalry, mistrust, and hostility just short of overt military action. The tearing down of the Berlin Wall in November 1989 is generally considered as the (symbolic) end of the cold-war period.

***Columbia* accident**   While gliding back to Earth on February 1, 2003, after a successful 16-day scientific research mission in low Earth orbit,

NASA's space shuttle *Columbia* experienced a catastrophic reentry accident and broke apart at an altitude of about 39 miles (63 km) over Texas. The STS-107 mission disaster claimed the lives of six American astronauts (Rick D. Husband, William C. McCool, Michael P. Anderson, Kalpana Chawla, Laurel Blair Salton Clark, and David M. Brown) and the first Israeli astronaut (Ilan Ramon).

continuously crewed spacecraft   A spacecraft that has accommodations for continuous habitation (human occupancy) during its mission. The *International Space Station* is an example. Sometimes (though not preferred) called a continuously manned spacecraft.

cooperative target   A three-axis, stabilized, orbiting object that has signaling devices to support rendezvous and docking/capture operations by a chaser spacecraft.

co-orbital   Sharing the same or similar orbit; for example, during a rendezvous operation, the chaser spacecraft and its cooperative target are said to be co-orbital.

coronal mass ejection (CME)   A high-speed (six to 620 miles per second [10 to 1,000 km/s]) ejection of matter from the Sun's corona. A CME travels through space disturbing the solar wind and giving rise to geomagnetic storms when the disturbance reaches Earth.

cosmic   Of or pertaining to the universe, especially that part outside Earth's atmosphere. This term frequently appears in the Russian (former Soviet Union) space program as the equivalent to space or astro-, such as cosmic station (versus space station) or cosmonaut (versus astronaut).

cosmic rays   Extremely energetic particles (usually bare atomic nuclei) that move through outer space at speeds just below the speed of light and bombard Earth from all directions.

cosmonaut   The title given by Russia (formerly the Soviet Union) to its space travelers or "astronauts."

countdown   The step-by-step process that leads to the launch of a rocket or aerospace vehicle. A countdown takes place in accordance with a specific schedule, with zero being the go, or activate, time.

crew-tended spacecraft   A spacecraft that is visited and/or serviced by astronauts but can only provide temporary accommodations for human habitation during its overall mission. Sometimes referred to as a man-tended spacecraft.

deboost   A retrograde (opposite-direction) burn of one or more low-thrust rockets or an aerobraking maneuver that lowers the altitude of an orbiting spacecraft.

debris   Jettisoned human-made materials, discarded launch vehicle components, and derelict or nonfunctioning spacecraft in orbit around Earth.

decay (orbital)   The gradual lessening of both the apogee and perigee of an orbiting object from its primary body. The orbital decay process for abandoned spacecraft, artificial satellites, and space debris often results in their ultimate fiery plunge into the denser regions of the Earth's atmosphere.

de-orbit burn   A retrograde (opposite direction) rocket engine firing by which a space vehicle's velocity is reduced to less than that required to remain in orbit around a celestial body.

Destiny   The American-built laboratory module delivered to the *International Space Station* by the space shuttle *Atlantis* during the STS-98 mission (February 2001). *Destiny* is the primary research laboratory for U.S. payloads.

diurnal   Having a period of, occurring in, or related to a day; daily.

docking   The act of physically joining two orbiting spacecraft. This is usually accomplished by independently maneuvering one spacecraft (the chaser spacecraft) into contact with the other (the target spacecraft) at a chosen physical interface. For spacecraft with human crews, a docking module assists in the process and often serves as a special passageway (airlock) that permits hatches to be opened and crewmembers to move from one spacecraft to the other without the use of a space suit and without losing cabin pressure.

docking module   A structural element that provides a support and attachment interface between a docking mechanism and a spacecraft. For example, the special component added to the U.S. Apollo spacecraft so that it could be joined with the Russian Soyuz spacecraft in the Apollo-Soyuz Test Project; or the component carried in the cargo bay of the U.S. space shuttle so that it could be joined with the Russian *Mir* space station.

doffing   The act of removing wearing apparel or other apparatus, such as a space suit.

donning   The act of putting on wearing apparel or other apparatus, such as a space suit.

dose   In radiation protection, a general term describing the amount of energy delivered to a given volume of matter, a particular body organ, or a person (i.e., a whole-body dose) by ionizing radiation.

dose equivalent (symbol: H)   In radiation protection, the product of absorbed dose and a suitable weighting factor or quality factor that characterizes and evaluates the biological effects of ionizing radiation doses received by human beings (or other living creatures). The traditional unit of dose equivalent is the rem, while the sievert is the special unit for dose equivalent in the international unit system; 100 rem = 1 sievert.

downlink   The telemetry signal received at a ground station from a spacecraft.

downrange   A location away from the launch site but along the intended flight path (trajectory) of a launch vehicle flown from a rocket range.

drogue parachute   A small parachute used specifically to pull a larger parachute out of stowage; a small parachute used to slow down a descending space capsule or aerospace vehicle.

dwarf planet   As defined by the International Astronomical Union in August 2006, a celestial body that is (a) in orbit around the Sun, (b) has sufficient mass for its self-gravity to overcome rigid body forces so that it assumes a nearly round shape, (c) has not cleared the cosmic neighborhood around its orbit, and (d) is not a satellite of another (larger) body. Included in this definition are: Pluto, Ceres (the largest asteroid), and Eris (a large, distant Kuiper belt object, also called 2003 UB313).

Dyna-Soar (Dynamic Soaring)   An early U.S. Air Force space project from 1958 to 1963 that involved a crewed boost-glide orbital vehicle that was to be sent into orbit by an expendable launch vehicle, perform its military mission, and return to Earth using wings to glide through the atmosphere during reentry (in a manner similar to NASA's space shuttle orbiter vehicle). The project was canceled in favor of the civilian (NASA) human spaceflight program, involving the Mercury Project, Gemini Project, and Apollo Project. Also called the X-20 Project.

dysbarism   A general aerospace medicine term describing a variety of symptoms within the human body caused by the existence of a pressure differential between the total ambient pressure and the total pressure of dissolved and free gases within the body tissues, fluids, and cavities. For example, increased ambient pressure, as accompanies a descent from higher altitudes, might cause painful distention of the eardrums.

Earth's trapped radiation belts   Two major belts (or zones) of energetic atomic particles (mainly electrons and protons) that are trapped by Earth's magnetic field hundreds of miles above the atmosphere. Also called the Van Allen belts after the American physicist James Alfred Van Allen (1914–2006), who discovered them in 1958.

ejection capsule   In a crewed spacecraft or human-rated launch vehicle, a detachable compartment that may be ejected as a unit and parachuted to the ground during an emergency.

electromagnetic radiation (EMR)   Radiation made up of oscillating electric and magnetic fields and propagated with the speed of light. Includes (in order of decreasing frequency) gamma rays, X-rays, ultraviolet radiation, visible radiation, infrared radiation, radar, and radio waves.

electron volt (symbol: eV)   A unit of energy equivalent to the energy gained by an electron when it experiences a potential difference of one volt. Larger multiple units of the electron volt are often encountered frequently—as, for example, *keV* for a thousand (or kilo-) electron volts (10^3 eV); *MeV* for a million (or mega-) electron volts (10^6 eV); and *GeV* for a billion (or giga-) electron volts (10^9 eV). One electron volt is equivalent to 1.519×10^{-22} Btu (1.602×10^{-19} J).

encounter   The close flyby or rendezvous of a spacecraft with a target body. The target of an encounter can be a natural celestial body (such as a planet, asteroid, or comet) or a human-made object (such as another spacecraft).

Enos   The primate (astrochimp) used by NASA to test the Mercury Project space capsule during a successful orbital flight test on November 29, 1961. Enos's test flight qualified the space capsule for use by human beings during subsequent orbital missions.

ergometer   A bicycle-like instrument for measuring muscular work and for exercising in place. Astronauts and cosmonauts use specially designed ergometers to exercise while on extended orbital flights.

escape rocket   A small rocket engine attached to the leading end of an escape tower, which is used to provide additional thrust to the crew capsule to obtain separation of this capsule from an expendable launch vehicle in the event of a launch pad abort or emergency.

escape tower   A trestle tower placed on top of a crew (space) capsule, which during liftoff connects the capsule to the escape rocket. After a successful liftoff and ascent, the escape tower and escape rocket are separated from the capsule.

escape velocity (common symbol: V_e)   The minimum velocity that an object must acquire to overcome the gravitational attraction of a celestial body. The escape velocity for an object launched from the surface of Earth is approximately seven miles per second (11.2 km/s), while the escape velocity from the surface of Mars is about three miles per second (5.0 km/s).

European Space Agency (ESA)   An international organization that promotes the peaceful use of outer space and cooperation among the European member states in space research and applications.

exoatmospheric   Occurring outside Earth's atmosphere; events and actions that take place at altitudes above about 62 miles (100 km).

explosive decompression   A rapid reduction of air pressure inside the pressurized portion (i.e., crew compartment) of an aircraft, aerospace vehicle, or spacecraft. For example, collision with a large piece of space debris might puncture the wall of one of the pressurized modules on a space station, causing an explosive decompression situation within that module. Air locks would activate, sealing off the stricken portion of the pressurized space habitat.

external tank (ET)   The large tank that contains the cryogenic liquid propellants for the three space shuttle main engines. This tank forms the structural backbone of NASA's space shuttle vehicle.

extraterrestrial contamination   The contamination of one world by life-forms, especially microorganisms, from another world. Taking Earth's biosphere as the reference, planetary contamination is called forward contamination when an alien world is contaminated by contact with terrestrial organisms, and it is called back contamination when alien organisms are released into Earth's biosphere.

extraterrestrial life   Life-forms that may have evolved independent of and now exist beyond the terrestrial biosphere.

extravehicular activity (EVA)   Activities conducted by an astronaut or cosmonaut in outer space or on the surface of another planet (or moon), outside of the protective environment of his/her aerospace vehicle, space-craft, or lander. Astronauts and cosmonauts must put on space suits that contain portable life-support systems to perform EVA tasks.

eyeballs-in, eyeballs-out   Early American space program expression used to describing the acceleration-related sensations experienced by an astro-naut at liftoff or when retrorockets fired. The experience at liftoff is eyeballs-in (due to positive g-forces on the human body when the launch vehicle accelerates). The experience when the retrorockets fire is eyeballs-out (due to negative g-forces on the human body as a spacecraft decelerates).

farside   The side of the Moon that never faces Earth.

ferry flight   An in-the-atmosphere flight of NASA's space shuttle orbiter vehicle while mated on top of a specially configured Boeing 747 shuttle carrier aircraft.

flare (solar)   A bright eruption from the Sun's corona. An intense flare represents a major ionizing radiation hazard to astronauts traveling beyond Earth's magnetosphere through interplanetary space or while exploring the surface of the Moon or Mars.

flight crew   Personnel assigned to an aerospace vehicle (like NASA's space shuttle), a space station, or an interplanetary spacecraft for a specific flight or mission. The space shuttle flight crew usually consists of astronauts serving as the commander, the pilot, and one to several mission specialists.

free fall   The unimpeded fall of an object in a gravitational field. For example, all the astronauts and objects inside an Earth-orbiting spacecraft experience a continuous state of free fall and appear weightless as the force of inertia counterbalances the force of Earth's gravity.

free-flying spacecraft ("free-flyer")   Any spacecraft or payload that can be detached from NASA's space shuttle or the *International Space Station* and then operate independently on orbit.

frequency (common symbol: f or ν)   The rate of repetition of a recur-ring or regular event; the number of cycles of a wave per second. For electromagnetic radiation, the frequency is equal to the speed of light divided by the wavelength. *See also* HERTZ.

fuel cell   A direct-conversion device that transforms chemical energy directly into electrical energy by reacting continuously supplied chemicals. In a modern fuel cell, an electrochemical catalyst (such as platinum) promotes a noncombustible reaction between a fuel (such as hydrogen) and an oxidant (such as oxygen).

fuselage   The central part of an aerospace vehicle or aircraft that accommodates crew, passengers, payload, or cargo.

g   The symbol used for the acceleration due to gravity. At sea level on Earth, g is approximately 32.2 feet per second, squared (ft/s^2) ($9.8 m/s^2$)—that is, "one g." This term is used as a unit of stress for bodies experiencing acceleration. When a rocket accelerates during launch, everything inside it (including astronauts and cosmonauts) experiences a g-force that can be as high as several g's.

Gagarin, Yuri A. (1934–68)   The Russian cosmonaut Yuri A. Gagarin became the first human being to travel in outer space. He accomplished this feat on April 12, 1961, with an historic orbit of Earth mission in the *Vostok 1* spacecraft. A popular hero of the Soviet Union, he died in an aircraft training flight near Moscow on March 27, 1968.

gamma ray (symbol: γ)   Very-short-wavelength, high-frequency packets (or quanta) of electromagnetic radiation. Gamma-ray photons are similar to X-rays, except that they originate within the atomic nucleus and have energies between 10,000 electron volts (10 keV) and 10 million electron volts (10 MeV) or more.

Gemini Project   The second U.S. crewed space project (1964–66) and the start of more sophisticated missions by pairs of American astronauts in each Gemini space capsule. Through this project, NASA expanded the results of the Mercury Project and prepared for the ambitious lunar-landing missions of the Apollo Project.

geocentric   Relative to Earth as the center; measured from the center of Earth.

geomagnetic storm   Sudden, often global fluctuations in Earth's magnetic field, associated with the shock waves from solar flares that arrive at Earth within about 24 to 36 hours after violent activity on the Sun.

Glenn, John Herschel, Jr. (b. 1921)   An American astronaut, U.S. Marine Corps officer, and U.S. senator, John H. Glenn, Jr., was the first American to orbit Earth in a spacecraft. He accomplished this historic feat on February 20, 1962, as part of NASA's Mercury Project. Launched from Cape Canaveral by an Atlas rocket, Glenn flew into space inside the *Friendship 7* Mercury space capsule and made three orbits of Earth. In 1998, he became the oldest human being to travel in space when he served as a member of the space shuttle *Discovery* crew during the STS-95 orbital mission.

gravitation   The force of attraction between two masses. From Sir Isaac Newton's law of gravitation, this attractive force operates along a line joining the centers of mass, and its magnitude is inversely proportional to the square of the distance between the two masses. From Albert Einstein's general relativity theory, gravitation is viewed as a distortion of the space-time continuum.

gravity   The attraction of a celestial body for any nearby mass. For example, the downward force imparted by Earth on a mass near Earth or on the planet's surface.

Greenwich mean time (GMT)   Mean solar time at the meridian of Greenwich, England, used as the basis for standard time throughout the world. It is normally expressed in four numerals, 0001 to 2400. Also called *universal time (UT)*.

ground elapsed time (GET)   The time expired since launch.

ground track   The path followed by a spacecraft over Earth's surface.

G suit   A suit that exerts pressure on the abdomen and lower parts of the body to prevent or retard the collection of blood below the chest under conditions of positive acceleration.

habitable payload   A payload with a pressurized compartment suitable for supporting a crewperson in a shirtsleeve environment.

Hadley Rille   A long, ancient lava channel on the Moon that was the landing site for the Apollo 15 mission during NASA's Apollo Project.

half-life (radioactive)   The time required for one-half of the atoms of a particular radioactive isotope population to disintegrate to another nuclear form. Measured half-lives vary from millionths of a second to billions of years.

Ham   The primate (astrochimp) used by NASA on January 31, 1961, to test the Mercury Project space capsule in a suborbital flight. Ham's successful test flight qualified the space capsule for use of by human beings during subsequent suborbital flights.

hangfire   A faulty condition in the ignition system of a rocket engine.

hard-landing   A relatively high-velocity impact of a lander spacecraft on a solid planetary surface. The impact usually destroys all equipment, except perhaps a rugged instrument package or payload container.

hatch   A tightly sealed access door in the pressure hull of an aerospace vehicle, spacecraft, or space station.

hertz (symbol: Hz)   The SI unit of frequency. One hertz is equal to one cycle per second. Named in honor of the German physicist Heinrich Rudolf Hertz (1857–94), who produced and detected radio waves for the first time in 1888.

high Earth orbit (HEO)   An orbit around Earth at an altitude greater than 3,475 miles (5,600 km).

highlands   Oldest-exposed areas on the surface of the Moon; extensively cratered and chemically distinct from the maria.

Hohmann transfer orbit   The most efficient orbit transfer path between two coplanar circular orbits. The maneuver consists of two impulsive high-thrust burns (or firings) of a spacecraft's propulsion system. The technique was suggested in 1925 by the German engineer Walter Hohmann (1880–1945).

hold   To stop the sequence of events during a countdown until an impediment has been removed so that the countdown to launch can be resumed.

"housekeeping" (spacecraft)   The collection of routine tasks that must be performed to keep a spacecraft functioning properly during an orbital flight or interplanetary mission.

Hubble Space Telescope (HST)   A cooperative European Space Agency and NASA program to operate a long-lived space-based optical observatory. Launched on April 25, 1990, by NASA's space shuttle *Discovery* (STS-31 mission), subsequent on-orbit repair and refurbishment missions by shuttle-based astronauts have allowed this powerful Earth-orbiting optical observatory to revolutionize scientific knowledge of the size, structure, and makeup of the universe. Named in honor of the American astronomer Edwin Powell Hubble (1889–1953).

human-factor engineering   The branch of engineering involved in the design, development, testing, and construction of devices, equipment, and artificial living environments to the anthropometric, physiological, and/or psychological requirements of the human beings who will use them. One aerospace example is the design of a functional microgravity toilet that is suitable for use by both male and female crewpersons.

hyperoxia   An aerospace medicine term used to describe a condition in which the total oxygen content of the body is increased above that normally existing at sea level.

hypoxia   An aerospace medicine term used to describe an oxygen deficiency in the blood, cells, or tissues of the body in such degree as to cause psychological and physiological disturbances.

HZE particles   The most potentially damaging cosmic rays, with high atomic number and high kinetic energy.

Imbrium basin   Large (about 810 miles [1,300 km] across), ancient impact crater on the Moon.

International Space Station (ISS)   A major human spaceflight project headed by NASA. Russia, Canada, Europe, Japan, and Brazil are also contributing key elements to this large, modular space station in low Earth orbit that represents a permanent human outpost in outer space for microgravity research and advanced space technology demonstrations. On-orbit assembly began in December 1998.

international system of units   *See* SI UNITS.

interplanetary   Between the planets; within the solar system.

interstellar   Between or among the stars.

intravehicular activity (IVA)   Astronaut or cosmonaut activities performed inside an orbiting spacecraft or aerospace vehicle. *Compare with* EXTRAVEHICULAR ACTIVITY.

ionizing radiation   Any type of nuclear radiation that displaces electrons from atoms or molecules, thereby producing ions within the irradiated material. Examples include: alpha radiation, beta radiation, gamma radiation, protons, neutrons, and X-rays.

jettison   To discard or toss away.

Kennedy Space Center (KSC)   Sprawling NASA spaceport on the east-central coast of Florida adjacent to Cape Canaveral Air Force Station. Launch site (Complex 39) and primary landing/recovery site for the space shuttle.

Komarov, Vladimir M. (1927–67)   Russian cosmonaut who was the first person to make two trips into outer space. He was also the first person to die while engaged in space travel. On April 23, 1967, he flew the new Soviet *Soyuz 1* spacecraft into orbit. This flight encountered many difficulties, eventually forcing him to execute an emergency reentry maneuver on April 24. During the final stage of reentry, the spacecraft's recovery parachute became entangled, and the *Soyuz 1* impacted the ground at high speed—instantly killing the cosmonaut.

launch site   The extensive, well-defined area used to launch rocket vehicles for operational or for test purposes. Also called the launch complex.

launch vehicle (LV)   An expendable or reusable rocket-propelled vehicle that provides sufficient thrust to place a spacecraft in orbit around Earth or to send a payload on an interplanetary trajectory to another celestial body. Sometimes called a booster or space-lift vehicle.

launch window   An interval of time during which a launch may be made to satisfy some mission objective. Sometimes it is just a short period each day for a certain number of days.

life support system (LSS)   The system that maintains life throughout the entire aerospace flight environment, including (as appropriate) travel in outer space, activities on the surface of another world (e.g., the lunar surface), and ascent and descent through Earth's atmosphere. The LSS must reliably satisfy a human crew's daily needs for clean air, potable water, food, and effective waste removal.

liftoff   The action of a rocket or aerospace vehicle as it separates from its launch pad in a vertical ascent.

lithium hydroxide (LiOH)   A white crystalline compound used for removing carbon dioxide from a closed atmosphere, such as found on a crewed spacecraft, aerospace vehicle, or space station. Space suit life support systems also use lithium hydroxide canisters to purge the suit's closed atmosphere of the carbon dioxide exhaled by the astronaut occupant.

Liwei, Yang (b. 1965)   On October 15, 2003, Yang Liwei became the first taikonaut (astronaut) from the People's Republic of China. He rode on board the *Shenzhou 5* spacecraft as a Long March 2F rocket carried it into orbit from the Jiuquan Satellite Launch Center. After Liwei made 14 orbits of Earth, the *Shenzhou 5* spacecraft made a successful reentry and soft-landing on October 16. He was then recovered safely in the Chinese portion of Inner Mongolia.

lunar   Of or pertaining to Earth's natural satellite, the Moon.

lunar base   A permanently inhabited complex on the surface of the Moon. It is the next logical step after brief human exploration expeditions, such as NASA's Apollo Project.

lunar excursion module (LEM)   The lander spacecraft used by NASA to deliver astronauts to the surface of the Moon during the Apollo Project.

lunar highlands   The light-colored, heavily cratered mountainous part of the Moon's surface.

lunar rover vehicle (LRV)   The electrically powered "Moon car" used by Apollo Project astronauts on the lunar surface during the Apollo 15, 16, and 17 expeditions.

magnetosphere   The region around a planet in which charged atomic particles are influenced (and often trapped) by the planet's own magnetic field rather than the magnetic field of the Sun as projected by the solar wind.

manned   An aerospace vehicle or system that is occupied by one or more persons, male or female. The terms *crewed, human,* or *personed* are preferred today in the aerospace literature. For example, a "manned mission to Mars" should be called a "human mission to Mars."

manned vehicle   An older aerospace term describing a rocket or spacecraft that carried one or more human beings, male or female. Used to distinguish that craft from a robot (i.e., pilotless) aircraft, a ballistic missile, or an automated (and uncrewed) satellite or planetary probe. The expression "crewed vehicle" or "personed vehicle" is now preferred.

man-rated   A launch vehicle, spacecraft, aerospace system, or component considered safe and reliable enough to be used by human crew members. The term *human-rated* is now preferred.

maria (singular: mare)   Latin word for "seas." Originally used by the Italian astronomer Galileo Galilei (1564–1642) to describe the large, dark, ancient lava flows on the lunar surface, since he and other astronomers thought these features were bodies of water on the Moon's surface. Following tradition, this term is still used by modern astronomers.

Mars base   The surface base needed to support human explorers during a Mars expedition later this century.

Mars expedition   The first crewed mission to visit Mars in this century. Current concepts suggest a 600- to 1,000-day mission (starting from Earth orbit), a total crew size of up to 15 astronauts, and about 30 days for surface excursion activities on Mars.

Martian   Of or relating to the planet Mars.

mating   The act of fitting together two major components of an aerospace system, such as the mating of a launch vehicle and its payload—a scientific spacecraft. Also the physical joining of two orbiting spacecraft either through a docking or a berthing process.

Mercury Project   The initial United States astronaut program (1958–63) in which NASA selected seven military test pilots to become the first Americans to fly in outer space. They flew in cramped, one-person space capsules, such as John Herschel Glenn, Jr.'s *Friendship 7* Mercury capsule.

metric system   *See* SI UNITS.

microgravity (common symbol: μg)   Because its inertial trajectory compensates for the force of gravity, a spacecraft in orbit around Earth travels in a state of continual free fall. All objects inside appear weightless—as if they were in a zero gravity environment. However, the venting of gases, the minuscule drag exerted by Earth's residual atmosphere (at low orbital altitudes), and crew motions tend to create nearly imperceptible forces on objects inside the orbiting vehicle. These tiny forces are collectively called microgravity.

Mir ("Peace")   A third-generation Russian space station of modular design that was assembled on-orbit around a core module launched in February 1986. Although used extensively by many cosmonauts and guest researchers (including American astronauts), the massive station was eventually abandoned because of economics and was safely de-orbited into a remote area of the Pacific Ocean in March 2001.

mission specialist   The space shuttle crewmember and NASA career astronaut responsible for coordinating payload/Space Transportation System (STS) interaction. During the payload operation phase of a space shuttle flight, the mission specialist directs the allocation of STS and crew resources to accomplish payload-related mission objectives.

Moon   Earth's only natural satellite and closest celestial neighbor. It has an equatorial diameter of 2,159 miles (3,476 km), keeps the same side (nearside) toward Earth, and orbits at an average distance (center to center) of 238,758 miles (384,400 km).

nadir   The direction from a spacecraft directly down toward the center of a planet. It is the opposite of the ZENITH.

NASA   The National Aeronautics and Space Administration, the civilian space agency of the United States. Created in 1958 by an act of Congress, NASA's overall mission is to plan, direct, and conduct civilian (including scientific) aeronautical and space activities for peaceful purposes.

nearside   The side of the Moon that always faces Earth.

nuclear radiation   Ionizing radiation consisting of particles (such as alpha particles, beta particles, and neutrons) and very energetic electromagnetic radiation (that is, gamma rays). Atomic nuclei emit this type of radiation during a variety of energetic nuclear reaction processes, including radioactive decay, fission, and fusion.

nuclear rocket   A rocket vehicle that derives its propulsive thrust from nuclear energy. For example, the nuclear thermal rocket uses a nuclear reactor to heat hydrogen to extremely high temperatures before expelling it through a thrust-producing nozzle.

one-g   The downward acceleration of gravity at Earth's surface (approximately 32.2 ft/s^2 [9.8 m/s^2]).

orbit   The path followed by a body in space, generally under the influence of gravity—as for example a satellite around a planet.

orbital injection   The process of providing a space vehicle or a satellite with sufficient velocity to establish an orbit.

orbital period   The interval between successive passages of a satellite or spacecraft through the same point in its orbit. Often called period.

orbiter (spacecraft)   A spacecraft especially designed to travel through interplanetary space, achieve a stable orbit around the target planet (or other celestial body), and conduct a program of detailed scientific investigation.

Orbiter (space shuttle)   The winged aerospace vehicle portion of NASA's space shuttle. It carries astronauts and payload into orbit and returns from outer space by gliding and landing like an airplane. The operational orbiter vehicle (OV) fleet includes *Discovery* (OV-103), *Atlantis* (OV-104), and *Endeavour* (OV-105).

Orbiting Quarantine Facility (OQF)   A proposed Earth-orbiting, crew-tended laboratory in which soil and rock samples from Mars and other worlds in the solar system would first be tested for potentially harmful alien microorganisms—before these materials are allowed to enter Earth's biosphere.

Orion   The name given by NASA to the agency's new crew exploration vehicle, which is being designed to carry astronauts back to the Moon (ca. 2020) and later to Mars (ca. 2030). By 2015, the *Orion* spacecraft—launched by the Ares I rocket vehicle—will succeed the space shuttle as NASA's primary vehicle for human space exploration.

outer space   Any region beyond Earth's atmospheric envelope—usually considered to begin at between 62 and 125 miles (100 and 200 km) altitude.

pad   The platform from which a rocket vehicle is launched.

parking orbit   The temporary (but stable) orbit of a spacecraft around a celestial body. It is used for assembly and/or transfer of crew or equipment, as well as to wait for conditions favorable for departure from that orbit.

payload bay   The large (15 feet [4.6 m] in diameter) and long (60 feet [18.3 m]) enclosed volume within NASA's space shuttle orbiter vehicle designed to carry a wide variety of payloads including upper stage vehicles, deployable spacecraft, and attached equipment. Also called cargo bay.

payload specialist   The noncareer astronaut who flies as a space shuttle passenger and is responsible for achieving the payload/experiment objectives. He/she is the onboard scientific expert in charge of the operation of a particular payload or experiment.

peri-   A prefix meaning "near."

perigee   The point at which a satellite's orbit is the closest to its primary body; the minimum altitude attained by an Earth-orbiting object. *Compare with* APOGEE.

perilune   The point in an elliptical orbit around the Moon that is nearest to the lunar surface.

permanently crewed capability (PCC)   A space station or planetary surface base that can be continuously occupied and operated by a human crew.

pitch   The rotation (angular motion) of an aerospace vehicle or spacecraft about its lateral axis. *See also* ROLL; YAW.

pitchover   The programmed turn from the vertical that a launch vehicle (under power) takes as it describes an arc and points in a direction other than vertical.

planet   A nonluminous celestial body that orbits around the Sun or some other star. The name "planet" comes from the ancient Greek *planetes* ("wanderers")—since early astronomers identified the planets as the wandering points of light relative to the fixed stars. There are eight major planets in this solar system and numerous minor planets (or asteroids). The distinction between a planet and a large satellite is not always precise. The Moon is nearly the size of Mercury and is very large in comparison to Earth—suggesting the Earth-Moon system might easily be treated as a double-planet system. In August 2006, the International Astronomical Union, clarified the difference between a planet and a dwarf planet. A planet is defined as a celestial body that (a) is in orbit around the Sun, (b) has sufficient mass for its self-gravity to overcome rigid body forces so as to assume a nearly round shape, and (c) has cleared the cosmic neighborhood around its orbit. Within this definition, there are eight major planets in the solar system: Mercury, Venus, Earth, Mars, Jupiter, Saturn, Uranus and Neptune. Pluto is now regarded as a DWARF PLANET.

planetary albedo The fraction of incident solar radiation that is reflected by a planet (and its atmosphere) and returned to outer space.

planet fall The act of landing a spacecraft or space vehicle on a planet or moon.

polar orbit An orbit around a planet (or primary body) that passes over or near its poles; an orbit with an inclination of about 90 degrees.

pressurized habitable environment Any module or enclosure in outer space in which an astronaut may perform activities in a shirtsleeve environment.

primary body The celestial body around which a satellite, moon, or other object orbits or from which it is escaping or toward which it is falling.

prograde orbit An orbit having an inclination of between 0 degrees and 90 degrees.

Progress An uncrewed Russian supply spacecraft configured to perform automated rendezvous and docking operations with space stations and other orbiting spacecraft.

rad In radiation protection, the traditional unit for an absorbed dose of ionizing radiation. A dose of one rad means the absorption of 100 ergs of ionizing radiation energy per gram of absorbing material (or 0.01 joule per kilogram in SI units). The term is an acronym derived from *r*adiation *a*bsorbed *d*ose.

radiation belt The region(s) in a planet's magnetosphere where there is a high density of trapped atomic particles from the solar wind. *See also* EARTH'S TRAPPED RADIATION BELTS.

radiation sickness A potentially fatal illness resulting from excessive exposure to ionizing radiation. *See also* ACUTE RADIATION SYNDROME.

radioactivity   The spontaneous decay or disintegration of an unstable (atomic) nucleus accompanied by the emission of nuclear radiation, such as alpha particles, beta particles, or gamma rays.

radio frequency (RF)   The portion of the electromagnetic spectrum useful for telecommunications with a frequency range between 10,000 and 3×10^{11} hertz.

Red Planet   The planet Mars—so named because of its distinctive reddish soil.

reentry   The return of objects, originally launched from Earth, back into the sensible atmosphere; the action involved in this event. The major types of reentry are: ballistic, gliding, and skip. When a piece of space debris undergoes an uncontrolled ballistic reentry, it usually burns up in the atmosphere due to excessive aerodynamic heating. An aerospace vehicle, like NASA's space shuttle, is designed to make a controlled atmospheric reentry by using a gliding trajectory, which carefully dissipates the vehicle's kinetic energy and potential energy prior to landing.

regenerative life support system (RLSS)   A controlled ecological life support system in which biological and physiochemical subsystems produce plants for food and process solid, liquid, and gaseous wastes for reuse in the system.

regolith (lunar)   The unconsolidated mass of surface debris that overlies the Moon's bedrock. This blanket of pulverized lunar dust and soil was created by millions of years of meteoric and cometary impacts.

rem   In radiation protection, the traditional unit for dose equivalent (symbol: H). The dose equivalent in rem is the product of the absorbed dose in rad and a suitable weighting factor or quality factor—as well as any other modifying factors considered necessary to characterize and evaluate the biological effects of an ionizing radiation dose received by a human being or other living creature. The term is an acronym derived from the expression: *roentgen equivalent man*. The rem is related to the sievert (the SI unit of dose equivalent) as follows: 100 rem = 1 sievert.

remote manipulator system (RMS)   The dexterous, Canadian-built, 50-foot- (15.2-m-) long articulated arm that is remotely controlled by astronauts from the aft flight deck of NASA's space shuttle orbiter vehicle.

rendezvous   The close approach of two or more spacecraft in the same orbit, so that docking can take place. These objects meet at a preplanned location and time with essentially zero relative velocity.

robot spacecraft   A semiautomated or fully automated spacecraft capable of executing its primary exploration mission with minimal or no human supervision.

rocket   A completely self-contained projectile or flying vehicle propelled by a reaction engine. Since a rocket carries all of its required propellant, it can function in the vacuum of outer space and represents the key to space travel. There are chemical rockets, nuclear rockets, and electric propulsion rockets. Chemical rockets are further divided into solid-propellant rockets and liquid-propellant rockets.

roll   The rotational or oscillatory movement of an aerospace vehicle or spacecraft about its longitudinal (lengthwise) axis. *See also* PITCH; YAW.

rover   A crewed or robot space vehicle used to explore a planetary surface.

Salyut (**"Salute"**)   An evolutionary series of early space stations placed in orbit around Earth in the 1970s and 1980s by the former Soviet Union to support a variety of military and civilian missions.

satellite   A secondary (smaller) celestial body in orbit around a larger primary body. For example, Earth is a natural satellite of the Sun, while the Moon is a natural satellite of Earth. A human-made spacecraft placed in orbit around Earth is called an artificial satellite—or more commonly just a satellite.

Saturn (launch vehicle)   Family of powerful expendable launch vehicles developed for NASA by Wernher von Braun (1912–77) to carry astronauts to the Moon during the Apollo Project.

scientific air lock   A special opening in a crewed spacecraft or space station through which experiments and research equipment can be extended outside (into outer space) without violating the atmospheric integrity of the pressurized interior of the space vehicle.

scrub   To cancel or postpone a rocket firing either before or during the countdown.

sensible atmosphere　That portion of a planet's atmosphere that offers resistance to a body passing through it.

sensor　The portion of a scientific instrument that detects and/or measures some physical phenomenon.

Shenzhou 5 **spacecraft**　On October 15, 2003, the People's Republic of China became the third nation—following Russia (former Soviet Union) and the United States—to place a human being in orbit around Earth using a national launch vehicle. On that date, a Chinese Long March 2F rocket lifted off from the Jiuquan Satellite Launch Center and placed the *Shenzhou 5* spacecraft with taikonaut Yang Liwei on board into orbit around Earth. After 14 orbits around Earth, the spacecraft reentered the atmosphere on October 16, and Yang Liwei was safely recovered in the Chinese portion of Inner Mongolia.

Shepard, Alan B. (1923–98)　Selected as one of the original seven Mercury Project astronauts, Alan B. Shepard, Jr., became the first American to travel in outer space on May 5, 1961, when he rode inside the *Freedom 7* space capsule on a suborbital flight from Cape Canaveral. In February 1971, he served as the commander of NASA's Apollo 14 lunar-landing mission. Along with astronaut Edgar Dean Mitchell (b. 1930), Shepard explored the Moon's Fra Mauro region.

shirtsleeve environment　A space station module or spacecraft cabin in which the atmosphere is similar to that found on the surface of Earth, that is, it does not require a pressure suit.

sievert (symbol: Sv)　In radiation protection, the dose equivalent in sieverts is the product of the absorbed dose in grays and the radiation weighting factor or (previously) the quality factor, as well as any other modifying factors considered necessary to characterize and evaluate the biological effects of the ionizing radiation received by a human being or other living creatures. The sievert is the special SI unit for dose equivalent and is related to the traditional dose equivalent unit (the rem) as follows: 1 sievert = 100 rem. *See also* ACUTE RADIATION SYNDROME.

SI units　The international system of units (the metric system) that uses the meter (m), kilogram (kg), and second (s) as its basic units of length, mass, and time, respectively.

Skylab　The first U.S. space station that NASA placed in orbit in 1973 and was visited by three astronaut crews between 1973 and 1974. It reentered the atmosphere on July 11, 1979, as a large, abandoned derelict—with surviving space debris impacting in the Indian Ocean and remote portions of Australia.

soft-landing   The act of landing on the surface of a planet without damaging any portion of a spacecraft or its payload, except possibly an expendable landing gear structure. *Compare with* HARD-LANDING.

solar flare   A highly concentrated, explosive release of electromagnetic radiation and nuclear particles within the Sun's atmosphere near an active sunspot.

solar panel   The winglike assembly of solar cells used by a spacecraft to convert sunlight (solar energy) directly into electrical energy. Also called a solar array.

solar storm   A major disturbance in the space environment triggered by an intense solar flare (or flares) that produces bursts of electromagnetic radiation and charged particles, threatening unprotected spacecraft and astronauts alike.

solar system   In general, any star and its gravitationally bound collection of nonluminous objects, such as planets, asteroids, and comets; specifically, humans' home solar system, consisting of the Sun and all the objects bound to it by gravitation—including eight major planets, three dwarf planets with more than 60 known moons, over 2,000 asteroids (minor planets), and a large number of comets. Except for the comets, all the other celestial objects travel around the Sun in the same direction.

solar wind   The variable stream of plasma (that is, electrons, protons, alpha particles, and other atomic nuclei) that flows continuously outward from the Sun into interplanetary space.

solid rocket booster (SRB)   The two large solid-propellant rockets that operate in parallel to augment the thrust of the space shuttle's three main engines. After burning for about 120 seconds, the depleted SRBs are jettisoned from the space shuttle flight vehicle and recovered in the Atlantic Ocean downrange of Cape Canaveral for refurbishment and propellant reloading.

Soyuz (**"Union"**) **spacecraft**   The evolutionary family of crewed Russian spacecraft used by cosmonauts on a wide variety of Earth-orbiting missions since 1967.

space base   A large, permanently inhabited space facility located in orbit around a celestial body or on its surface that would serve as the center of future human operations in some particular region of the solar system.

space capsule   The family of small, container-like, tear-shaped spacecraft used to carry American astronauts into outer space and return them to Earth as part of NASA's Mercury Project, Gemini Project, and Apollo Project.

space colony   An earlier term used to describe a large, permanent space habitat and industrial complex occupied by up to 10,000 persons. Currently, the term *space settlement* is preferred.

spacecraft   A platform that can function, move, and operate in outer space or on a planetary surface. Spacecraft can be human-occupied or uncrewed (robot) platforms. They can operate in orbit around Earth or while on an interplanetary trajectory to another celestial body. Some spacecraft travel through space and orbit another planet, while others descend to a planet's surface making a hard-landing (collision impact) or a (survivable) soft-landing. Exploration spacecraft are often categorized as either flyby, orbiter, atmospheric probe, lander, or rover spacecraft.

spacecraft clock   The time-keeping component within a spacecraft's command and data-handling system. It meters the passing time during a mission and regulates nearly all activity within the spacecraft.

space debris   Space junk; abandoned or discarded human-made objects in orbit around Earth. It includes operational debris (items discarded during spacecraft deployment), used or failed rockets, inactive or broken satellites, and fragments from collisions and space object breakup. When a spacecraft collides with an object or a discarded rocket spontaneously explodes, thousands of debris fragments become part of the orbital debris population.

Spacelab (SL)   An orbiting laboratory facility delivered into space and sustained while in orbit within the huge cargo bay of NASA's space shuttle orbiter. Developed by the European Space Agency in cooperation with NASA, *Spacelab* featured several interchangeable elements that were arranged in various configurations to meet the particular needs of a given flight.

space launch vehicle (SLV)   The expendable or reusable rocket-propelled vehicle used to lift a payload or spacecraft from the surface of Earth and place it in orbit around the planet or on an interplanetary trajectory.

spaceman   A person, male or female, who travels in outer space. The term *astronaut* is preferred.

space medicine   The branch of aerospace medicine concerned specifically with the health of persons who make, or expect to make, flights beyond Earth's sensible (measurable) atmosphere into outer space.

spaceport   A facility that serves as both a doorway to outer space from the surface of a planet and a port of entry for aerospace vehicles returning from space to the planet's surface. NASA's Kennedy Space Center with its space shuttle launch site and landing complex is an example.

space radiation environment   One of the major concerns associated with the development of a permanent human presence in outer space is the ionizing radiation environment, both natural and human-made. The natural portion of the space radiation environment consists primarily of Earth's trapped radiation belts (also called the Van Allen belts), solar particle events (SPEs), and galactic cosmic rays (GCRs).

space resources   The resources available in outer space that could be used to support an extended human presence and eventually become the physical basis for a thriving solar system–level civilization. These resources include unlimited solar energy; minerals on the Moon, asteroids, Mars, and numerous outer planet moons; lunar (water) ice; and special environmental conditions like access to high vacuum and physical isolation from terrestrial biosphere.

space settlement   A proposed large, human-made habitat in outer space within which from 1,000 to 10,000 people would live, work, and play while supporting various research and commercial activities, such as the construction of satellite power systems.

spaceship   An interplanetary spacecraft that carries a human crew.

space shuttle   The major spaceflight component of NASA's Space Transportation System. It consists of a winged orbiter vehicle, three space shuttle main engines, the giant external tank—which feeds liquid hydrogen and liquid oxygen to the shuttle's three main liquid-propellant rocket engines—and the two solid rocket boosters.

space sickness   The Space Age form of motion sickness whose symptoms include nausea, vomiting, and general malaise. This temporary condition lasts no more than a day or so but affects 50 percent of the astronauts or cosmonauts when they encounter the microgravity environment (weight-lessness) of an orbiting spacecraft after a launch. Also called space adaptation syndrome.

space station An Earth-orbiting facility designed to support long-term human habitation in outer space. *See also* INTERNATIONAL SPACE STATION.

space suit The flexible, outer, garmentlike structure (including visored helmet) that protects an astronaut in the hostile environment of outer space. It provides portable life support functions, supports communications, and accommodates some level of movement and flexibility so the astronaut can perform useful tasks during an extravehicular activity or while exploring the surface of another world.

Space Transportation System (STS) The official name for NASA's space shuttle.

space vehicle The general term describing a crewed or robot vehicle capable of traveling through outer space. An aerospace vehicle can operate both in outer space and in Earth's atmosphere.

space walk The popular term for an extravehicular activity.

splashdown That portion of a human space mission in which the space capsule (reentry craft) containing the crew lands in the ocean—quite literally, "splashing down." A team of helicopters, aircraft, and/or surface ships then recovers the astronauts. This term was used during NASA's Mercury, Gemini, Apollo, Skylab, and Apollo-Soyuz Projects (1961–75). In the space shuttle era, the orbiter vehicle returns to Earth by landing much like an aircraft. Consequently, the orbiter vehicle and its crew are said to "touch down."

starship A space vehicle capable of traveling the great distances between star systems. Even the closest stars in the Milky Way Galaxy are light-years apart. The term *starship* is generally used to describe an interstellar spaceship capable of carrying intelligent beings to other star systems; robot interstellar spaceships are often referred to as interstellar probes.

stationkeeping The sequence of maneuvers that maintains a space vehicle or spacecraft in a predetermined orbit.

Surveyor Project   The NASA Moon exploration effort in which five lander spacecraft softly touched down on the lunar surface between 1966 and 1968—the robot precursor to the Apollo Project human expeditions.

taikonaut   The suggested Chinese equivalent to astronaut and cosmonaut. *Taikong* is the Chinese word for space or cosmos; so the prefix "taiko-" assumes the same concept and significance as the use of "astro-" or "cosmo-" to form the words *astronaut* and *cosmonaut*.

telecommunications   The transmission of information over great distances using radio waves or other portions of the electromagnetic spectrum.

telemetry   The process of making measurements at one point and transmitting the information via radio waves over some distance to another location for evaluation and use. Telemetered data on a spacecraft's communications downlink often include scientific data as well as spacecraft state-of-health data.

teleoperation   The technique by which a human controller operates a versatile robot system that is at a distant, often hazardous, location. High-resolution vision and tactile sensors on the robot, reliable telecommunications links, and computer-generated virtual reality displays enable the human worker to experience telepresence.

telepresence   The process, supported by an information-rich control station environment, that enables a human controller to manipulate a distant robot through teleoperation and almost feel physically present in the robot's remote location.

telescope   An instrument that collects electromagnetic radiation from a distant object so as to form an image of the object or to permit the radiation signal to be analyzed. Optical (astronomical) telescopes are divided into two general classes: refracting telescopes and reflecting telescopes. Earth-based astronomers also use large radio telescopes, while orbiting observatories use optical, infrared, ultraviolet, X-ray, and gamma-ray telescopes to study the universe.

Tereshkova, Valentina (b. 1937)   Cosmonaut Valentina Tereshkova holds the honor of being the first woman to travel in outer space. She accomplished this feat on June 16, 1963, by riding the *Vostok 6* spacecraft into orbit. During this historic mission, she completed 48 orbits of Earth.

terminator   The distinctive boundary line separating the illuminated (that is, sunlit) and dark portions of a nonluminous celestial body such as the Moon.

terrestrial   Of or relating to Earth.

terrestrial planets   In addition to Earth, the planets Mercury, Venus, and Mars—all of which are relatively small, high-density celestial bodies composed of metals and silicates with shallow or no atmospheres in comparison to the Jovian planets.

Titov, Gherman S. (1935–2000)   In August 1961, Russian cosmonaut Gherman S. Titov became the second person to travel in orbit around Earth. His *Vostok 2* spacecraft made 17 orbits, during which he became the first of many space travelers to experience space sickness.

tracking   Following the movement of a satellite, rocket, or aerospace vehicle. It is usually performed with optical, infrared, radar, or radio wave systems.

trajectory   The three-dimensional path traced by any object moving because of an externally applied force; the flight path of a space vehicle.

transfer orbit   An elliptical, interplanetary trajectory tangent to the orbits of both the departure planet and target planet (or moon). *See also* HOHMANN TRANSFER ORBIT.

Unity   The first U.S.-built component of the *International Space Station*. A six-sided connecting module and passageway (node), *Unity* was the primary cargo of the space shuttle *Endeavour* during the STS-88 mission in early December 1998. Once delivered into orbit, astronauts mated *Unity* to the Russian-built *Zarya* module—delivered earlier into orbit by a Russian Proton rocket that lifted off from the Baikonur Cosmodrome.

universal time coordinated (UTC)   The worldwide scientific standard of timekeeping, based on carefully maintained atomic clocks. Its reference point is Greenwich, England.

uplink   The telemetry signal sent from a ground station to a spacecraft or planetary probe.

upper stage   The second, third, or later rocket stage of a multistage rocket vehicle. Once lifted into low Earth orbit, a spacecraft often uses an attached upper stage to reach its final destination—a higher-altitude orbit around Earth or an interplanetary trajectory.

Van Allen radiation belts *See* EARTH'S TRAPPED RADIATION BELTS.

Voskhod ("Sunrise") An early Russian three-person spacecraft that evolved from the Vostok spacecraft. *Voskhod 1* was launched on October 12, 1964, and carried the first three-person crew into space. *Voskhod 2* was launched on March 18, 1965, and carried a crew of two cosmonauts, including Alexei Arkhipovich Leonov (b. 1934), who performed the world's first extravehicular activity, or space walk (about 10 minutes in duration), during the orbital mission.

Vostok ("East") The first Russian crewed spacecraft, with room for just a single cosmonaut. *Vostok 1* was launched on April 12, 1961, carrying cosmonaut Yuri A. Gagarin (1934–68), the first human to fly in space. Gagarin's flight made one orbit of Earth and lasted about 108 minutes.

weightlessness The condition of free fall (or zero-g) in which objects inside an Earth-orbiting, unaccelerated spacecraft appear weightless even though the objects and the spacecraft are still under the influence of Earth's gravity. It is the condition in which no acceleration, whether of gravity or another force, can be detected by an observer within the system in question.

X-1 The rocket-powered research aircraft, patterned on the lines of a 50-caliber machine-gun bullet, that was the first human-crewed vehicle to fly faster than the speed of sound. On October 14, 1947, the *Bell X-1,* named "Glamorous Glennis" and piloted by Captain Charles "Chuck" Yeager, was carried aloft by a Boeing B-29 mother ship and then released. Yeager ignited the aircraft's rocket engine, climbed, and accelerated, reaching 700 miles per hour (1,127 km/h), or Mach 1.06, as he flew over Edwards Air Force Base in California at an altitude of approximately eight miles (13 km).

X-15 The North American X-15 rocket-powered experimental aircraft helped bridge the gap between human flight within the atmosphere and human flight in space. It was developed and flown in the 1960s to provide in-flight information and data on aerodynamics, structures, flight controls, and the physiological aspects of high-speed, high-altitude flight.

X-ray A penetrating form of electromagnetic radiation of very short wavelength (approximately 0.01 to 10 nanometers) and high photon energy (approximately 100 electron volts to some 100 kiloelectron volts).

yaw   The rotation or oscillation of a missile or aerospace vehicle about its vertical axis so as to cause the longitudinal axis of the vehicle to deviate from the flight line or heading in its horizontal plane. *See also* PITCH; ROLL.

Zarya (**"Dawn"**)   The Russian-built and American-financed module that was the first-launched of numerous modules that make up the *International Space Station*—a large, habitable spacecraft being assembled in low Earth orbit. The first assembly step of the *ISS* occurred in late November and early December 1998. During a NASA space shuttle–supported orbital assembly operation, astronauts linked *Zarya,* the initial control module, together with *Unity,* the American six-port habitable connection module. *Zarya* is also known as the *Functional Cargo Block,* or *FGM*—when the Russian equivalent acronym is transliterated.

zenith   The point on the celestial sphere vertically overhead. *Compare with* NADIR, the point 180 degrees from the zenith.

zero-g   Common (but imprecise) term for the condition of continuous free fall and apparent weightlessness experienced by astronauts and objects in an Earth-orbiting spacecraft. *See also* MICROGRAVITY.

zero-gravity (zero-g) aircraft   An aircraft that flies a special parabolic trajectory to create low-gravity conditions (typically 0.01 g) for short periods of time (10–30 seconds). This type of aircraft accommodates a variety of experiments and often is used to support astronaut training and to refine spaceflight experiment techniques and equipment.

Zvezda (**"Star"**)   The Russian service module for the *International Space Station* (*ISS*). The 20-ton module has three docking hatches and 14 windows. Launched by a Proton rocket from the Baikonur Cosmodrome on July 12, 2000, the module automatically docked with the *Zarya* module of the orbiting *ISS* complex on July 26, 2000.

Further Reading

RECOMMENDED BOOKS

Angelo, Joseph A., Jr. *The Dictionary of Space Technology.* Rev. ed. New York: Facts On File, Inc., 2004.

———. *Encyclopedia of Space Exploration.* New York: Facts On File, Inc., 2000.

Angelo, Joseph A., Jr., and Irving W. Ginsberg, eds. *Earth Observations and Global Change Decision Making, 1989: A National Partnership.* Malabar, Fla.: Krieger Publishing, 1990.

Brown, Robert A., ed. *Endeavour Views the Earth.* New York: Cambridge University Press, 1996.

Burrows, William E., and Walter Cronkite. *The Infinite Journey: Eyewitness Accounts of NASA and the Age of Space.* Discovery Books, 2000.

Chaisson, Eric, and Steve McMillian. *Astronomy Today.* 5th ed. Upper Saddle River, N.J.: Pearson Prentice Hall, 2005.

Cole, Michael D. *International Space Station. A Space Mission.* Springfield, N.J.: Enslow Publishers, 1999.

Collins, Michael. *Carrying the Fire.* New York: Cooper Square Publishers, 2001.

Consolmagno, Guy J., et al. *Turn Left at Orion: A Hundred Night Objects to See in a Small Telescope—And How to Find Them.* New York: Cambridge University Press, 2000.

Damon, Thomas D. *Introduction to Space: The Science of Spaceflight.* 3d ed. Malabar, Fla.: Krieger Publishing Co., 2000.

Dickinson, Terence. *The Universe and Beyond.* 3d ed. Willowdater, Ont.: Firefly Books Ltd., 1999.

Heppenheimer, Thomas A. *Countdown: A History of Space Flight.* New York: John Wiley and Sons, 1997.

Kluger, Jeffrey. *Journey beyond Selene: Remarkable Expeditions Past Our Moon and to the Ends of the Solar System.* New York: Simon & Schuster, 1999.

Kraemer, Robert S. *Beyond the Moon: A Golden Age of Planetary Exploration, 1971–1978.* Smithsonian History of Aviation and Spaceflight Series. Washington, D.C.: Smithsonian Institution Press, 2000.

Lewis, John S. *Rain of Iron and Ice: The Very Real Threat of Comet and Asteroid Bombardment.* Reading, Mass.: Addison-Wesley, 1996.

Logsdon, John M. *Together in Orbit: The Origins of International Participation in the Space Station.* NASA History Division, Monographs in Aerospace History 11, Washington, D.C.: Office of Policy and Plans, November 1998.

Matloff, Gregory L. *The Urban Astronomer: A Practical Guide for Observers in Cities and Suburbs.* New York: John Wiley and Sons, 1991.

Neal, Valerie, Cathleen S. Lewis, and Frank H. Winter. *Spaceflight: A Smithsonian Guide.* New York: Macmillan, 1995.

Pebbles, Curtis L. *The Corona Project: America's First Spy Satellites.* Annapolis, Md.: Naval Institute Press, 1997.

Seeds, Michael A. *Horizons: Exploring the Universe.* 6th ed. Pacific Grove, Calif.: Brooks/Cole Publishing, 1999.

Sutton, George Paul. *Rocket Propulsion Elements.* 7th ed. New York: John Wiley & Sons, 2000.

Todd, Deborah, and Joseph A. Angelo, Jr. *A to Z of Scientists in Space and Astronomy.* New York: Facts On File, Inc., 2005.

EXPLORING CYBERSPACE

In recent years, numerous Web sites dealing with astronomy, astrophysics, cosmology, space exploration, and the search for life beyond Earth have appeared on the Internet. Visits to such sites can provide information about the status of ongoing missions, such as NASA's *Cassini* spacecraft as it explores the Saturn system. This book can serve as an important companion, as you explore a new Web site and encounter a person, technology phrase, or physical concept unfamiliar to you and not fully discussed within the particular site. To help enrich the content of this book and to make your astronomy and/or space technology–related travels in cyberspace more enjoyable and productive, the following is a selected list of Web sites that are recommended for your viewing. From these sites you will be able to link to many other astronomy or space-related locations on the Internet. Please note that this is obviously just a partial list of the many astronomy and space-related Web sites now available. Every effort has been made at the time of publication to ensure the accuracy of the information provided. However, due to the dynamic nature of the Internet, URL changes do occur and any inconvenience you might experience is regretted.

Selected Organizational Home Pages

European Space Agency (ESA) is an international organization whose task is to provide for and promote, exclusively for peaceful purposes, cooperation among European states in space research and technology and their applications. URL: http://www.esrin.esa.it. Accessed April 12, 2005.

National Aeronautics and Space Administration (NASA) is the civilian space agency of the United States government and was created in 1958 by an act of Congress. NASA's overall mission is to plan, direct, and conduct American civilian (including scientific) aeronautical and space activities for peaceful purposes. URL: http://www.nasa.gov. Accessed April 12, 2005.

National Oceanic and Atmospheric Administration (NOAA) was established in 1970 as an agency within the U.S. Department of Commerce to ensure the safety of the general public from atmospheric phenomena and to provide the public with an understanding of Earth's environment and resources. URL: http://www.noaa.gov. Accessed April 12, 2005.

National Reconnaissance Office (NRO) is the organization within the Department of Defense that designs, builds, and operates U.S. reconnaissance satellites. URL: http://www.nro.gov. Accessed April 12, 2005.

United States Air Force (USAF) serves as the primary agent for the space defense needs of the United States. All military satellites are launched from Cape Canaveral Air Force Station, Florida or Vandenberg Air Force Base, California. URL: http://www.af.mil. Accessed April 14, 2005.

United States Strategic Command (USSTRATCOM) is the strategic forces organization within the Department of Defense, which commands and controls U.S. nuclear forces and military space operations. URL: http://www.stratcom.mil. Accessed April 14, 2005.

Selected NASA Centers

Ames Research Center (ARC) in Mountain View, California, is NASA's primary center for exobiology, information technology, and aeronautics. URL: http://www.arc.nasa.gov. Accessed April 12, 2005.

Dryden Flight Research Center (DFRC) in Edwards, California, is NASA's center for atmospheric flight operations and aeronautical flight research. URL: http://www.dfrc.nasa.gov. Accessed April 12, 2005.

Glenn Research Center (GRC) in Cleveland, Ohio, develops aerospace propulsion, power, and communications technology for NASA. URL: http://www.grc.nasa.gov. Accessed April 12, 2005.

Goddard Space Flight Center (GSFC) in Greenbelt, Maryland, has a diverse range of responsibilities within NASA, including Earth system science, astrophysics, and operation of the *Hubble Space Telescope* and other Earth-orbiting spacecraft. URL: http://www.nasa.gov/goddard. Accessed April 14, 2005.

Jet Propulsion Laboratory (JPL) in Pasadena, California, is a government-owned facility operated for NASA by Caltech. JPL manages and operates NASA's deep-space scientific missions, as well as the NASA's Deep Space Network, which communicates with solar system exploration spacecraft. URL: http://www.jpl.nasa.gov. Accessed April 12, 2005.

Johnson Space Center (JSC) in Houston, Texas, is NASA's primary center for design, development, and testing of spacecraft and associated systems for human space flight, including astronaut selection and training. URL: http://www.jsc.nasa.gov. Accessed April 12, 2005.

Kennedy Space Center (KSC) in Florida is the NASA center responsible for ground turnaround and support operations, prelaunch checkout, and launch of the space shuttle. This center is also responsible for NASA launch facilities at Vandenberg Air Force Base, California. URL: http://www.ksc.nasa.gov. Accessed April 12, 2005.

Langley Research Center (LaRC) in Hampton, Virginia, is NASA's center for structures and materials, as well as hypersonic flight research and aircraft safety. URL: http://www.larc.nasa.gov. Accessed April 15, 2005.

Marshall Space Flight Center (MSFC) in Huntsville, Alabama, serves as NASA's main research center for space propulsion, including contemporary rocket engine development as well as advanced space transportation system concepts. URL: http://www.msfc.nasa.gov. Accessed April 12, 2005.

Stennis Space Center (SSC) in Mississippi is the main NASA center for large rocket engine testing, including space shuttle engines as well as future generations of space launch vehicles. URL: http://www.ssc.nasa.gov. Accessed April 14, 2005.

Wallops Flight Facility (WFF) in Wallops Island, Virginia, manages NASA's suborbital sounding rocket program and scientific balloon flights to Earth's upper atmosphere. URL: http://www.wff.nasa.gov. Accessed April 14, 2005.

White Sands Test Facility (WSTF) in White Sands, New Mexico, supports the space shuttle and space station programs by performing tests on and evaluating potentially hazardous materials, space flight components, and rocket propulsion systems. URL: http://www.wstf.nasa.gov. Accessed April 12, 2005.

Selected Space Missions

Cassini Mission is an ongoing scientific exploration of the planet Saturn. URL: http://saturn.jpl.nasa.gov. Accessed April 14, 2005.

Chandra X-ray Observatory (CXO) is a space-based astronomical observatory that is part of NASA's Great Observatories Program. *CXO* observes the universe in the X-ray portion of the electromagnetic spectrum. URL: http://www.chandra.harvard.edu. Accessed April 14, 2005.

Exploration of Mars is the focus of this Web site, which features the results of numerous contemporary and previous flyby, orbiter, and lander robotic spacecraft. URL: http://mars.jpl.nasa.gov. Accessed April 14, 2005.

National Space Science Data Center (NSSDC) provides a worldwide compilation of space missions and scientific spacecraft. URL: http://nssdc.gsfc.nasa.gov/planetary. Accessed April 14, 2005.

Voyager (Deep Space/Interstellar) updates the status of NASA's *Voyager 1* and *2* spacecraft as they travel beyond the solar system. URL: http://voyager.jpl.nasa.gov. Accessed April 14, 2005.

Other Interesting Astronomy and Space Sites

Arecibo Observatory in the tropical jungle of Puerto Rico is the world's largest radio/radar telescope. URL: http://www.naic.edu. Accessed April 14, 2005.

Astrogeology (USGS) describes the USGS Astrogeology Research Program, which has a rich history of participation in space exploration efforts and planetary mapping. URL: http://planetarynames.wr.usgs.gov. Accessed April 14, 2005.

Hubble Space Telescope (**HST**) is an orbiting NASA Great Observatory that is studying the universe primarily in the visible portions of the electromagnetic spectrum. URL: http://hubblesite.org. Accessed April 14, 2005.

NASA's Deep Space Network (DSN) is a global network of antennas that provide telecommunications support to distant interplanetary spacecraft and probes. URL: http://deepspace.jpl.nasa.gov/dsn. Accessed April 14, 2005.

NASA's Space Science News provides contemporary information about ongoing space science activities. URL: http://science.nasa.gov. Accessed April 14, 2005.

National Air and Space Museum (NASM) of the Smithsonian Institution in Washington, D.C., maintains the largest collection of historic aircraft and spacecraft in the world. URL: http://www.nasm.si.edu. Accessed April 14, 2005.

Planetary Photojournal is a NASA-/JPL- sponsored Web site that provides an extensive collection of images of celestial objects within and beyond the solar system, historic and contemporary spacecraft used in space exploration, and advanced aerospace technologies. URL: http://photojournal.jpl.nasa.gov. Accessed April 14, 2005.

Planetary Society is the nonprofit organization founded in 1980 by Carl Sagan and other scientists that encourages all spacefaring nations to explore other worlds. URL: http://planetary.org. Accessed April 14, 2005.

Search for Extraterrestrial Intelligence (SETI) Projects at UC Berkeley is a Web site that involves contemporary activities in the search for extraterrestrial intelligence (SETI), especially a radio SETI project that lets anyone with a computer and an Internet connection participate. URL: http://www.setiathome.ssl.berkeley.edu. Accessed April 14, 2005.

Solar System Exploration is a NASA-sponsored and -maintained Web site that presents the last events, discoveries and missions involving the exploration of the solar system. URL: http://solarsystem.nasa.gov. Accessed April 14, 2005.

Space Flight History is a gateway Web site sponsored and maintained by the NASA Johnson Space Center. It provides access to a wide variety of interesting data and historic reports dealing with (primarily U.S.) human space flight. URL: http://www11.jsc.nasa.gov/history. Accessed April 14, 2005.

Space Flight Information (NASA) is a NASA-maintained and -sponsored gateway Web site that provides the latest information about human spaceflight activities, including the *International Space Station* and the space shuttle. URL: http://spaceflight.nasa.gov/home/index.html. Accessed April 14, 2005.

Index

Gemini 11 mission
115, 116
Gemini 12 mission
117, 118
Gemini B spacecraft
121
Gemini Project 39, 93,
95–121, *98, 102, 106*
geostationary Earth
orbit (GEO) 47
Gibson, Edward G. 176,
177
Gibson, Robert L. 209,
217
Gidzenko, Yuri 190,
233, 236
GIRD (Gruppa
Isutcheniya
Reaktivnovo
Dvisheniya; Group
for Investigation of
Reactive Motion) 22
Glazhkov, Yuri 184
Glenn, John Herschel,
Jr. 20, 83, *84,* 89–91
gloves, for Apollo space
suits 40
Goddard, Robert 8,
10–12
Gorbatko, Victor 184
Gordo (squirrel
monkey) 63–64
Gordon, Richard F., Jr.
114–117, 150, 151, 167
Gordy (rhesus monkey)
79
Grabe, Ronald J. 211
gravity, artificial. *See*
artificial gravity
Great Pyramid (Giza,
Egypt) 2
Grechko, Geogi 183
Greece, ancient 3
Greek mythology, space
project names and
123–124
greenhouse, lunar 50
Griffin, Michael 224

Grissom, Virgil I. "Gus"
84, 137
Apollo 1 tragedy
136–138
Gemini 3 mission
100, 101
Mercury Project 83
Mercury-Redstone 4
mission 86–87
Grumman Aerospace
Corporation 132
Gubarev, Alexi 183
Gumdrop CSM 141–143

H
Haber, Heinz 15
Haise, Fred W., Jr. 153–
156, *155,* 167
Haisheng, Nie 6
Hale, Edward Everett 12
Hall, Asaph 266
HAM. *See* Holloman
Aero-Medical
Research Laboratory
Ham (astrochimp) xii,
61, 67–71, *70*
Harbaugh, Gregory J.
217
Hart, Terry J. 210
Hartsfield, Henry W.
210, 212
Hauck, Frederick H.
207, 214
Hawley, Steven A. 210,
215
hazards
concerns during
Mercury Project
84–85
from external tank
insulation debris
200–201
to space travelers/
workers 43–48, *46*
heart, effect of
microgravity on 37
heat, for space suit 42,
43

heliocentric universe 7
helmets, for space suits
40
Helms, Susan J. 34, *35,*
233, 236
Hermes Project 58
Hieb, Richard J. 216
Hilmers, David C. 211–
212, 214
Hoffman, Jeffrey A. 216
Holloman Aero-Medical
(HAM) Research
Laboratory 60, 61, 68,
71, 73
*Hubble Space Telescope
(HST)* 215–217
human spaceflight, first.
See Gagarin, Yuri A.
Husband, Rick 219,
220
hygiene, personal, in
microgravity 33
HZE particles 53

I
ICBM. *See*
intercontinental
ballistic missile
IGY. *See* International
Geophysical Year
in situ resource
utilization (ISRU)
271–272
insulation debris
200–201, 220–221
intercontinental
ballistic missile
(ICBM) 20, 21
intermediate-moisture
food 48
intermediate-range
ballistic missile
(IRBM) 19
International Council of
Scientific Unions 18
International
Geophysical Year
(IGY) 15, 18